# In the Tradition of Ahlfors–Bers, VII

CONTEMPORARY
MATHEMATICS

696

# In the Tradition of Ahlfors–Bers, VII

6th Ahlfors–Bers Colloquium
October 23–26, 2014
Yale University, New Haven, CT

Ara S. Basmajian
Yair N. Minsky
Alan W. Reid
Editors

American Mathematical Society
Providence, Rhode Island

2010 *Mathematics Subject Classification.* Primary 20-XX, 30-XX, 31-XX, 32-XX, 51-XX, 53-XX, 57-XX, 58-XX, 60-XX.

---

**Library of Congress Cataloging-in-Publication Data**

Names: Ahlfors–Bers Colloquium (6th : 2014 : Yale University) | Basmajian, Ara S., 1958– editor. | Minsky, Yair Nathan, editor. | Reid, Alan W., editor.

Title: In the tradition of Ahlfors–Bers, VII : 6th Ahlfors–Bers Colloquium, October 23–26, 2014, Yale University, New Haven, Connecticut / Ara S. Basmajian, Yair N. Minsky, Alan W. Reid, editors.

Description: Providence, Rhode Island : American Mathematical Society, [2017] | Series: Contemporary mathematics ; volume 696 | Includes bibliographical references.

Identifiers: LCCN 2017015094 | ISBN 9781470426514 (alk. paper)

Subjects: LCSH: Functions–Congresses. | Riemann surfaces–Congresses. | Mappings (Mathematics)–Congresses. | Teichmüller spaces–Congresses. | Manifolds (Mathematics)–Congresses. | AMS: Group theory and generalizations. msc | Functions of a complex variable. msc | Potential theory. msc | Several complex variables and analytic spaces. msc | Geometry. msc | Differential geometry. msc | Manifolds and cell complexes. msc | Global analysis, analysis on manifolds. msc | Probability theory and stochastic processes. msc

Classification: LCC QA331 .A462 2014 | DDC 510–dc23 LC record available at https://lccn.loc.gov/2017015094

Contemporary Mathematics ISSN: 0271-4132 (print); ISSN: 1098-3627 (online)

DOI: http://dx.doi.org/10.1090/conm/696

---



∞ The paper used in this book is acid-free and falls within the guidelines
established to ensure permanence and durability.
Visit the AMS home page at http://www.ams.org/

10 9 8 7 6 5 4 3 2 1 22 21 20 19 18 17

# Contents

# Preface

The Ahlfors–Bers Colloquia commemorate the mathematical legacy of Lars Ahlfors and Lipman Bers. The core of this legacy lies in the fields of geometric function theory, Teichmüller theory, hyperbolic geometry, and partial differential equations. Today we see the influence of Ahlfors and Bers on algebraic geometry, mathematical physics, dynamics, probability, geometric group theory, number theory and topology. Recent years have seen a flowering of this legacy with an increased interest in their work.

The sixth Ahlfors–Bers Colloquium took place at Yale University in October 2014. There were ten plenary speakers, 39 contributed talks, and 124 registered participants. The organizing committee members were Kari Astala, Ara Basmajian, Mario Bonk, William Goldman, Ursula Hamenstadt, Michael Kapovich, Steve Kerckhoff, Irwin Kra, Vlad Markovic, Yair Minsky, Alan Reid, Rubi Rodriguez, Steffen Rohde, and Anna Wienhard.

The conference website

https://sites.google.com/site/ahlforsberscolloquium/

contains the full schedule and videos of all the plenary talks.

This current volume contains articles on a wide variety of subjects that are central to this legacy. These include papers in Kleinian groups, classical Riemann surface theory, Teichmüller theory, mapping class groups, geometric group theory and statistical mechanics. The majority of the papers present new research, but there are survey articles as well.

There have been six volumes of these colloquia previously published by the AMS in the Contemporary Mathematics Series: volumes 211, 256, 355, 432, 510, and 590.

As this volume was going to press we were informed of the death of Clifford Earle on June 12, 2017. Cliff Earle was a treasured member of the mathematics community in general and of the Ahlfors–Bers family in particular. His contributions and his company will be missed.

Ara Basmajian
Yair N. Minsky
Alan W. Reid

## List of Plenary Lectures

Stergios Antonakoudis (Cambridge University): The complex geometry of Teichmüller space and symmetric domains

Mladen Bestvina (University of Utah): Teichmüller theory in outer space

Jon Chaika (University of Utah): On isomorphism and disjointness of interval exchanges and flows on flat surfaces.

David Dumas (UIC): Computing the image of Thurston's skinning map

Peter Haïssinsky (Toulouse): Quasi-isometric rigidity of the class of convex-cocompact Kleinian groups

Jeremy Kahn (CUNY Graduate Center): Surface subgroups, cube complexes, and the virtual Haken theorem

Sarah Koch (University of Michigan): Roots of polynomials and parameter spaces

Nick Makarov (Caltech): The circle and the cardioid

Kasra Rafi (University of Toronto): Rigidity of Teichmüller space

Wendelin Werner (ETH Zürich): Conformal invariance and critical behavior within critical fractal carpets

Contemporary Mathematics
Volume **696**, 2017
http://dx.doi.org/10.1090/conm/696/14013

# The complex geometry of Teichmüller spaces and bounded symmetric domains

Stergios M. Antonakoudis

## 1. Introduction

We study isometric maps between Teichmüller spaces $\mathcal{T}_{g,n} \subset \mathbb{C}^{3g-3+n}$ and bounded symmetric domains $\mathcal{B} \subset \mathbb{C}^N$ in their intrinsic Kobayashi metric. From a complex analytic perspective, these two important classes of geometric spaces have several features in common, but also exhibit many differences. The focus here is on recent results proved by the author; we give a list of open questions at the end.

In a nutshell, we will see that Teichmüller spaces, equipped with their intrinsic Kobayashi metric, exhibit a remarkable rigidity property reminiscent of rank one bounded symmetric domains; in particular, we will show that isometric disks are Teichmüller disks. However, we will see that Teichmüller spaces and bounded symmetric domains *do not mix* isometrically, so long as both have dimension two or more.

The proofs of these results, although technically different, use the common theme of *complexification* and *realification*; they also involve ideas from geometric topology.

## 2. The setting

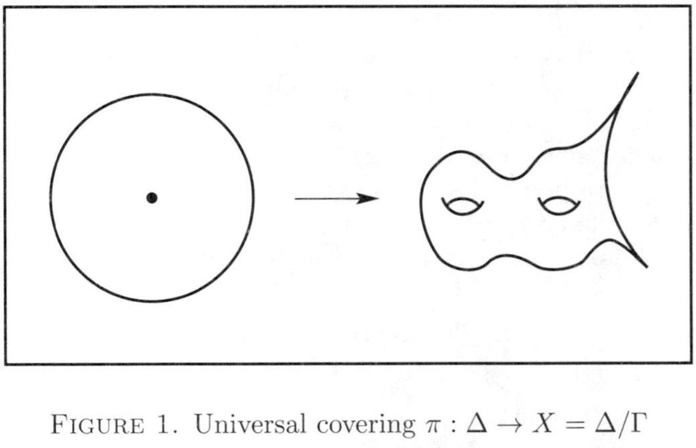

FIGURE 1. Universal covering $\pi : \Delta \to X = \Delta/\Gamma$

Let $X$ be a hyperbolic Riemann surface of finite type homeomorphic to a fixed oriented topological surface $\Sigma_{g,n}$ of genus $g$ with $n$ punctures. More concretely, the Riemann surface $X$ can be presented as a quotient space $X = \Delta/\Gamma$, where $\Gamma \leq \mathrm{Aut}(\Delta)$ is discrete group of automorphisms of the unit disk $\Delta \cong \{ z \in \mathbb{C} : |z| < 1 \}$ and $\pi : \Delta \to X = \Delta/\Gamma$ is the universal covering map.

The unit disk $\Delta$ is equipped with a metric $|dz|/(1-|z|^2)$ of constant curvature, known as the Poincaré metric, which we shall denote by $\mathbb{CH}^1$ and refer to as the complex hyperbolic line. The group $\mathrm{Aut}(\Delta)$ can be identified with the group $\mathrm{Isom}^+(\mathbb{CH}^1)$ of orientation preserving isometries of $\mathbb{CH}^1$; hence, we can endow $X = \Delta/\Gamma$ with a finite-volume metric of constant curvature.

The moduli space $\mathcal{M}_{g,n}$ parametrizing isomorphism classes of Riemann surfaces $X$ has a similar description: it is a complex quasi-projective variety, that can be presented as a quotient space $\mathcal{M}_{g,n} = \mathcal{T}_{g,n}/\mathrm{Mod}_{g,n}$, where $\mathrm{Mod}_{g,n} \leq \mathrm{Aut}(\mathcal{T}_{g,n})$ is a discrete group of automorphisms of a contractible, bounded domain of holomorphy $\mathcal{T}_{g,n} \subset \mathbb{C}^{3g-3+n}$ [**Bers**].

Teichmüller space $\mathcal{T}_{g,n}$, which parametrizes isomorphism classes of marked Riemann surfaces, is therefore the orbifold universal cover of the moduli space of Riemann surfaces $\mathcal{M}_{g,n}$ and it is naturally a complex manifold of dimension $3g-3+n$. It is equipped with a complete intrinsic metric, known as the Teichmüller metric, which endows $\mathcal{M}_{g,n}$ with the structure of a finite-volume complex orbifold.

Classically, another class of complex spaces admitting a similar description is that of locally symmetric varieties $\mathcal{V}$ of non-compact type, which can be presented as a quotient $\mathcal{V} = \mathcal{B}/\Gamma$, where $\Gamma \leq \mathrm{Aut}(\mathcal{B})$ is a lattice, a discrete group of automorphisms of a bounded symmetric domain $\mathcal{B} \subset \mathbb{C}^N$ [**Hel**, §VIII].

Let $\mathcal{B} \subset \mathbb{C}^N$ be a bounded domain; we call $\mathcal{B}$ a bounded symmetric domain if every point $p \in \mathcal{B}$ is an *isolated* fixed point of a holomorphic involution $\sigma_p : \mathcal{B} \to \mathcal{B}$, with $\sigma_p^2 = \mathrm{id}_{\mathcal{B}}$. Bounded symmetric domains are contractible and homogeneous as complex manifolds. It is classically known that all Hermitian symmetric spaces of non-compact type can be realized as bounded symmetric domains $\mathcal{B} \subset \mathbb{C}^N$ by the Harish-Chandra embeddings. The simplest example is given by the unit disk $\Delta \cong \mathbb{CH}^1$, which is in fact the unique (up to isomorphism) contractible bounded domain of complex dimension one.

A feature that Teichmüller spaces and bounded symmetric domains have in common is that they contain holomorphic isometric copies of $\mathbb{CH}^1$ through every point and complex direction; in particular, in complex dimension one, Teichmüller spaces and bounded symmetric domains coincide. However, in higher dimensions, the situation is quite different. [**Roy**] Royden proved that, when $\dim_{\mathbb{C}} \mathcal{T}_{g,n} \geq 2$, $\mathrm{Aut}(\mathcal{T}_{g,n})$ is a discrete group and, therefore, $\mathcal{T}_{g,n}$ is not a symmetric space. Central to Royden's work was the use of the intrinsic Kobayashi metric of $\mathcal{T}_{g,n}$.

## 3. The Kobayashi metric

Let $\mathcal{B} \subset \mathbb{C}^N$ be a bounded domain. Its intrinsic Kobayashi metric is the *largest* complex Finsler metric such that every holomorphic map $f : \mathbb{CH}^1 \to \mathcal{B}$ is non-expanding: $||df||_{\mathcal{B}} \leq 1$. [**Ko**] It determines both a family of norms $||\cdot||_{\mathcal{B}}$ on the tangent bundle $T\mathcal{B}$ and a distance $d_{\mathcal{B}}(\cdot,\cdot)$ on pairs of points.

We recall that the Schwarz-Pick lemma [**Ko**, §2] shows that every holomorphic map $f : \mathbb{CH}^1 \to \mathbb{CH}^1$ is non-expanding. The Kobayashi metric provides a natural generalisation - it has the fundamental property that every holomorphic map between complex domains is non-expanding and therefore every holomorphic automorphism is an isometry. In particular, the Kobayashi metric of complex domain depends only on its structure as a complex manifold.

**Examples.**

(1) $\mathbb{CH}^1 \cong \{ z \in \mathbb{C} : |z| < 1 \}$ with its Poincaré metric $|dz|/(1-|z|^2)$ coincides with the Kobayashi metric, by Schwarz's lemma. More generally, the Kobayashi metric of the ball $\mathbb{CH}^N \cong \{ (z_i)_{i=1}^N \in \mathbb{C}^N : \sum_{i=1}^N |z_i|^2 < 1 \}$ coincides with its complete invariant (Kähler) metric of constant holomorphic curvature -4. See [**Ko**, Example 3.1.24].

(2) The Kobayashi metric of the bi-disk $\mathbb{CH}^1 \times \overline{\mathbb{CH}^1}$ is the maximum metric of the two factors. It is a complex Finsler metric, which is not Hermitian. The distance function is given by $d_{\mathbb{CH}^1 \times \overline{\mathbb{CH}^1}}((z_1, z_2), (w_1, w_2)) = \max\{d_{\mathbb{CH}^1}(z_1, w_1), d_{\mathbb{CH}^1}(z_2, w_2)\}$, for points $(z_1, z_2), (w_1, w_2) \in \mathbb{CH}^1 \times \overline{\mathbb{CH}^1}$. See [**Ko**, Corollary 3.1.10].

(3) The Kobayashi metric of $\mathcal{T}_{g,n}$ coincides with its classical Teichmüller metric [**Roy**]; it endows $\mathcal{T}_{g,n}$ with the structure of a complete geodesic metric space. We will discuss this example in more detail in §5.

Incidentally, examples (1) and (2) describe all bounded symmetric domains, up to isomorphism, in complex dimensions one and two.

## 4. Main results

An important feature of the Kobayashi metric of Teichmüller space is that every holomorphic map $f : \mathbb{CH}^1 \hookrightarrow \mathcal{T}_{g,n}$ such that $df$ is an isometry on tangent spaces is *totally geodesic*: it sends real geodesics to real geodesics preserving their length. Moreover, there are such holomorphic isometries, known as Teichmüller disks, through every point in every complex direction.

**Holomorphic rigidity.** Our first result is the following: [1]

THEOREM 4.1. *Every totally geodesic isometry $f : \mathbb{CH}^1 \hookrightarrow \mathcal{T}_{g,n}$ for the Kobayashi metric is either holomorphic or anti-holomorphic; in particular, it is a Teichmüller disk.*

Informally, our theorem shows that the intrinsic Teichmüller-Kobayashi metric of $\mathcal{T}_{g,n}$ determines its natural structure as a complex manifold.

It is classically known that this result is also true for bounded symmetric domains with rank one; however, it is not true for bounded symmetric domains with rank two or more. In [**SMA2**, Theorem 4.2] we show that Theorem 4.1 is true for a large class of *disk-rigid* domains, including strictly convex bounded domains, as well as Teichmüller spaces.

[1] Theorem 4.1 solves problem 5.3 from [**FM**].

The following corollary follows easily from the theorem above.

COROLLARY 4.2. *Every totally geodesic isometry $f : \mathcal{T}_{g,n} \hookrightarrow \mathcal{T}_{h,m}$ is either holomorphic or anti-holomorphic.*

We note that there are many holomorphic isometries $f : \mathcal{T}_{g,n} \hookrightarrow \mathcal{T}_{h,m}$ between Teichmüller spaces $\mathcal{T}_{g,n}, \mathcal{T}_{h,m}$ in their Kobayashi metric, induced by pulling back complex structures from a fixed topological covering map $\psi : \Sigma_{h,m} \to \Sigma_{g,n}$ between the underlying topological surfaces $\Sigma_{g,n}, \Sigma_{h,m}$ [**FaMa**, See discussion in §13.2.1]. In a coming paper [**SMA3**], we prove that all such isometries arise in this way.

**Symmetric spaces vs Teichmüller spaces.** [**SMA1**, §6] Like Teichmüller spaces, there are also many holomorphic isometries $f : \mathcal{B} \hookrightarrow \widetilde{\mathcal{B}}$ between bounded symmetric domains $\mathcal{B}, \widetilde{\mathcal{B}}$ in their Kobayashi metric. However, in dimension two or more, Teichmüller spaces and bounded symmetric domains *do not mix* isometrically. More precisely, we have [**SMA5**]:

THEOREM 4.3. *Let $\mathcal{B}$ be a bounded symmetric domain and $\mathcal{T}_{g,n}$ be a Teichmüller space with $dim_{\mathbb{C}}\mathcal{B}, dim_{\mathbb{C}}\mathcal{T}_{g,n} \geq 2$. There are no holomorphic isometric immersions*

$$\mathcal{B} \xhookrightarrow{f} \mathcal{T}_{g,n} \quad or \quad \mathcal{T}_{g,n} \xhookrightarrow{f} \mathcal{B}$$

*equivalently, there are no holomorphic maps $f$ such that $df$ is an isometry for the Kobayashi norms on tangent spaces.*

We record the following special case, which we shall prove in §7.

THEOREM 4.4. *There is no holomorphic isometry $f : \mathbb{CH}^2 \hookrightarrow \mathcal{T}_{g,n}$ for the Kobayashi metric.*

We also have a similar result for submersions [**SMA1**, §6], [**SMA4**]:

THEOREM 4.5. *Let $\mathcal{B}$ and $\mathcal{T}_{g,n}$ be as in Theorem 4.3. There are no holomorphic isometric submersions*

$$\mathcal{B} \xrightarrow{g} \mathcal{T}_{g,n} \quad or \quad \mathcal{T}_{g,n} \xrightarrow{g} \mathcal{B}$$

*equivalently, there are no holomorphic maps $g$ such that $dg^*$ is an isometry for the dual Kobayashi norms on cotangent spaces.*

REMARK 4.6. The existence of isometrically immersed curves in $\mathcal{M}_{g,n}$, known as Teichmüller curves, has far-reaching applications to the dynamics of billiards in rational polygons [**V**], [**Mc2**]. The following immediate Corollary of Theorem 4.3 shows that there are no higher dimensional, locally symmetric, analogues of Teichmüller curves.

COROLLARY 4.7. *There is no locally symmetric variety $\mathcal{V}$ isometrically immersed in the moduli space of curves $\mathcal{M}_{g,n}$, nor is there an isometric copy of $\mathcal{M}_{g,n}$ in $\mathcal{V}$, for the Kobayashi metrics, so long as both have dimension two or more.*

REMARKS 4.8. The Torelli map, associating to a Riemann surface its Jacobian, is a holomorphic map $\mathcal{T}_{g,n} \xrightarrow{\tau} \mathcal{H}_g$ of Teichmüller space to a bounded symmetric domain (Siegel space). It is known that Torelli maps are isometric, for the Kobayashi metric, in some directions [**Kra**], but strictly contracting in most directions [**Mc4**].

It is known that there are holomorphic isometric submersions $\mathcal{T}_{g,n} \xrightarrow{g} \mathbb{CH}^1$, which are of the form $g = \rho \circ \tau$, where $\tau$ is the Torelli map $\mathcal{T}_{g,n} \xrightarrow{\tau} \mathcal{H}_g$ and $\mathcal{H}_g \xrightarrow{\rho} \mathbb{CH}^1$ is a holomorphic isometric submersion; see [**Mc3**, Theorem 5.1].

Further references, details and proofs of the results we mention in §4 appear in [**SMA1**],[**SMA2**],[**SMA4**] and [**SMA5**]. In this paper, we focus on explaining the proofs of Theorem 4.1 and Theorem 4.4 using the common theme of *complexification* and *realification*. We start with some preliminaries on Teichmüller spaces and their complex and real geodesics in the intrinsic Teichmüller-Kobayashi metric.

## 5. Preliminaries in Teichmüller theory

**Teichmüller space.** [**GL**], [**Hub**] Let $\Sigma_{g,n}$ be a connected, oriented surface of genus $g$ and $n$ punctures and $\mathcal{T}_{g,n}$ denote the Teichmüller space of Riemann surfaces marked by $\Sigma_{g,n}$. A point in $\mathcal{T}_{g,n}$ is specified by an equivalence class of orientation preserving homeorphisms $\phi : \Sigma_{g,n} \to X$ to a Riemann surface of finite type, where we consider two marked Riemann surfaces $\phi : \Sigma_{g,n} \to X$, $\psi : \Sigma_{g,n} \to Y$ equivalent if $\psi \circ \phi^{-1} : X \to Y$ is isotopic to a holomorphic bijection.

Teichmüller space $\mathcal{T}_{g,n}$ is naturally a complex manifold of dimension $3g-3+n$ and forgetting the marking realises $\mathcal{T}_{g,n}$ as the complex orbifold universal cover of moduli space $\mathcal{M}_{g,n}$. When it is clear from the context we often denote a point specified by $\phi : \Sigma_{g,n} \to X$ simply by $X$.

For each $X \in \mathcal{T}_{g,n}$, we let $Q(X)$ denote the space of holomorphic quadratic differentials $q = q(z)(dz)^2$ on $X$ with finite total mass: $||q||_1 = \int_X |q(z)||dz|^2 < +\infty$, which is equivalent to $q$ having at worse simple poles at the punctures of $X$.

The tangent and cotangent spaces to Teichmüller space at $X \in \mathcal{T}_{g,n}$ are described in terms of the natural pairing $(q, \mu) \mapsto \int_X q\mu$ between the space $Q(X)$ and the space $M(X)$ of $L^\infty$-measurable Beltrami differentials on $X$; in particular, the tangent $T_X\mathcal{T}_{g,n}$ and cotangent $T^*_X\mathcal{T}_{g,n}$ spaces are naturally isomorphic to $M(X)/Q(X)^\perp$ and $Q(X)$, respectively.

The Teichmüller-Kobayashi metric on $\mathcal{T}_{g,n}$ is given by norm duality on the tangent space $T_X\mathcal{T}_{g,n}$ from the norm $||q||_1 = \int_X |q|$ on the cotangent space $Q(X)$ at $X$. The corresponding distance function is given by the formula $d_{\mathcal{T}_{g,n}}(X,Y) = \inf \frac{1}{2}\log K(\phi)$ and measures the minimal dilatation $K(\phi)$ of a quasiconformal map $\phi : X \to Y$ respecting their markings.

**Measured foliations.** Let $\mathcal{MF}_{g,n}$ denote the space of equivalent classes[2] of nonzero (singular) measured foliations on $\Sigma_{g,n}$. [**FLP**] It is known that $\mathcal{MF}_{g,n}$ has the structure of a *piecewise linear* manifold homeomorphic to $\mathbb{R}^{6g-6+2n} \setminus \{0\}$.

We denote the geometric intersection number of a pair of measured foliations $\mathcal{F}, \mathcal{G}$ by $i(\mathcal{F}, \mathcal{G})$. [**Bon**] It induces a continuous map $i(\cdot,\cdot) : \mathcal{MF}_{g,n} \times \mathcal{MF}_{g,n} \to \mathbb{R}_{\geq 0}$, which extends the geometric intersection pairing on pairs of (isotopy classes of) simple closed curves on $\Sigma_{g,n}$.

[2] Two measured foliations $\mathcal{F}, \mathcal{G}$ are equivalent $\mathcal{F} \sim \mathcal{G}$ if they differ by a finite sequence of Whitehead moves followed by an isotopy of $\Sigma_{g,n}$, preserving their transverse measures.

Given $\mathcal{F} \in \mathcal{MF}_{g,n}$ and $X \in \mathcal{T}_{g,n}$, we denote the *extremal length* of $\mathcal{F}$ on the Riemann surface $X$ by $\lambda(\mathcal{F}, X)$; it is given by $\lambda(\mathcal{F}, X) = \sup \frac{\ell_\rho(\mathcal{F})^2}{\text{area}(\rho)}$, where $\ell_\rho(\mathcal{F})$ denotes the $\rho$-length of $\mathcal{F}$ and the supremum is over all (Borel-measurable) conformal metrics $\rho$ of finite area on $X$.

Every nonzero quadratic differential $q \in Q(X)$ induces a conformal metric $|q|$ on $X$, which is non-singular of zero curvature away from the zeros of $q$, and a measured foliation $\mathcal{F}(q)$ tangent to vectors $v = v(z)\frac{\partial}{\partial z}$ with $q(v) = q(z)(v(z))^2 < 0$. The transverse measure of the foliation $\mathcal{F}(q)$ is (locally) given by integrating $|\text{Re}(\sqrt{q})|$ along arcs transverse to its leaves.

We refer to $\mathcal{F}(q)$ as the vertical measured foliation induced by the pair $(X, q)$. In local coordinates where $q = dz^2$ (such coordinates exist away from the zeros of $q$), the metric $|q|$ coincides with the Euclidean metric $|dz|^2$ and the measured foliation $\mathcal{F}(q)$ has leaves given by vertical lines and transverse measure by the total horizontal variation $|\text{Re}(dz)|$. We note that the (horizontal) measured foliation $\mathcal{F}(-q)$ has leaves orthogonal to $\mathcal{F}(q)$ and the product of their transverse measures is just the area form of the conformal metric $|q|$.

When it is clear from the context we often identify the measured foliation $\mathcal{F}(q)$ with its equivalence class in $\mathcal{MF}_{g,n}$. The following fundamental theorem relates quadratic differentials and measured foliations on a fixed Riemann surface.

THEOREM 5.1. *([**HM**];Hubbard-Masur) Let $X \in \mathcal{T}_{g,n}$; the map $q \mapsto \mathcal{F}(q)$ induces a homeomorphism $Q(X) \setminus \{0\} \cong \mathcal{MF}_{g,n}$. Moreover, $|q|$ is the unique extremal metric for $\mathcal{F}(q)$ on $X$ and its extremal length is given by the formula $\lambda(\mathcal{F}(q), X) = ||q||_1$.*

**Complex geodesics.** We denote by $Q\mathcal{T}_{g,n} \cong T^*\mathcal{T}_{g,n}$ the complex vector-bundle of holomorphic quadratic differentials over $\mathcal{T}_{g,n}$ and by $Q_1\mathcal{T}_{g,n}$ the associated sphere-bundle of quadratic differentials with unit mass. There is a natural norm-preserving action of $\text{SL}_2(\mathbb{R})$ on $Q\mathcal{T}_{g,n}$, with the diagonal subgroup $\{\text{diag}(e^t, e^{-t})\}_{t\in\mathbb{R}}$ giving the co-geodesic flow of the Teichmüller-Kobayashi metric of $\mathcal{T}_{g,n}$.

For every $(X, q) \in Q_1\mathcal{T}_{g,n}$, the orbit $\text{SL}_2(\mathbb{R}) \cdot (X, q) \subset Q_1\mathcal{T}_{g,n}$ induces a holomorphic map $\mathbb{CH}^1 \cong \text{SO}_2(\mathbb{R}) \setminus \text{SL}_2(\mathbb{R}) \hookrightarrow \mathcal{T}_{g,n}$ that is a totally geodesic isometry for the Kobayashi metric. These *complex geodesics* are known as *Teichmüller disks* and they are abundant in Teichmüller space $\mathcal{T}_{g,n}$.

**Real geodesics.** Let $\gamma : [0, \infty) \to \mathcal{T}_{g,n}$ be a Teichmüller geodesic ray with unit speed; it has a unique lift $\tilde{\gamma}(t) = (X_t, q_t) \in Q_1\mathcal{T}_{g,n}$ such that $\gamma(t) = X_t$ and $\tilde{\gamma}(t) = \text{diag}(e^t, e^{-t}) \cdot (X_0, q_0)$ for $t \in \mathbb{R}_{\geq 0}$. The map $q \mapsto (\mathcal{F}(q), \mathcal{F}(-q))$ gives an embedding $Q\mathcal{T}_{g,n} \hookrightarrow \mathcal{MF}_{g,n} \times \mathcal{MF}_{g,n}$, that satisfies $||q||_1 = i(\mathcal{F}(q), \mathcal{F}(-q))$ and sends the canonical lift $\tilde{\gamma}(t) = (X_t, q_t)$ of $\gamma$ to a path of the form $(e^t\mathcal{F}(q), e^{-t}\mathcal{F}(-q))$.

Let $(X, q) \in Q_1\mathcal{T}_{g,n}$ generate the real Teichmüller geodesic $\gamma : \mathbb{R} \to \mathcal{T}_{g,n}$ with $\gamma(0) = X$. The real geodesic $\gamma$ extends uniquely to a holomorphic totally geodesic isometry $\gamma_{\mathbb{C}} : \Delta \cong \mathbb{CH}^1 \hookrightarrow \mathcal{T}_{g,n}$ satisfying $\gamma(t) = \gamma_{\mathbb{C}}(\tanh(t))$ for $t \in \mathbb{R}$; the real Teichmüller geodesic generated by the quadratic differential $e^{i\theta}q \in Q(X)$, for $\theta \in \mathbb{R}/2\pi\mathbb{Z}$, is given by $t \mapsto \gamma_{\mathbb{C}}(e^{-i\theta}\tanh(t))$, for $t \in \mathbb{R}$.

**Holomorphic disks.** The following two important results about holomorphic disks in Teichmüller space will be used in the proofs of Theorems 6.1 and 7.1 below.

The first result characterises holomorphic disks in Teichmüller space that are complex geodesics for the Kobayashi metric; see [**EKK**] for a simple proof, based on Slodkowski's theorem [**Sl**].

THEOREM 5.2. *Let* $f : \Delta \cong \mathbb{CH}^1 \to \mathcal{T}_{g,n}$ *be a holomorphic map that satisfies* $||f'(0)||_{\mathcal{T}_{g,n}} = 1$, *then* $f$ *is a totally geodesic isometry for the Kobayashi metric; in particular, it is a Teichmüller disk.*

The second result, which is a consequence of Sullivan's rigidity theorem [**Sul**], shows that there are no non-trivial holomorphic families of *essentially proper* holomorphic disks in Teichmüller space. See [**Tan**, Theorem 1] for a simple proof and [**Mc1**] for further applications and related ideas.

THEOREM 5.3. *Let* $\{f_t\}_{t\in\Delta}$ *be a holomorphic family of holomorphic maps* $f_t : \Delta \cong \mathbb{CH}^1 \to \mathcal{T}_{g,n}$ *and* $B_{f_0} \subset \partial\Delta$ *denote the set of bounded rays of* $f_0$, *i.e.* $B_{f_0} = \{ e^{i\theta} \in \partial\Delta \ : \ \sup_{t\in[0,1)} d_{\mathcal{T}_{g,n}}(f_0(0), f_0(te^{i\theta})) < +\infty \}$. *If* $\partial\Delta \setminus B_{f_0}$ *has positive (Lebesgue) measure, then the family is trivial:* $f_t = f_0$ *for all* $t \in \Delta$.

## 6. Holomorphic rigidity

In this section we prove:

THEOREM 6.1. *A totally geodesic isometry, for the Kobayashi metric,*

$$f : \mathbb{CH}^1 \hookrightarrow \mathcal{T}_{g,n}$$

*is either holomorphic or anti-holomorphic. In particular, it is a Teichmüller disk.*

The proof of the theorem uses the idea of *complexification* and leverages the following two facts: firstly, a complete real geodesic in $\mathcal{T}_{g,n}$ is contained in a unique Teichmüller disk and secondly, a holomorphic family $\{f_t\}_{t\in\Delta}$ of *essentially proper* holomorphic maps $f_t : \mathbb{CH}^1 \to \mathcal{T}_{g,n}$ is *trivial* i.e. $f_t = f_0$ for $t \in \Delta$.

**Outline of the proof.** Let $\gamma \subset \mathbb{CH}^1$ be a complete (real) geodesic and denote by $\gamma_{\mathbb{C}} \subset \mathbb{CH}^1 \times \overline{\mathbb{CH}^1}$ the *maximal* holomorphic extension of the diagonal embedding of $\gamma$ to the bi-disk. We note that $\gamma_{\mathbb{C}} \cong \mathbb{CH}^1$ and we define $F|_{\gamma_{\mathbb{C}}}$ to be the *unique* holomorphic extension of $f|_\gamma$ which is a Teichmüller disk.

Applying this construction to all real geodesics in $\mathbb{CH}^1$, we deduce that the isometry $f : \mathbb{CH}^1 \to \mathcal{T}_{g,n}$ extends to a map $F : \mathbb{CH}^1 \times \overline{\mathbb{CH}^1} \to \mathcal{T}_{g,n}$ such that $f(z) = F(z, z)$, for $z \in \Delta \cong \mathbb{CH}^1$, and prove that $F$ is holomorphic. Using that $f$ is a totally geodesic isometry, we show that the map $F$ is *essentially proper* and conclude by Theorem 5.3 that either $F(z, w) = F(z, z)$ or $F(z, w) = F(w, w)$ for $(z, w) \in \mathbb{CH}^1 \times \overline{\mathbb{CH}^1}$. □

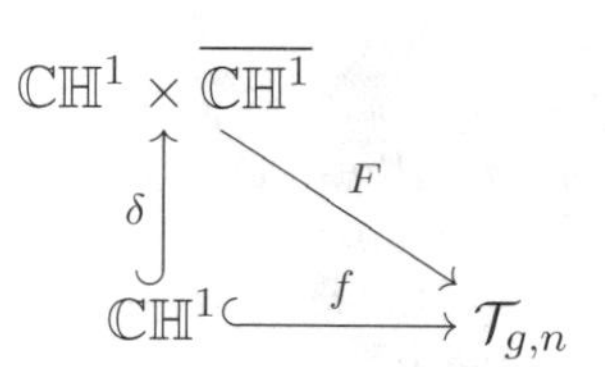


We start with some preliminary constructions.

**The totally real diagonal.** Let $\overline{\mathbb{CH}^1}$ be the complex hyperbolic line with its conjugate complex structure. The identity map is a *canonical* anti-holomorphic isomorphism $\mathbb{CH}^1 \cong \overline{\mathbb{CH}^1}$ and its graph is a totally real embedding $\delta : \mathbb{CH}^1 \hookrightarrow \mathbb{CH}^1 \times \overline{\mathbb{CH}^1}$, given by $\delta(z) = (z, z)$ for $z \in \Delta \cong \mathbb{CH}^1$. We call $\delta(\mathbb{CH}^1)$ the *totally real diagonal.*

**Geodesics and graphs of reflections.** Let $\mathcal{G}$ denote the set of all real, unoriented, complete geodesics $\gamma \subset \mathbb{CH}^1$. In order to describe their *maximal* holomorphic extensions $\gamma_{\mathbb{C}} \subset \mathbb{CH}^1 \times \overline{\mathbb{CH}^1}$, such that $\gamma_{\mathbb{C}} \cap \delta(\mathbb{CH}^1) = \delta(\gamma)$, it is convenient to parametrize $\mathcal{G}$ in terms of the set $\mathcal{R}$ of hyperbolic reflections of $\mathbb{CH}^1$ - or equivalently, the set of anti-holomorphic involutions of $\mathbb{CH}^1$. The map that associates a reflection $r \in \mathcal{R}$ with the set $\gamma = \mathrm{Fix}(r) \subset \mathbb{CH}^1$ of its fixed points gives a bijection between $\mathcal{R}$ and $\mathcal{G}$.

Let $r \in \mathcal{R}$ and denote its graph by $\Gamma_r \subset \mathbb{CH}^1 \times \overline{\mathbb{CH}^1}$; there is a natural holomorphic isomorphism $\mathbb{CH}^1 \cong \Gamma_r$, given by $z \mapsto (z, r(z))$ for $z \in \Delta \cong \mathbb{CH}^1$. We note that $\Gamma_r$ is the *maximal* holomorphic extension $\gamma_{\mathbb{C}}$ of the geodesic $\gamma = \mathrm{Fix}(r)$ to the bi-disk and it is *uniquely* determined by the property $\gamma_{\mathbb{C}} \cap \delta(\mathbb{CH}^1) = \delta(\gamma)$.

**The foliation by graphs of reflections.** The union of the graphs of reflections $\bigcup_{r\in\mathcal{R}} \Gamma_r$ gives rise to a (singular) foliation of $\mathbb{CH}^1 \times \overline{\mathbb{CH}^1}$ with holomorphic leaves $\Gamma_r$ parametrized by the set $\mathcal{R}$. We have $\Gamma_r \cap \delta(\mathbb{CH}^1) = \delta(\mathrm{Fix}(r))$ for all $r \in \mathcal{R}$, and

$$\Gamma_r \cap \Gamma_s = \delta(\mathrm{Fix}(r) \cap \mathrm{Fix}(s)) \tag{6.1}$$

which is either empty or a single point for all $r, s \in \mathcal{R}$ with $r \neq s$. In particular, the foliation is smooth in the complement of the totally real diagonal $\delta(\mathbb{CH}^1)$.

We emphasize that the following simple observation plays a key role in the proof of the theorem. For all $r \in \mathcal{R}$:

$$(z, w) \in \Gamma_r \iff (w, z) \in \Gamma_r \tag{6.2}$$

**Geodesics and the Klein model.** The Klein model gives a real-analytic identification $\mathbb{CH}^1 \cong \mathbb{RH}^2 \subset \mathbb{R}^2$ with an open disk in $\mathbb{R}^2$. It has the nice property that the hyperbolic geodesics are affine straight lines intersecting the disk. [**Rat**]

REMARK 6.2. The holomorphic foliation by graphs of reflections defines a *canonical* complex structure in a neighborhood of the zero section of the tangent bundle of $\mathbb{RH}^2$.

The description of geodesics in the Klein model is convenient in the light of the following theorem.

THEOREM 6.3. *(S. Bernstein) Let $f : [0,1]^2 \to \mathbb{C}$ be a function on $[0,1]^2 \subset \mathbb{R}^2 \subset \mathbb{C}^2$ and let $E \subset \mathbb{C}$ denote an ellipse with foci at 0 and 1. If for every vertical and every horizontal unit line segment $\ell \cong [0,1]$, there is a holomorphic function $F_\ell : E_\ell \to \mathbb{C}$, defined on the ellipse $E_\ell$ with foci at the end-points of $\ell$ and congruent to $E$, that agrees with $f$ along the line segment $\ell$: $f|_\ell \cong F_\ell|_\ell$, then the function $f$ has a (unique germ of) holomorphic extension in a neighborhood of $[0,1]^2$ in $\mathbb{C}^2$.*

See [**AhRo**] for a simple proof of the theorem [3]; and [**JP**, Theorem 5.6.5], [**BZ**, Theorem 3.4 & Corollary 3.6] for a more general result and discussion of related ideas.

We use Theorem 6.3 to prove:

LEMMA 6.4. *Every totally geodesic isometry* $f : \mathbb{CH}^1 \hookrightarrow \mathcal{T}_{g,n}$ *admits a unique holomorphic extension in a neighborhood of the totally real diagonal* $\delta(\mathbb{CH}^1) \subset \mathbb{CH}^1 \times \overline{\mathbb{CH}^1}$.

PROOF OF 6.4. Using the fact that analyticity is a local property and the description of geodesics in the Klein model of $\mathbb{RH}^2$, we can assume - without loss of generality - that the map $f$ is defined in a neighborhood of the unit square $[0,1]^2$ in $\mathbb{R}^2$ and has the property that its restriction on every horizontal and vertical line segment $\ell \cong [0,1]$ is a real-analytic parametrization of a Teichmüller geodesic segment. Moreover, we can also assume that the lengths of all these segments, measured in the Teichmüller metric, are uniformly bounded from above and from below away from zero.

Since every segment of a Teichmüller geodesic extends to a holomorphic Teichmüller disk in $\mathcal{T}_{g,n}$, there exists an ellipse $E \subset \mathbb{C}$ with foci at 0,1 such that the restrictions $f|_\ell$ extend to holomorphic maps $F_\ell : E \to \mathcal{T}_{g,n}$ for all horizontal and vertical line segments $\ell \cong [0,1]$ of $[0,1]^2$. Hence, the proof of the lemma follows from Theorem 6.3.

□

**Proof of Theorem 6.1.**

Let $f : \mathbb{CH}^1 \hookrightarrow \mathcal{T}_{g,n}$ be a totally geodesic isometry. Applying Lemma 6.4, we deduce that $f$ has a *unique* holomorphic extension in a neighborhood of the totally real diagonal $\delta(\mathbb{CH}^1) \subset \mathbb{CH}^1 \times \overline{\mathbb{CH}^1}$. We will show that $f$ extends to a holomorphic map from $\mathbb{CH}^1 \times \overline{\mathbb{CH}^1}$ to $\mathcal{T}_{g,n}$.

We start by defining a *new* map $F : \mathbb{CH}^1 \times \overline{\mathbb{CH}^1} \to \mathcal{T}_{g,n}$, satisfying:

1. $F(z,z) = f(z)$ for all $z \in \Delta \cong \mathbb{CH}^1$.
2. $F|_{\Gamma_r}$ is the *unique* holomorphic extension of $f|_{\mathrm{Fix}(r)}$ for all $r \in \mathcal{R}$.

Let $r \in \mathcal{R}$ be a reflection. There is a *unique* (holomorphic) Teichmüller disk $\phi_r : \mathbb{CH}^1 \hookrightarrow \mathcal{T}_{g,n}$ such that the intersection $\phi_r(\mathbb{CH}^1) \cap f(\mathbb{CH}^1) \subset \mathcal{T}_{g,n}$ contains the Teichmüller geodesic $f(\mathrm{Fix}(r))$ and $\phi_r(z) = f(z)$ for all $z \in \mathrm{Fix}(r)$. We define $F$ by $F(z, r(z)) = \phi_r(z)$ for $z \in \mathbb{CH}^1$ and $r \in \mathcal{R}$; equation (6.1) shows that $F$ is well-defined and satisfies conditions (1) and (2) above.

We *claim* that $F : \mathbb{CH}^1 \times \overline{\mathbb{CH}^1} \to \mathcal{T}_{g,n}$ is the *unique* holomorphic extension of the map $f : \mathbb{CH}^1 \hookrightarrow \mathcal{T}_{g,n}$ such that $F(z,z) = f(z)$ for $z \in \mathbb{CH}^1$.

*Proof of claim.* We note that the restriction of $F$ on the totally real diagonal $\delta(\mathbb{CH}^1)$ agrees with $f$ and that there is a *unique* germ of holomorphic maps near $\delta(\mathbb{CH}^1)$ whose restriction on $\delta(\mathbb{CH}^1)$ coincides with $f$. Let us fix an element of this

[3] S. Bernstein's clever proof was to consider this as problem on the torus obtained from a branched covering $S^1 \times S^1 \to [0,1] \times [0,1]$ with Galois group $\mathbb{Z}/2\mathbb{Z} \times \mathbb{Z}/2\mathbb{Z}$, and use the symmetries of the pull back function to show that its Fourier coefficients vanish outside the first quadrant.

germ $\tilde{F}$ defined on a neighborhood $U \subset \mathbb{CH}^1 \times \overline{\mathbb{CH}^1}$ of $\delta(\mathbb{CH}^1)$. For every $r \in \mathcal{R}$, the restrictions of $F$ and $\tilde{F}$ on the intersection $U_r = U \cap \Gamma_r$ are holomorphic and equal along the real-analytic arc $U_r \cap \delta(\mathbb{CH}^1) \subset U_r$; hence they are equal on $U_r$. Since $\mathbb{CH}^1 \times \overline{\mathbb{CH}^1} = \bigcup_{r \in \mathcal{R}} \Gamma_r$, we conclude that $F|_U = \tilde{F}$ and, in particular, $F$ is holomorphic near the totally real diagonal $\delta(\mathbb{CH}^1)$. Since, in addition to this, $F$ is holomorphic along all the leaves $\Gamma_r$ of the foliation, we deduce [4] that it is holomorphic at all points of $\mathbb{CH}^1 \times \overline{\mathbb{CH}^1}$. □

In order to complete the proof, we recall the *key* observation (6.2) as follows: the points $(z, w)$ and $(w, z)$ are always contained in the same leaf $\Gamma_r$ of the foliation; and since the restriction of $F$ on every leaf $\Gamma_r$ is a Teichmüller disk, we conclude that: $d_{\mathcal{T}_{g,n}}(F(z,w), F(w,z)) = d_{\mathbb{CH}^1}(z,w)$ for all $z, w \in \Delta \cong \mathbb{CH}^1$. In particular, at least one of $F(\rho e^{i\theta}, 0)$ and $F(0, \rho e^{i\theta})$ diverges in $\mathcal{T}_{g,n}$, as $\rho \to 1$, for every $\theta \in \mathbb{R}/2\pi\mathbb{Z}$.

It follows that there is a subset $I \subset \mathbb{R}/2\pi\mathbb{Z}$ with (positive) Lebesgue measure at least $1/2$, so that either $F(\rho e^{i\theta}, 0)$ diverges as $\rho \to 1$, for all $\theta \in I$, or $F(0, \rho e^{i\theta})$ diverges as $\rho \to 1$, for all $\theta \in I$. If we assume that the former of the two is true, we deduce from Theorem 5.3 that the holomorphic family $\{F(z, \overline{w})\}_{w \in \Delta}$ of holomorphic maps $F(\cdot, \overline{w}) : \Delta \cong \mathbb{CH}^1 \to \mathcal{T}_{g,n}$ is *trivial*; therefore, $F(z, 0) = F(z, z) = f(z)$ for all $z \in \Delta$ and, in particular, $f$ is holomorphic. If we assume that the latter of the two is true, we similarly deduce that $F(0, z) = F(z, z) = f(z)$ for all $z \in \Delta$ and, in particular, $f$ is anti-holomorphic. □

## 7. Extremal length geometry

In this section we prove:

THEOREM 7.1. *There is no holomorphic isometry $f : \mathbb{CH}^2 \hookrightarrow \mathcal{T}_{g,n}$ for the Kobayashi metric.*

The proof of the theorem uses the idea of *realification* and leverages the fact that extremal length provides a link between the geometry of Teichmüller geodesics and the geometric intersection pairing for measured foliations.

**Outline of the proof.** Using Theorem 5.2, we deduce that such an isometry would be totally geodesic: it would send real geodesics in $\mathbb{CH}^2$ to real Teichmüller geodesics in $\mathcal{T}_{g,n}$ preserving their length. We can parametrize the set of Teichmüller geodesic rays from any base point $X \in \mathcal{T}_{g,n}$, using Theorem 5.1, by the subspace of measured foliations $\mathcal{F} \in \mathcal{MF}_{g,n}$ with extremal length $\lambda(\mathcal{F}, X) = 1$.

Assuming the existence of $f$, we consider pairs of measured foliations that parametrize orthogonal geodesic rays in the image of a *totally real* geodesic hyperbolic plane $\mathbb{RH}^2 \subset \mathbb{CH}^2$. We obtain a contradiction by computing their geometric intersection number in two different ways.

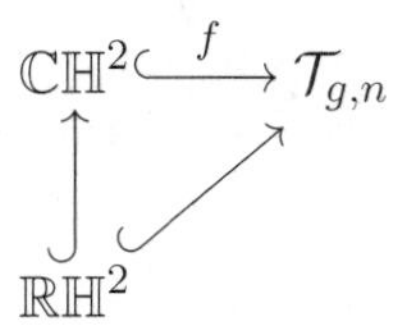


[4]For a simple proof of this, consider the power series expansion of $F$ at $(0,0) \in \mathbb{CH}^1 \times \overline{\mathbb{CH}^1}$; see [**Hör**, Lemma 2.2.11].

On the one hand, we use the geometry of complex hyperbolic horocycles and extremal length to show that the geometric intersection number does not depend on the choice of the totally real geodesic plane. On the other hand, by a direct geometric argument we show that this is impossible. More precisely, we prove:

PROPOSITION 7.2. *Let* $q \in Q_1\mathcal{T}_{g,n}$ *and* $\mathcal{G} \in \mathcal{MF}_{g,n}$, *there exist constants* $v_1, \ldots, v_N \in \mathbb{C}^*$ *such that* $i(\mathcal{F}(e^{i\theta}q), \mathcal{G}) = \sum_{i=1}^{N} |Re(e^{i\theta/2}v_i)|$ *for* $\theta \in \mathbb{R}/2\pi\mathbb{Z}$.

The proof of the proposition is given at the end of the section. □
We start with preliminaries on compex hyperbolic and extremal length horocycles.

**Complex hyperbolic horocycles.** Let $\gamma : [0, \infty) \to \mathbb{CH}^2$ be a geodesic ray with unit speed. Since $\mathbb{CH}^2$ is a homogeneous space, we have $\gamma = \alpha \circ \gamma_1$, where $\gamma_1(t) = (\tanh(t), 0)$, for $t \geq 0$, and $\alpha$ is a holomorphic isometry of $\mathbb{CH}^2$. Each geodesic ray $\gamma$ is contained in the image of unique holomorphic totally-geodesic isometry $\phi : \mathbb{CH}^1 \hookrightarrow \mathbb{CH}^2$, satisfying $\gamma(t) = \phi(\tanh(t))$, e.g. the ray $\gamma_1$ is contained in the image of $\phi_1(z) = (z, 0)$, for $z \in \Delta \cong \mathbb{CH}^1$. We also note that image of every complex geodesic $\phi : \mathbb{CH}^1 \hookrightarrow \mathbb{CH}^2$ arises uniquely as the intersection of the unit ball in $\mathbb{C}^2$ with a complex affine line.

Associated to each geodesic ray $\gamma : [0, \infty) \to \mathbb{CH}^2$ is a pair of transverse foliations of $\mathbb{CH}^2$, one by real geodesics asymptotic to $\gamma$ and another by complex hyperbolic horocycles asymptotic to $\gamma$. For each $p \in \mathbb{CH}^2$ there exists a *unique* geodesic $\gamma_p : \mathbb{R} \to \mathbb{CH}^2$ and a *unique* time $t_p \in \mathbb{R}$ such that $\gamma_p(t_p) = p$ and $\lim_{t\to\infty} d_{\mathbb{CH}^2}(\gamma(t), \gamma_p(t)) \to 0$. For each $s \in \mathbb{R}_+$, we define the set $H(\gamma, s) = \{ p \in \mathbb{CH}^2 \mid \exp(t_p) = s \}$. The collection of subsets $\{H(\gamma, s)\}_{s\in\mathbb{R}_+}$ defines the foliation of $\mathbb{CH}^2$ by *complex hyperbolic horocycles* asymptotic to $\gamma$.

**Extremal length horocycles.** Let $\gamma : [0, \infty) \to \mathcal{T}_{g,n}$ be a unit speed Teichmüller geodesic ray, with its unique lift $\tilde{\gamma}(t) = (X_t, q_t) \in Q_1\mathcal{T}_{g,n}$ such that $\gamma(t) = X_t$ and $\tilde{\gamma}(t) = \mathrm{diag}(e^t, e^{-t}) \cdot (X_0, q_0)$. We recall that $q \mapsto (\mathcal{F}(q), \mathcal{F}(-q))$ gives an embedding $Q\mathcal{T}_{g,n} \hookrightarrow \mathcal{MF}_{g,n} \times \mathcal{MF}_{g,n}$ that satisfies $||q||_1 = i(\mathcal{F}(q), \mathcal{F}(-q))$ and sends the lift $\tilde{\gamma}(t) = (X_t, q_t)$ of $\gamma$ to a path of the form $(e^t\mathcal{F}(q), e^{-t}\mathcal{F}(-q))$.

This description of Teichmüller geodesics and Theorem 5.1 show that the extremal length of $\mathcal{F}(q_t)$ along $\gamma$ satisfies $\lambda(\mathcal{F}(q_t), X_s) = e^{2(t-s)}$ for all $t, s \in \mathbb{R}_+$, which motivates the following definition: For each $\mathcal{F} \in \mathcal{MF}_{g,n}$ the *extremal length horocycles* asymptotic to $\mathcal{F}$ are the level-sets of extremal length $H(\mathcal{F}, s) = \{ X \in \mathcal{T}_{g,n} \mid \lambda(\mathcal{F}, X) = s \}$ for $s \in \mathbb{R}_+$. The collection of subsets $\{H(\mathcal{F}, s)\}_{s\in\mathbb{R}_+}$ defines the foliation of $\mathcal{T}_{g,n}$ by *extremal length horocycles* asymptotic to $\mathcal{F}$.

There is also a transverse foliation of $\mathcal{T}_{g,n}$ by real Teichmüller geodesics having canonical lifts $\{(X_t, q_t)\}_{t\in\mathbb{R}}$ that satisfy $\mathcal{F}(q_t) \in \mathbb{R}_+ \cdot \mathcal{F}$. We note that, although one might expect that this foliation of $\mathcal{T}_{g,n}$ is analogous to the foliation of $\mathbb{CH}^2$ by geodesics that are positively asymptotic to $\gamma$, this is not always true; however, it is true for *generic* measured foliations $\mathcal{F} \in \mathcal{MF}_{g,n}$.

THEOREM 7.3. *(*[**Mas**]*; H. Masur) Let* $(X_t, q_t)$ *and* $(Y_t, p_t)$ *be two Teichmüller geodesics and* $\mathcal{F}(q_0) \in \mathcal{MF}_{g,n}$ *be uniquely ergodic.* [5] *Then* $lim_{t\to\infty} d_{\mathcal{T}_{g,n}}(X_t, Y_t) \to 0$ *if and only if* $\mathcal{F}(q_0) = \mathcal{F}(p_0)$ *in* $\mathcal{MF}_{g,n}$ *and* $\lambda(\mathcal{F}(q_0), X_0) = \lambda(\mathcal{F}(p_0), Y_0)$.

---

[5] A measured foliation $\mathcal{F}$ is *uniquely ergodic* if it is minimal and admits a unique, up to scaling, transverse measure; in particular, $i(\gamma, \mathcal{F}) > 0$ for all simple closed curves $\gamma$. Compare with [**Mas**].

REMARK 7.4. It is known that this result is not true for measured foliations that are not uniquely ergodic.

**Proof of Theorem 7.1.**

Let $f : \mathbb{CH}^2 \hookrightarrow \mathcal{T}_{g,n}$ be a holomorphic isometry for the Kobayashi metric. We summarize the proof in the following three steps:

**1.** *Asymptotic behavior of geodesics determines the extremal length horocycles.*
**2.** *The geometry of horocycles determines the geometric intersection pairing.*
**3.** *Get a contradiction by a direct computation of the geometric intersection pairing.*

**Step 1.** Let $X = f((0,0)) \in \mathcal{T}_{g,n}$ and $q, p \in Q_1(X)$ unit area quadratic differentials generating the two Teichmüller geodesic rays $f(\gamma_1)$,$f(\gamma_2)$, where $\gamma_1$,$\gamma_2$ are two orthogonal geodesic rays in $\mathbb{CH}^2$ contained in the image of a totally real geodesic hyperbolic plane $\mathbb{RH}^2 \subset \mathbb{CH}^2$; more explicitly, we consider the two rays in $\mathbb{CH}^2$ given by $\gamma_1(t) = (\tanh(t), 0)$, $\gamma_2(t) = (0, \tanh(t))$, for $t \geq 0$.

For every $(X, q) \in Q_1\mathcal{T}_{g,n}$ there is a dense set of $\theta \in \mathbb{R}/2\pi\mathbb{Z}$ such that the measured foliation $\mathcal{F}(e^{i\theta}q)$ is uniquely ergodic [**KMS**, Theorem 2]; hence, we can assume without loss of generality (up to a holomorphic automorphism of $\mathbb{CH}^2$) that both $\mathcal{F}(q)$ and $\mathcal{F}(p)$ are (minimal) uniquely ergodic measured foliations. In particular, we can apply Theorem 7.3 to study the extremal length horocycles asymptotic to $\mathcal{F}(q)$ and $\mathcal{F}(p)$ respectively.

The complex hyperbolic horocycle $H(\gamma_1, 1)$ is characterized by the property that for the points $P \in H(\gamma_1, 1)$ the geodesic distance between $\gamma_P(t)$ and $\gamma_1(t)$ tends to zero as $t \to +\infty$, where $\gamma_P(t)$ is the unique geodesic with unit speed through $P$ that is positively asymptotic to $\gamma_1$. Using Theorem 7.3, we obtain that:

$$f(\mathbb{CH}^2) \cap H(\mathcal{F}(q), 1) = f(H(\gamma_1, 1)) \tag{7.1}$$

$$f(\mathbb{CH}^2) \cap H(\mathcal{F}(p), 1) = f(H(\gamma_2, 1)) \tag{7.2}$$

**Step 2.** Let $\delta$ be the complete real geodesic in $\mathbb{CH}^2$, which is asymptotic to $\gamma_1$ in the positive direction and to $\gamma_2$ in the negative direction, i.e. its two endpoints are $(1, 0), (0, 1) \in \mathbb{C}^2$ in the boundary of the unit ball. Let $P_1$ and $P_2$ be the two points where $\delta$ intersects the horocycles $H(\gamma_1, 1)$ and $H(\gamma_2, 1)$, respectively; see Figure 2.

The image of $\delta$ under the map $f$ is a Teichmüller geodesic which is determined by a pair of measured foliations $\mathcal{F}, \mathcal{G} \in \mathcal{MF}_{g,n}$ with $i(\mathcal{F}, \mathcal{G}) = 1$ and its unique lift to $Q_1\mathcal{T}_{g,n}$ is given by $(e^t\mathcal{F}, e^{-t}\mathcal{G})$, for $t \in \mathbb{R}$. Let $\widetilde{P}_i = (e^{t_i}\mathcal{F}, e^{-t_i}\mathcal{G})$, for $i = 1, 2$, denote the lifts of $P_1, P_2$ along the geodesic $\delta$. Then, the distance between the two points is given by $d_{\mathbb{CH}^2}(P_1, P_2) = t_2 - t_1$. From Step 1, we conclude that $e^{t_1}\mathcal{F} = \mathcal{F}(q)$ (7.1) and $e^{-t_2}\mathcal{G} = \mathcal{F}(p)$ ((7.2). Therefore we have $i(\mathcal{F}(q), \mathcal{F}(p)) = e^{t_1 - t_2}$.

REMARK 7.5. A simple calculation shows that $t_2 - t_1 = \log(2)$; hence

$$i(\mathcal{F}(q), \mathcal{F}(p)) = \frac{1}{2}.$$

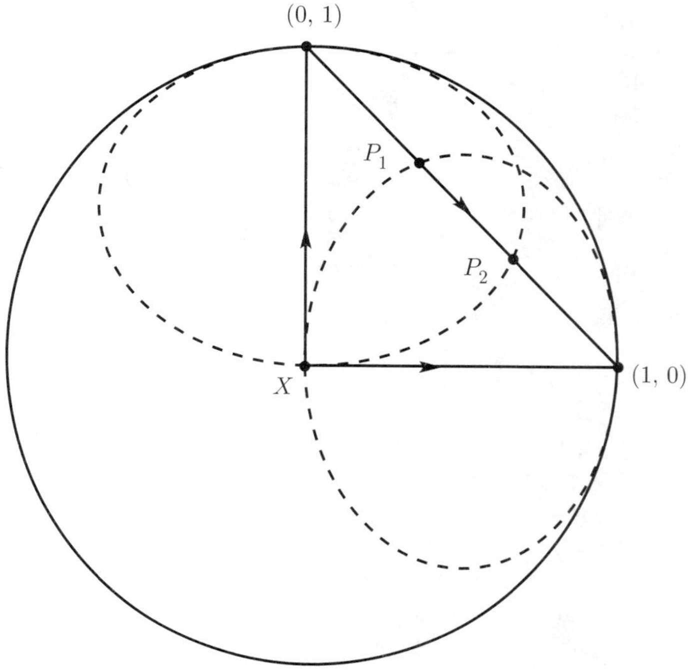


FIGURE 2. The real slice of $\mathbb{CH}^2 \subset \mathbb{C}^2$ coincides with the Klein model of the real hyperbolic plane $\mathbb{RH}^2 \subset \mathbb{R}^2$ with curvature -1.

**Step 3.** The holomorphic automorphism given by $\phi(z,w) = (e^{-i\theta}z, w)$, for $(z,w) \in \mathbb{CH}^2$, is an isometry of $\mathbb{CH}^2$ and sends the two horocycles $H(\gamma_i, 1)$ to the horocycles $H(\phi(\gamma_i), 1)$, for $i = 1, 2$. The Teichmüller geodesic ray $f(\phi(\gamma_1))$ is now generated by $e^{i\theta}q$, whereas the Teichmüller geodesic ray $f(\phi(\gamma_2))$ is still generated by $p \in Q(X)$. Since the distance between $P_1$ and $P_2$ is equal to the distance between $\phi(P_1)$ and $\phi(P_2)$, using Step 2 and the continuity of the geometric intersection pairing we conclude that $i(\mathcal{F}(e^{i\theta}q), \mathcal{G}) = \frac{1}{2}$ for all $\theta \in \mathbb{R}/2\pi\mathbb{Z}$. However, this contradicts the following Proposition 7.2. □

PROPOSITION 7.2. *Let $q \in Q_1\mathcal{T}_{g,n}$ and $\mathcal{G} \in \mathcal{MF}_{g,n}$, there exist constants $v_1, \ldots, v_N \in \mathbb{C}^*$ such that $i(\mathcal{F}(e^{i\theta}q), \mathcal{G}) = \sum_{i=1}^N |Re(e^{i\theta/2}v_i)|$ for $\theta \in \mathbb{R}/2\pi\mathbb{Z}$.*

PROOF OF PROPOSITION 7.2. Let $q \in Q(X)$ be a unit area quadratic differential. We assume first that $q$ has no poles and that $\mathcal{G}$ is an isotopy class of simple closed curves. The metric given by $|q|$ is flat with conical singularities of negative curvature at its set of zeros and hence the isotopy class of simple closed curves $\mathcal{G}$ has a unique geodesic representative, which is a finite union of saddle connections of $q$. In particular, we can readily compute $i(\mathcal{F}(e^{i\theta}q), \mathcal{G})$ by integrating $|\mathrm{Re}(\sqrt{e^{i\theta}q})|$ along the union of these saddle connections. It follows that:

$$i(\mathcal{F}(e^{i\theta}q), \mathcal{G}) = \sum_{i=1}^{N} |\mathrm{Re}(e^{i\theta/2}v_i)| \quad \text{for all} \quad \theta \in \mathbb{R}/2\pi\mathbb{Z}, \tag{7.3}$$

where $N$ is the number of the saddle connections and $\{v_i\}_{i=1}^N \subset \mathbb{C}^*$ their associated holonomy vectors. We note that when $q$ has simple poles, there need not be a geodesic representative in $\mathcal{G}$ anymore. Nevertheless, equation (7.3) is still true by applying the argument to a sequence of length minimizing representatives.

Finally, we observe that the number of saddle connections $N$ is bounded from above by a constant that depends only on the topology of the surface. Combining this observation with the fact that any $\mathcal{G} \in \mathcal{MF}_{g,n}$ is a limit of simple closed curves and that the geometric intersection pairing $i(\cdot,\cdot) : \mathcal{MF}_{g,n} \times \mathcal{MF}_{g,n} \to \mathbb{R}$ is continuous, we conclude that equation (7.3) is true in general. □

## 8. Final remarks

We conclude this discussion with a few open questions and further results.

**Questions.**

(1) Is Theorem 4.1 true for $f : \mathbb{CH}^1 \hookrightarrow \mathcal{T}_{g,n}$ a (real) $C^1$-smooth *local* isometry?

(2) Is there a *round* complex two-dimensional linear slice in $T_X\mathcal{T}_{g,n}$?

(3) Is there a holomorphic isometric immersion $f : (\mathcal{M}, g) \hookrightarrow \mathcal{T}_{g,n}$ from a Hermitian manifold with $\dim_{\mathbb{C}}\mathcal{M} \geq 2$?

(4) Is there a holomorphic retraction $\mathcal{T}_{g,n} \xrightarrow{g} \mathbb{CH}^1$ onto every Teichmüller disk $\mathbb{CH}^1 \xhookrightarrow{f} \mathcal{T}_{g,n}$ such that $g \circ f = id_{\mathbb{CH}^1}$? Equivalently, does the Caratheodory metric equal to the Kobayashi metric for every complex direction of $\mathcal{T}_{g,n}$?

**Further results.**

The following two results suggest that the answers to questions 2 & 3 are *no*.

THEOREM 8.1 ([**SMA4**]). *There is no complex linear isometry*

$$P : (\mathbb{C}^2, ||\cdot||_2) \hookrightarrow (Q(X), ||\cdot||_1).$$

REMARK 8.2. This result is used in the proof of Theorem 4.5.

As an application of Theorem 4.3, we prove:

THEOREM 8.3. *Let $(\mathcal{M}, g)$ be a complete Kähler manifold with $\dim_{\mathbb{C}}\mathcal{M} \geq 2$ and holomorphic sectional curvature at least $-4$. There is no holomorphic map $f : \mathcal{M} \to \mathcal{T}_{g,n}$ such that $df$ is an isometry on tangent spaces.*

PROOF. The monotonicity of holomorphic sectional curvature under holomorphic maps and the existence of (totally geodesic) holomorphic isometries $\mathbb{CH}^1 \hookrightarrow \mathcal{T}_{g,n}$ through every complex direction imply that $\mathcal{M}$ has constant holomorphic curvature -4. [**Roy**] Since $\mathcal{M}$ is a complete Kähler manifold, we have $\mathcal{M} \cong \mathbb{CH}^N$, which is impossible when $N \geq 2$ by Theorem 4.3. □

We also mention the following immediate corollaries of Theorem 8.1 and Theorem 8.3, respectively.

COROLLARY 8.4. *Let $(\mathcal{M}, g)$ be a Hermitian manifold with $\dim_{\mathbb{C}}\mathcal{M} \geq 2$. There is no holomorphic isometric submersion $g : \mathcal{T}_{g,n} \twoheadrightarrow \mathcal{M}$.*

COROLLARY 8.5. *There is no holomorphic, totally geodesic isometry from a Kähler manifold $\mathcal{M}$ into a Teichmüller space $\mathcal{T}_{g,n}$, so long as $\mathcal{M}$ has dimension two or more.*

Finally, we refer to [**Kra**], [**Mc4**] and [**FM**] for partial results and references towards question 4.

## References


[AhRo] N. I. Ahiezer and L. I. Ronkin, *Separately analytic functions of several variables, and "edge of the wedge" theorems* (Russian), Uspehi Mat. Nauk **28** (1973), no. 3(171), 27–42. MR0419829

[SMA1] S. M. Antonakoudis, *The complex geometry of Teichmüller space*, ProQuest LLC, Ann Arbor, MI, 2014. Thesis (Ph.D.)–Harvard University. MR3251342

[SMA2] S. M. Antonakoudis, *Isometric disks are holomorphic*, Invent. Math. **207** (2017), no. 3, 1289–1299, DOI 10.1007/s00222-016-0685-1. MR3608291

[SMA3] S. M. Antonakoudis. Isometric maps between Teichmüller spaces are geometric. *In preparation.*

[SMA4] S. M. Antonakoudis. Royden's theorem and birational geometry. *Preprint.*

[SMA5] S. M. Antonakoudis. Teichmüller spaces and bounded symmetric domains do not mix isometrically. *Preprint.*

[BZ] L. Baracco and G. Zampieri, *Separate real analyticity and CR extendibility*, Forum Math. **21** (2009), no. 3, 519–531, DOI 10.1515/FORUM.2009.025. MR2526798

[Bers] L. Bers, *Finite-dimensional Teichmüller spaces and generalizations*, Bull. Amer. Math. Soc. (N.S.) **5** (1981), no. 2, 131–172, DOI 10.1090/S0273-0979-1981-14933-8. MR621883

[Bon] F. Bonahon, *The geometry of Teichmüller space via geodesic currents*, Invent. Math. **92** (1988), no. 1, 139–162, DOI 10.1007/BF01393996. MR931208

[EKK] C. J. Earle, I. Kra, and S. L. Krushkalʹ, *Holomorphic motions and Teichmüller spaces*, Trans. Amer. Math. Soc. **343** (1994), no. 2, 927–948, DOI 10.2307/2154750. MR1214783

[FaMa] B. Farb and D. Margalit, *A primer on mapping class groups*, Princeton Mathematical Series, vol. 49, Princeton University Press, Princeton, NJ, 2012. MR2850125

[FLP] A. Fathi, F. Laudenbach, and V. Poénaru, *Travaux de Thurston sur les surfaces* (French), Astérisque, vol. 66, Société Mathématique de France, Paris, 1979. Séminaire Orsay; With an English summary. MR568308

[FM] A. Fletcher and V. Markovic, *Infinite dimensional Teichmüller spaces*, Handbook of Teichmüller theory. Vol. II, IRMA Lect. Math. Theor. Phys., vol. 13, Eur. Math. Soc., Zürich, 2009, pp. 65–91, DOI 10.4171/055-1/3. MR2497790

[GL] F. P. Gardiner and N. Lakic, *Quasiconformal Teichmüller theory*, Mathematical Surveys and Monographs, vol. 76, American Mathematical Society, Providence, RI, 2000. MR1730906

[Hel] S. Helgason, *Differential geometry, Lie groups, and symmetric spaces*, Pure and Applied Mathematics, vol. 80, Academic Press, Inc. [Harcourt Brace Jovanovich, Publishers], New York-London, 1978. MR514561

[Hör] L. Hörmander, *An introduction to complex analysis in several variables*, 3rd ed., North-Holland Mathematical Library, vol. 7, North-Holland Publishing Co., Amsterdam, 1990. MR1045639

[HM] J. Hubbard and H. Masur, *Quadratic differentials and foliations*, Acta Math. **142** (1979), no. 3-4, 221–274, DOI 10.1007/BF02395062. MR523212

[Hub] J. H. Hubbard. *Teichmüller Theory, vol. I.* Matrix Editions, 2006.

[JP] M. Jarnicki and P. Pflug, *Separately analytic functions*, EMS Tracts in Mathematics, vol. 16, European Mathematical Society (EMS), Zürich, 2011. MR2840986

[KMS] S. Kerckhoff, H. Masur, and J. Smillie, *Ergodicity of billiard flows and quadratic differentials*, Ann. of Math. (2) **124** (1986), no. 2, 293–311, DOI 10.2307/1971280. MR855297

[Ko] S. Kobayashi, *Hyperbolic complex spaces*, Grundlehren der Mathematischen Wissenschaften [Fundamental Principles of Mathematical Sciences], vol. 318, Springer-Verlag, Berlin, 1998. MR1635983

[Kra] I. Kra, *The Carathéodory metric on abelian Teichmüller disks*, J. Analyse Math. **40** (1981), 129–143 (1982), DOI 10.1007/BF02790158. MR659787

[Mas] H. Masur, *Uniquely ergodic quadratic differentials*, Comment. Math. Helv. **55** (1980), no. 2, 255–266, DOI 10.1007/BF02566685. MR576605

[Mc1] C. T. McMullen, *From dynamics on surfaces to rational points on curves*, Bull. Amer. Math. Soc. (N.S.) **37** (2000), no. 2, 119–140, DOI 10.1090/S0273-0979-99-00856-3. MR1713286

[Mc2] C. T. McMullen, *Billiards and Teichmüller curves on Hilbert modular surfaces*, J. Amer. Math. Soc. **16** (2003), no. 4, 857–885, DOI 10.1090/S0894-0347-03-00432-6. MR1992827

[Mc3] C. T. McMullen, *Foliations of Hilbert modular surfaces*, Amer. J. Math. **129** (2007), no. 1, 183–215, DOI 10.1353/ajm.2007.0002. MR2288740

[Mc4] C. T. McMullen, *Entropy on Riemann surfaces and the Jacobians of finite covers*, Comment. Math. Helv. **88** (2013), no. 4, 953–964, DOI 10.4171/CMH/308. MR3134416

[Rat] J. G. Ratcliffe, *Foundations of hyperbolic manifolds*, Graduate Texts in Mathematics, vol. 149, Springer-Verlag, New York, 1994. MR1299730

[Roy] H. L. Royden, *Automorphisms and isometries of Teichmüller space*, Advances in the Theory of Riemann Surfaces (Proc. Conf., Stony Brook, N.Y., 1969), Ann. of Math. Studies, No. 66. Princeton Univ. Press, Princeton, N.J., 1971, pp. 369–383. MR0288254

[Sl] Z. Slodkowski, *Holomorphic motions and polynomial hulls*, Proc. Amer. Math. Soc. **111** (1991), no. 2, 347–355, DOI 10.2307/2048323. MR1037218

[Sul] D. Sullivan, *On the ergodic theory at infinity of an arbitrary discrete group of hyperbolic motions*, Riemann surfaces and related topics: Proceedings of the 1978 Stony Brook Conference (State Univ. New York, Stony Brook, N.Y., 1978), Ann. of Math. Stud., vol. 97, Princeton Univ. Press, Princeton, N.J., 1981, pp. 465–496. MR624833

[Tan] H. Tanigawa, *Holomorphic mappings into Teichmüller spaces*, Proc. Amer. Math. Soc. **117** (1993), no. 1, 71–78, DOI 10.2307/2159700. MR1113649

[V] W. A. Veech, *Teichmüller curves in moduli space, Eisenstein series and an application to triangular billiards*, Invent. Math. **97** (1989), no. 3, 553–583, DOI 10.1007/BF01388890. MR1005006

Department of Mathematics, Harvard University, Cambridge, Massachusetts 02138

*Current address*: Department of Pure Mathematics and Mathematical Statistics, Centre for Mathematical Sciences, University of Cambridge, Wilberforce Road, Cambridge, CB3 0WB, United Kingdom

*E-mail address*: stergios@dpmms.cam.ac.uk

Contemporary Mathematics
Volume **696**, 2017
http://dx.doi.org/10.1090/conm/696/14014

# Pseudo-Anosovs optimizing the ratio of Teichmüller to curve graph translation length

Tarik Aougab and Samuel J. Taylor

ABSTRACT. Given $\phi$ a pseudo-Anosov map, let $\ell_{\mathcal{T}}(\phi)$ denote the translation length of $\phi$ in the Teichmüller space, and let $\ell_{\mathcal{C}}(\phi)$ denote the stable translation length of $\phi$ in the curve graph. Gadre–Hironaka–Kent–Leininger showed that, as a function of Euler characteristic $\chi(S)$, the minimal possible ratio $\tau(\phi) = \frac{\ell_{\mathcal{T}}(\phi)}{\ell_{\mathcal{C}}(\phi)}$ is $\log(|\chi(S)|)$, up to uniform additive and multiplicative constants. In this short note, we introduce a new construction of such *ratio optimizers* and demonstrate their abundance in the mapping class group. Further, we show that ratio optimizers can be found arbitrarily deep into the Johnson filtration as well as in the point pushing subgroup.

## 1. Introduction

Let $S = S_{g,p}$ denote the orientable surface of genus $g$ with $p$ punctures, and let $\omega(g,p) = \omega(S) = 3g + p - 4$ be its *complexity*. Let $\text{Mod}(S)$ denote the mapping class group of $S$, $\text{Teich}(S)$ the Teichmüller space equipped with the Teichmüller metric, and $\mathcal{C}(S)$ the curve graph of $S$.

Consider the coarsely-defined map $\pi_{g,p} : \text{Teich}(S) \to \mathcal{C}^0(S)$, which sends a marked hyperbolic surface to the simple closed curve(s) of shortest length. The map $\pi_{g,p}$ was originally studied by Masur-Minsky, who, as part of the proof of the $\delta$-hyperbolicity of $\mathcal{C}(S)$, demonstrated the existence of a constant $K = K(g,p)$ such that $\pi_{g,p}$ is coarsely $K$– Lipschitz [**MM99**]. Recall that a map $f : X \to Y$ between metric spaces is *coarsely $K$–Lipschitz* if there is an $L \geq 0$ such that $d_Y(f(a), f(b)) \leq K \cdot d_X(a,b) + L$ for all $a, b \in X$.

Let $K(g,p)$ denote the optimal possible value of the Lipschitz constant for $\pi_{g,p}$ as a function of $S_{g,p}$; that is,

$$K(g,p) = \inf\{c \in \mathbb{R} : \pi_{g,p} \text{ is coarsely } c\text{-Lipschitz}\}.$$

Gadre–Hironaka–Leininger–Kent showed that $K(S_{g,0}) \sim \frac{1}{\log(g)}$ [**GHKL13**]. Thus, not only is $\pi_{g,p}$ coarsely Lipschitz, but it is coarsely contracting, and the optimal contraction factor approaches 0 as $g \to \infty$. Following this work, Valdivia showed that for $r \in \mathbb{Q}$ a fixed rational number, the optimal Lipschitz constant for a sequence

2010 *Mathematics Subject Classification.* Primary 46L55.
*Key words and phrases.* Curve complex, mapping class group.
The first author was supported in part by NSF Grant DMS- 1502623.
The second author was supported in part by NSF Grant DMS-1400498.

of surfaces $S_{g_i,p_i}$ with $g_i/p_i = r$, also decays logarithmically in complexity (relative to constants that a priori depend on $r$) [**Val14**].

In one direction, [**GHKL13**] follow Masur and Minsky's original proof, while controlling the portions of the argument that a priori grow with the complexity of $S$. Conversely, to show that a Lipschitz constant on the order of $1/\log(g)$ is optimal, they construct, for each $g$, a pseudo-Anosov map $\psi \in \mathrm{Mod}(S_g)$ such that the ratio $\tau(\psi) = \ell_\mathcal{T}(\psi)/\ell_\mathcal{C}(\psi)$ of its translation length in Teich$(S)$, denoted $\ell_\mathcal{T}(\psi)$, to its stable translation length in $\mathcal{C}(S)$, denoted $\ell_\mathcal{C}(\psi)$, is on the order of $\log(g)$.

The purpose of this short note is to give a new construction of pseudo-Anosov maps for which $\tau(\psi)$ is optimal, i.e. on the order of $\log(\omega(S))$. We call such pseudo-Anosovs *ratio optimizers* and our construction shows their abundance in the mapping class group:

THEOREM 1.1. *There exists a function $f(\omega) = O(\log(\omega))$, and a Teichmüller disk $\mathcal{D} \subset$ Teich$(S_{g,p})$ such that there are infinitely many conjugacy classes of primitive pseudo-Anosovs $\psi$ with $\tau(\psi) = \frac{\ell_\mathcal{T}(\psi)}{\ell_\mathcal{C}(\psi)} < f(\omega(g,p)))$, and the invariant axis of $\psi$ is contained in $\mathcal{D}$.*

We will see in Corollary 3.5 that the function $f(\omega)$ can be taken to be $\log(2D\cdot\omega)$ where $D \geq 1$ is a constant not depending on $\omega$.

In addition to establishing the abundance of ratio optimizers, our methods show that ratio optimizers can be constructed in subgroups of mapping class groups which are well-known not to contain pseudo-Anosov mapping classes that minimize Teichmüller space translation length alone. In particular, we build ratio optimizers arbitrarily deep into the Johnson filtration as well as in the point pushing subgroup for a mapping class group of a surface with a single puncture.

For a group $\Gamma$, let $\Gamma^{(k)}$ denote the $k$th term of its lower central series. That is $\Gamma^{(1)} = [\Gamma,\Gamma]$ is the commutator subgroup and $\Gamma^{(k+1)} = [\Gamma^{(k)},\Gamma]$. For any $k \geq 0$ there is a surjective homomorphism

$$\mathrm{Mod}(S) \to \mathrm{Out}(\pi_1(S)/\pi_1(S)^{(k)}),$$

whose kernel, denoted $J_k$, is the *$k$th term of the Johnson filtration.* These subgroups were introduced by Johnson in [**Joh83**]. Note that $J_1$ is the Torelli subgroup of $\mathrm{Mod}(S)$ and $J_2$ is the so-called Johnson kernel.

THEOREM 1.2. *There exists a uniform constant $C_J \geq 0$ satisfying the following. Let $S = S_{g,p}$, with $g \geq 2$ and $p = 0$ or $p = 1$, and denote by $J_k(S)$ the $k$th term of the Johnson filtration of* $\mathrm{Mod}(S)$. *Then there exists $\phi_k \in J_k(S)$ with*

$$\tau(\phi_k) = \frac{\ell_\mathcal{T}(\phi)}{\ell_\mathcal{C}(\phi)} \leq C_J \log\omega(S).$$

*That is, there are ratio optimizers arbitrarily deep into the Johnson filtration.*

We remark that Theorem 1.2 is entirely different from the situation of minimizing $\ell_\mathcal{T}(\phi)$ alone. In fact, Farb–Leininger–Margalit [**FLM08**] have shown that the minimal Teichmüller space translation length among pseudo-Anosov mapping class in $J_1$ is uniformly bounded from above and below, independent of genus. This is in contrast to work of Penner who shows that among all pseudo-Anosov homeomorphisms this quantity is on the order of $1/g$ [**Pen91**].

Finally, for a surface $S_{g,1}$ with $g \geq 2$, we denote the kernel of the natural map

$$\mathrm{Mod}(S_{g,1}) \to \mathrm{Mod}(S_{g,0})$$

by $PP_g$. This is the point pushing subgroup of $\text{Mod}(S_{g,1})$; it consists of mapping classes which are isotopic to the identity after ignoring the puncture. Similar to the situation discussed above, it is known that pseudo-Anosov mapping classes in $PP_g$ cannot minimize Teichmüller space translation length [**Dow11**]. However, this is not an issue for ratio optimizers:

THEOREM 1.3. *There exists a uniform constant $C_P \geq 0$ satisfying the following. Let $S = S_{g,1}$ with $g \geq 2$ and let $PP_g \leq \text{Mod}(S)$ be the point pushing subgroup of its mapping class group. Then there is $\phi \in PP_g$ with*

$$\tau(\phi) = \frac{\ell_{\mathcal{T}}(\phi)}{\ell_{\mathcal{C}}(\phi)} \leq C_P \log \omega(S).$$

**Acknowledgments:** We thank the anonymous referee for useful comments.

## 2. Background

**2.1. Curves, filling pairs and projections.** Let $S_{g,p}$ denote the genus $g$ surface with $p \geq 0$ punctures. The *complexity* of $S$ is defined as $\omega(S) = \omega(g,p) = 3g + p - 4$. For all surfaces in this paper we assume $\omega(S) > 0$. A simple closed curve $c$ on $S$ is *essential* if it is not homotopically trivial and if it is not homotopic into a neighborhood of a puncture. Given two essential simple closed curves $\alpha, \beta$, we denote their *geometric intersection number* by $i(\alpha,\beta)$. If $|\alpha \cap \beta| = i(\alpha,\beta)$, we say $\alpha$ and $\beta$ are in *minimal position.*

A collection of curves $\Gamma = \{\gamma_1, \dots, \gamma_n\}$ in pairwise minimal position is said to *fill* a surface $S$ if the complement of their union consists of a disjoint union of topological disks and once-punctured disks. Equivalently, $\Gamma$ fills $S$ so long as every essential simple closed curve $\alpha$ has positive geometric intersection number with at least one curve in $\Gamma$. Let $i_{g,p}$ denote the minimum possible geometric intersection number for a filling pair $\alpha, \beta$ on $S_{g,p}$. A simple Euler characteristic argument shows that $i_{g,p}$ must grow linearly in $\omega(g,p)$. The quantities $i_{g,p}$ were determined in [**AH15**], for $p = 0$, and in [**AT14**], for $p > 0$.

LEMMA 2.1 (Minimally intersecting filling pairs). *Minimally intersecting filling pairs intersect as follows:*

(1) *If $g \neq 2, 0$ and $p = 0$, $i_{g,p} = 2g - 1$.*
(2) *If $g \neq 2, 0$ and $p \geq 1$, $i_{g,p} = 2g + p - 2$.*
(3) *If $g = 0$ and $p \geq 6$ is even, then $i_{g,p} = p - 2$. On the other hand if $p$ is odd, $i_{g,p} = p - 1$.*
(4) *If $g = 2$ and $p \leq 2$, $i_{g,p} = 4$.*
(5) *If $g = 2$ and $p \geq 2$ is even, then $i_{g,p} = 2g + p - 2$; and if $p \geq 3$ is odd, $i_{g,p} \leq 2g + p - 1$.*

In our application to pseudo-Anosov mapping classes in the Johnson filtration, we also require information about filling pairs of separating curves. Let $i^{sep}_{g,p}$ denote the minimum geometric intersection number taken over all filling pairs $\alpha, \beta$ where both $\alpha, \beta$ are separating curves. Then we have the following:

LEMMA 2.2 (Separating filling pairs). *There exists a constant $C \geq 0$ such that if $g \geq 2$ and $p = 0$ or $1$, there is a filling pair $(\alpha, \beta)$ on $S_{g,p}$ with both $\alpha, \beta$ separating curves, satisfying $i(\alpha,\beta) \leq C \cdot \omega(g,p)$.*

PROOF. The proof is by induction. First suppose $p = 0$, and define $\alpha_2, \beta_2$ to be any pair of separating curves which fill $S_2$; similarly let $\alpha_3, \beta_3$ be any pair of separating curves filling $S_3$.

Now let $\rho, \gamma$ be a pair of simple separating arcs on $S_{2,1}$ (which we interpret as the genus 2 surface with one boundary component, as opposed to one puncture) having the property that any essential arc in $S_{2,1}$ intersects either $\rho$ or $\gamma$. Then given $\alpha_g, \beta_g$, we form $\alpha_{g+2}, \beta_{g+2}$ as follows: excise a small open disk centered at one of the points in $\alpha_g \cap \beta_g$. After excising, $\alpha_g$ and $\beta_g$ have become arcs which we denote by $\tilde{\alpha}_g, \tilde{\beta}_g$. We then glue on a copy of $S_{2,1}$, matching the endpoints of $\tilde{\alpha}_g$ to those of $\gamma$, and similarly matching the endpoints of $\tilde{\beta}_g$ to those of $\rho$. We obtain a pair of simple closed curves $\alpha_{g+2}, \beta_{g+2}$ on $S_{g+2}$, and we claim that these curves are both separating and that they fill.

Note first that $\alpha_{g+2}, \beta_{g+2}$ are in minimal position since no complementary region is a bigon, and therefore it suffices to prove that if $\kappa$ is any essential simple closed curve on $S_{g+2}$, $\kappa$ is not disjoint from $\alpha_{g+2} \cup \beta_{g+2}$. If $\kappa$ can be isotoped into the original copy of $S_g$, it must intersect either $\tilde{\alpha}_g$ or $\tilde{\beta}_g$ since $\alpha_g, \beta_g$ fill on $S_g$. Therefore, we can assume that $\kappa$ projects non-trivially to the copy of $S_{2,1}$; that is, that $\kappa$ intersects this copy of $S_{2,1}$ in at least one arc which is not boundary parallel. This arc must intersect either $\rho$ or $\gamma$ since any arc does so by construction. Therefore $\kappa$ intersects $\alpha_{g+2} \cup \beta_{g+2}$ and we conclude that the new pair fills $S_{g+2}$.

That $\alpha_{g+2}, \beta_{g+2}$ are both separating is immediate since both are obtained by concatenating a pair of separating arcs in disjoint subsurfaces. Finally, $i(\alpha_{g+2}, \beta_{g+2}) \leq i(\alpha_g, \beta_g) + i(\gamma, \rho) - 1$.

If $p = 1$, then by puncturing one complementary region of $S_{g,0} \setminus (\alpha_g \cup \beta_g)$, $(\alpha_g, \beta_g)$ is a filling pair on $S_{g,1}$ with the desired properties. □

**2.2. Annular projections and the bounded geodesic image theorem.** For an annulus $Y \subset S$ whose core curve $\alpha$ is essential, let $\tilde{Y}$ be the cover of $S$ associated to the conjugacy class of the cyclic subgroup of $\pi_1(S)$ represented by $\alpha$. Let $\overline{Y}$ be the compactification of $\tilde{Y}$ obtained by choosing a hyperbolic metric on $S$ and using the usual compactification of the hyperbolic plane $\tilde{S} = \mathbb{H}^2$. The curve graph $\mathcal{C}(Y)$ of the annulus $Y$ is the graph whose vertices are homotopy classes rel endpoints of properly embedded, simple arcs of $\overline{Y}$ whose endpoints lie on distinct boundary components. Two vertices $x$ and $y$ of $\mathcal{C}(Y)$ are joined by an edge of $\mathcal{C}(Y)$ if and only if $x$ and $y$ can be represented by arcs in $\overline{Y}$ with disjoint interiors. There is a projection $\pi_Y$ from the vertices of the curve graph of $S$ to arcs of $\mathcal{C}(Y)$, known as *subsurface projection* [**MM00**, Section 2]. Given $\beta \in \mathcal{C}^0(S)$ realize $\alpha$ and $\beta$ with minimal intersection in $S$. If $\beta$ is disjoint from $\alpha$ then define $\pi_Y(\beta) = \emptyset$. Otherwise, the preimage of $\beta$ in the cover $\tilde{Y}$ contains simple, properly embedded arcs with well-defined endpoints on distinct components of $\partial\overline{Y}$ and we define $\pi_Y(\beta) \subset \mathcal{C}^0(Y)$ to be this collection of arcs in $\overline{Y}$.

If $\alpha \in \mathcal{C}^0(S)$ is a curve which is the core of an annulus $Y$ then we also use the notation $\mathcal{C}(\alpha)$ for the curve complex $\mathcal{C}(Y)$ and we denote its path metric by $d_\alpha$. Let $\pi_\alpha : \mathcal{C}(S) \setminus N_1(\alpha) \to \mathcal{C}(\alpha)$ be the corresponding subsurface projection, where $N_1(\cdot)$ is the closed 1-neighborhood in $\mathcal{C}_1(S)$. Also, write $d_\alpha(\beta, \gamma)$ for $\operatorname{diam}_{\mathcal{C}(\alpha)}(\pi_\alpha(\beta) \bigcup \pi_\alpha(\gamma))$.

From [**MM00**], we recall the following:

LEMMA 2.3 (Masur–Minsky). *Let $S$ be a surface with $\omega(S) > 1$. For $\alpha \in \mathcal{C}^0(S)$ and any path $\gamma = \gamma_0, \gamma_1, \ldots, \gamma_n$ of curves in $\mathcal{C}(S)$ each intersecting $\alpha$ essentially, we have:*

(1) $\mathrm{diam}_{\mathcal{C}(\alpha)}\pi_\alpha(\gamma) \leq 1$
(2) $d_\alpha(\gamma_0, \gamma_n) \leq n+1$
(3) *If $T_\alpha$ is the Dehn twist about $\alpha$, then $d_\alpha(\gamma, T^N(\gamma)) \geq N-2$.*

Finally, we recall the bounded geodesic image theorem of Masur–Minsky [**MM00**]. The version we state here is due to Webb and gives a uniform, computable constant [**Web13**]. It is stated below for arbitrary subsurfaces $Y \subset S$, but we will use it only for annuli.

THEOREM 2.4 (Bounded geodesic image theorem). *There exists $M \geq 0$ so that for any surface $S$ and any geodesic $g$ in $\mathcal{C}(S)$, if each vertex of $g$ has nontrivial projection to the subsurface $Y$ then $\mathrm{diam}(\pi_Y(g)) \leq M$.*

**2.3. Mangahas' Lemma.** Let $C_A = N_1(\alpha)$ and $C_B = N_1(\beta)$ be 1-neighborhoods of the curves $\alpha$ and $\beta$ in $\mathcal{C}(S)$ and let $M$ be as in the bounded geodesic image theorem (Theorem 2.4). The following is a special case of a combination of Lemma 5.3 of [**Man13**] along with the claim used in its proof. Recall that for a reduced word $w$ in $\langle a, b\rangle$, a *syllable* of $w$ is a maximal subword of the form $a^k$ or $b^k$. The *syllable length* of $w$, denoted $|w|_s$, is the number of syllables in the word $w$. For example, for the reduced word $w = a^{k_1}b^{k_2}\ldots a^{k_l}$, $|w|_s = l$.

LEMMA 2.5 (Mangahas). *Let $a, b$ be powers of Dehn twists about curves $\alpha, \beta$, respectively, such that $d_S(\alpha, \beta) \geq 3$, i.e. $\alpha$ and $\beta$ fill $S$. Suppose that for all $k \neq 0$*

$$d_\alpha(C_B, a^k \cdot C_B) > 2M+4 \quad \text{and} \quad d_\beta(C_A, b^k \cdot C_A) > 2M+4.$$

*Then the subgroup $\langle a, b\rangle \leq \mathrm{Mod}(S)$ is isomorphic to a rank-two free group and for any word $w$ in $\langle a, b\rangle$, either*

$$d_S(w\cdot\alpha, \alpha) \geq |w|_s \quad \text{or} \quad d_S(w\cdot\beta, \beta) \geq |w|_s.$$

We remark that if $a = T_\alpha^{l_1}$ and $b = T_\beta^{l_2}$ then by Lemma 2.3 the hypotheses of Lemma 2.5 are satisfied so long as $|l_1|, |l_2| \geq 2M+7$.

**2.4. Pseudo-Anosovs and Teichmüller disks.** For curves $\alpha$ and $\beta$ which jointly fill the surface $S$ and have intersection number $i(\alpha, \beta) = n$, there is a representation $\Psi : \langle T_\alpha, T_\beta\rangle \to PSL_2(\mathbb{R})$ given by

$$\Psi(T_\alpha) = \begin{pmatrix} 1 & n \\ 0 & 1 \end{pmatrix}, \quad \Psi(T_\beta) = \begin{pmatrix} 1 & 0 \\ -n & 1 \end{pmatrix}. \tag{2.1}$$

Thurston showed that the pseudo-Anosov mapping classes of the subgroup $\langle T_\alpha, T_\beta\rangle \leq \mathrm{Mod}(S)$ are exactly the ones mapping to hyperbolic matrices in $PSL_2(\mathbb{R})$ (i.e. matrices with 2 distinct eigenvalues). Further, he showed that the dilatation of such a pseudo-Anosov is equal to the largest eigenvalue of its representative matrix [**Thu88**]. Since the Teichmüller space translation length of a pseudo-Anosov mapping class $\phi$ is equal to the logarithm of its dilatation, this allows a direct computation of $\ell_\mathcal{T}(\phi)$ for $\phi \in \langle T_\alpha, T_\beta\rangle$. For proofs of these facts see [**FM11**].

The representations $\Psi : \langle T_\alpha, T_\beta\rangle \to PSL_2(\mathbb{R})$ in (2.1) comes from the singular flat structure $S(\alpha, \beta)$ on $S$ associated to the pair $(\alpha, \beta)$. This is the structure induced by the quadratic differential $q$ on $S$ whose vertical foliation is equal to $\alpha$ and whose horizontal foliation is equal to $\beta$ as measured foliations. Alternatively, one can consider the dual square complex to the graph $\alpha \cup \beta$, which induces a complex structure on $S$ along with the quadratic differential $q$ obtained by taking $dz^2$ in the interior of each square. The Dehn twists $T_\alpha$ and $T_\beta$ can each be realized by an affine map with respect to the singular flat structure, and the "derivative" map induces the representation to $PSL_2(\mathbb{R})$. See [**FM11**] for details.

The image in $\mathcal{T}(S)$ of the $SL_2(\mathbb{R})$ orbit of a quadric differential $q$ is known as a *Teichmüller disk*. Given a filling pair $\alpha$, $\beta$, by $\mathbb{D}(\alpha, \beta)$ we mean the Teichmüller disk corresponding to the quadratic differential described above, determined by the dual square complex to $\alpha \cup \beta$. The subgroup of $\mathrm{Mod}(S)$ preserving $\mathbb{D}(\alpha, \beta)$ is known as the Veech group $V(\alpha, \beta)$ and equals the image in $\mathrm{Mod}(S)$ of the affine homeomorphisms of the singular flat surface $S(\alpha, \beta)$. In particular, we note that $\langle T_\alpha, T_\beta \rangle \le V(\alpha, \beta)$.

## 3. Ratio optimizers via QI trees

Let $S = S_{g,p}$. Choose simple closed curves $\alpha, \beta \in \mathcal{C}^0(S)$ which fill $S$, that is, for which $d_S(\alpha, \beta) \ge 3$. For notational convenience, set $i = i_{\alpha,\beta} = i(\alpha, \beta)$. We will use later that $\alpha$ and $\beta$ can be chosen so that $i(\alpha, \beta) \le i_{g,p}$ where $i_{g,p}$ is as in Section 2.1 and depends linearly on the complexity $\omega(S) = \omega(g, p)$ of $S$. Let $M$ be the bounded geodesic image constant of Theorem 2.4. Recall that $M$ is independent of the complexity of $S$. Let $D = 2M + 7$ and set $a = T_\alpha^D$ and $b = T_\beta^D$.

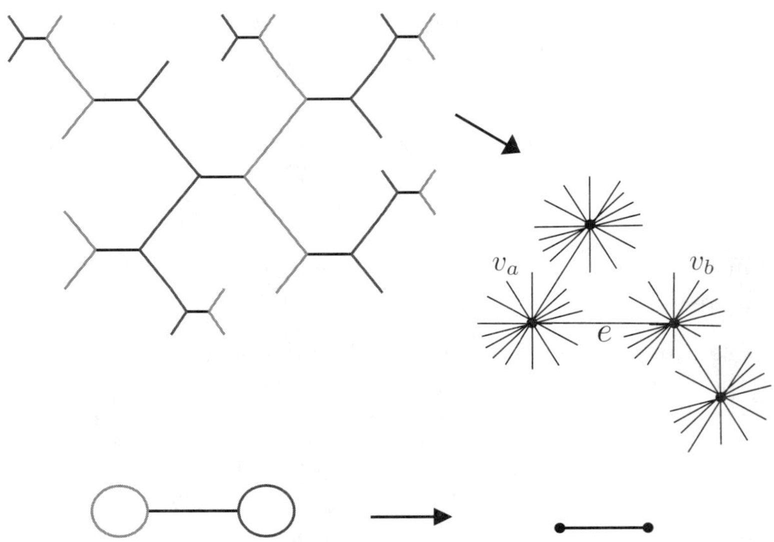


FIGURE 1. The tree $T_{\alpha,\beta}$ as a collapse of the trivalent tree.

It is well known that the subgroup of $\mathrm{Mod}(S)$ generated by $a, b$ is isomorphic to the free group of rank 2 (see, for example, [**Lei04**, Theorem 6.1]) and we make the identification $F_2 = \langle a, b \rangle \le \mathrm{Mod}(S)$. Let $T = T_{\alpha,\beta}$ be the Bass–Serre tree for the splitting $\langle a \rangle * \langle b \rangle$. In details, $T$ is the $F_2$–tree (tree with an action of $F_2$) obtained by taking the universal cover of the "barbell" graph whose loops are labeled by

$\{a, b\}$ and collapsing the lift of each $a$–edge and each $b$–edge. See Figure 1. Denote the image of the axis for $a$ by $v_a$ and note that this is the unique vertex of $T$ which is fixed by $a$. Similarly, denote the image of the axis for $b$ by $v_b$. Note that these vertices are joined by an edge, which we call $e$, and $e$ is a fundamental domain for the action $F_2 \curvearrowright T$.

We now define an equivariant map $\mathcal{O} : T \to \mathcal{C}(S)$. The vertex $v_a$ is mapped to $\alpha$, $v_b$ is mapped to $\beta$ and $e$ is mapped to a fixed geodesic from $\alpha$ to $\beta$ in $\mathcal{C}(S)$. Using the identification $F_2 \to \langle a, b\rangle \le \mathrm{Mod}(S)$, extend the map to all of $T$ by equivariance. This is well-defined since $a$ fixes $\alpha$ and $b$ fixes $\beta$. The main result of this section comes from the following proposition:

PROPOSITION 3.1 (The tree quasi-isometrically embeds). *With notation as above, the $\langle a, b\rangle$–equivariant map $\mathcal{O} : T_{\alpha,\beta} \to \mathcal{C}(S)$ is a $(3, 7)$–quasi-isometric embedding.*

We remark that by examining the proof of Lemma 2.5, the constants in Proposition 3.1 can be improved upon. This, however, will not be necessary for our application.

Before turning to the proof of Proposition 3.1, we first make a few remarks on distance in the (infinite valence) tree $T$. The syllable spelling of $w$ is exactly the normal form associated to the free product decomposition $F_2 = \langle a\rangle * \langle b\rangle$ and, hence, can be used to compute distance in $T$. In particular, let $x$ and $y$ be two vertices of $T$; there are four cases depending on whether $x$ and $y$ are in the orbit of $v_a$ or $v_b$. For example, suppose that $x = v_a$ and that $y \in F \cdot v_a \setminus \{v_a\}$. Then there is $w \in F$ with $y = w \cdot x$ and we can write $w = a^{l_1} w' a^{l_2}$ as a reduced syllable decomposition. Now it is easily seen that $d_T(x, y) = d_T(v_a, w'v_a) = |w'|_s + 1$. If $y \in F \cdot v_b$ then write $y = w \cdot v_b$ with $w = a^{l_1} w' b^{l_2}$ as a reduced syllable decomposition. Again we see that $d_T(x, y) = d_T(v_a, w'v_b) = |w'|_s + 1$. These elementary observations will be used in the proof of Proposition 3.1.

PROOF OF PROPOSITION 3.1. Let $x, y \in T$ and set $\gamma = \mathcal{O}(x)$ and $\delta = \mathcal{O}(y)$. Using equivariance and the fact that $d_S(\alpha, \beta) = 3$ we easily see that, $d_S(\gamma, \delta) \le 3 \cdot d_T(x, y)$.

For the other inequality, we may assume (by equivariance) that $x$ equals either $v_a$ or $v_b$; since the proofs in each case are identical we assume that $x = v_a$. First, suppose that $y$ is in the orbit of $v_a$, i.e. that $y = w \cdot v_a$ for $w \in F$. By the definition of $\mathcal{O}$ and the triangle inequality,

$$\begin{aligned} d_S(\gamma, \delta) &= d_S(\alpha, w \cdot \alpha) \\ &\ge d_S(\beta, w \cdot \beta) - 6, \end{aligned}$$

and thus by Lemma 2.5, $d_S(\gamma, \delta) \ge |w|_s - 6 \ge d_T(x, y) - 7$.

If on the other hand $y$ is in the obit of $v_b$, then we choose $w \in F$ so that $y = w \cdot v_b$. By the triangle inequality $d_S(\alpha, w \cdot \beta) \ge d_S(\alpha, w \cdot \alpha) - 3$ and $d_S(\alpha, w \cdot \beta) \ge d_S(\beta, w \cdot \beta) - 3$. Hence, we may apply Lemma 2.5 to conclude

$$\begin{aligned} d_S(\gamma, \delta) &= d_S(\alpha, w \cdot \beta) \\ &\ge |w|_s - 3 \\ &\ge d_T(x, y) - 4. \end{aligned}$$

□

We will say that $w \in \langle a, b\rangle$ is *cyclically reduced* if it has the smallest syllable length among any of its conjugates. The following is an immediate corollary of Proposition 3.1.

COROLLARY 3.2. *Let $w \in \langle a, b\rangle \leq \mathrm{Mod}(S)$ which is cyclically reduced. Then*

$$|w|_s \leq \ell_C(w) \leq 3|w|_s.$$

The next lemma is elementary and is used to bound the stretch factors of pseudo-Anosovs obtained by iterating composition.

LEMMA 3.3. *For any $i \geq 2$, let*

$$A = \begin{pmatrix} 1 & i \\ 0 & 1 \end{pmatrix}, \quad B = \begin{pmatrix} 1 & 0 \\ -i & 1 \end{pmatrix}$$

*and set $W = A^{\epsilon_1} B^{\delta_1} \ldots A^{\epsilon_k} B^{\delta_k}$, where $\epsilon_i, \delta_i \in \{\pm 1\}$. Then* $\mathrm{trace}(W) \leq (2i)^{2k}$.

PROOF. For a 2-by-2 matrix $A$, let $|A|_1$ denote its $l^1$-norm and $|A|_2$ its $l^2$-norm. For any such $A$, $|A|_1 \leq 2|A|_2$ . Moreover, for any matrices $A$ and $B$, $|AB|_2 \leq |A|_2|B|_2$. Hence,

$$\begin{aligned} \mathrm{trace}(W) &\leq |W|_1 \\ &\leq 2|W|_2 \\ &\leq 2\prod_i |A^{\epsilon_i} B^{\delta_i}|_2 \\ &\leq 2(i^2+1)^k < (2i)^{2k}. \end{aligned}$$

□

THEOREM 3.4 (Ratio bounds). *Let $\alpha, \beta$ be a filling pair of simple closed curves on $S$, and set $a = T_\alpha^D, b = T_\beta^D$. Let $w$ be a cyclically reduced word in $a, b$ satisfying $|w| = |w|_s$. Then*

$$\tau(w) = \frac{l_{\mathcal{T}}(w)}{l_{\mathcal{C}}(w)} \leq \log(2D \cdot i(\alpha, \beta)).$$

PROOF. Recall, as noted in Section 2.4, $\ell_{\mathcal{T}}(w)$ is equal to the logarithm of largest eigenvalue of the matrix $W = \Phi(w)$ corresponding to $w$. Applying Lemma 3.3 and Corollary 3.2, we compute

$$\begin{aligned} \tau(w) &\leq \frac{\log((2i)^{|w|})}{|w|_s} \\ &= \frac{|w|}{|w|_s} \log(2i). \end{aligned}$$

Since $|w| = |w|_s$ and $i = D \cdot i(\alpha, \beta)$, this completes the proof. □

The following corollary of Theorem 3.4 gives our construction of ratio optimizers.

COROLLARY 3.5 (Ratio optimizers). *Let $\alpha, \beta$ be a filling pair of simple closed curves satisfying $i(\alpha, \beta) = \eta_S$. Then for $w$ a cyclically reduced word as in Theorem 3.4,*

$$\begin{aligned} \tau(w) = \frac{l_{\mathcal{T}}(w)}{l_{\mathcal{C}}(w)} &\leq \log(D \cdot \eta_S) \\ &\leq \log(2D \cdot \omega(S)). \end{aligned}$$

## 4. Counting ratio optimizers in a Teichmüller Disk

In this section, we show that our construction yields infinitely many ratio optimizers whose maximal cyclic subgroups in $\mathrm{Mod}(S)$ are pairwise non-conjugate. That is, we will exhibit infinitely many ratio optimizers $\phi_1, \phi_2, \ldots$ such that for each $i \neq j$, no power of $\phi_i$ is conjugate to a power of $\phi_j$. Since each of our ratio optimizers is contained in the group generated by $T_\alpha, T_\beta$ for $\alpha, \beta$ a minimally intersecting filling pair, it will follow that the Teichmüller disk $\mathbb{D}(\alpha, \beta)$ is stabilized by infinitely many primitive, pairwise non-conjugate ratio optimizers. This will complete the proof of Theorem 1.1.

To begin, let $w_1, w_2, \ldots$ be an infinite collection of distinct cyclically reduced words in $a, b$, satisfying:

(1) $w_i \neq a^\pm, b^\pm$
(2) $|w_i| = |w_i|_s$.

Since each $w_i$ is cyclically reduced and all words in the collection are distinct, the words are pairwise non-conjugate. Furthermore, by property (1), each word $w_i$ has nonzero translation length on the tree $T_{\alpha,\beta}$, and therefore by Corollary 3.2, each corresponds to a pseudo-Anosov mapping class under the map sending the free group generated by $a, b$ to the subgroup generated by the Dehn twists $T_\alpha^D, T_\beta^D$.

We will refer to the pseudo-Anosov image of $w_i$ by $\mathcal{O}(w_i)$. By Proposition 3.1, $\mathcal{O}(w_j)$ admits a uniformly quasigeodesic axis $A_j$. By property (2) above, we may pass to a subsequence such that for each $k \geq 0$ the initial subword of $w_i$ of length $k$ is eventually constant as $i \to \infty$. Translating this fact to the tree $T$ and possibly passing to a further subsequence, we have the following property: the axis of $w_i$ in $T$ shares a segment centered around the origin of length at least $i$ with the axis for $w_{i-1}$. Thus, there exists a bi-infinite quasi-geodesic $\mathcal{R}$ in $\mathcal{C}(S)$ and a point $x \in \mathcal{R}$ so that $\mathcal{O}(w_i)$ admits an axis that shares a segment of length $i$ with $\mathcal{R}$, centered about $x$. Furthermore, as a consequence of Corollary 3.2 these axes do not fellow travel in $\mathcal{C}(S)$.

Now, let $\Gamma_1 \subset \{\mathcal{O}(w_2), \mathcal{O}(w_3), \ldots\}$ denote the set of words whose maximal cyclic subgroups are conjugate in $\mathrm{Mod}(S)$ to the maximal cyclic subgroup determined by $\mathcal{O}(w_1)$. By hyperbolicity of $\mathcal{C}(S)$, there is a constant $K > 0$ so that for $\mathcal{O}(w_i) \in \Gamma_1$, there is a conjugator $c_i \in \mathrm{Mod}(S)$ such that the quasigeodesic $c_i \cdot A_i$ $K$-fellow travels with $A_1$. Let $l$ denote the stable translation length of $\mathcal{O}(w_1)$.

It follows that there exists a uniform constant $r$ depending only on $l$, the hyperbolicity constant for $\mathcal{C}(S)$, and the quasigeodesic constants determined in Proposition 3.1, so that for any two points $t, s$ on $c_i \cdot A_i$, there exists a power of $\mathcal{O}(w_1)$ sending $t$ within $r$ of $s$. Hence the same is true for any two points on $A_i$, after replacing $\mathcal{O}(w_1)$ with its conjugate by $c_i^{-1}$.

We first show that $|\Gamma_1| < \infty$. Assume by contradiction that $\Gamma_1$ is infinite. Then there exists $\mathcal{O}(w_i) \in \Gamma_1$ with $i$ arbitrarily large. Choose such an $i \gg 1$, and let $y, z$ denote the endpoints of the segment of $\mathcal{R}$ that $A_i$ shares. Note that by construction $y, z \in A_j$ for all $j \geq i$.

Then for any $j > i$, $c_i^{-1} c_j$ sends $A_j$ to $A_i$, and post-composing this with some power $e_j$ of $c_i^{-1}\mathcal{O}(w_1)c_i$ sends each of $y$ and $z$ within $r$ of themselves. By acylindricity of the action of $\mathrm{Mod}(S)$ on $\mathcal{C}(S)$ [**Bow08**], there are at most finitely many mapping classes with this property. Hence,

$$\left\{(c_i^{-1}\mathcal{O}(w_1)^{e_j}c_i)c_i^{-1}c_j\right\}_{j>i} := \left\{c_j'\right\}_{j>i}$$

is a finite collection of mapping classes. We note that $c_i^{-1}c_j \neq c_i^{-1}c_k$ for any $j \neq k$, as the axes $A_j$ and $A_k$ do not fellow travel, and the inverse of $c_i^{-1}c_j$ sends the endpoints at infinity of $A_i$ to those of $A_j$. Moreover, $c'_j$ is obtained from $c_i^{-1}c_j$ by post-composition with a map that fixes the endpoints of $A_i$, and therefore $c'_j \neq c'_k$ for $k \neq j$. Thus $|\Gamma_1| < \infty$.

It follows that we may pass to a subsequence so that no corresponding pseudo-Anosov determines the same maximal cyclic subgroup up to conjugacy as $\mathcal{O}(w_1)$. Now we simply iterate this argument. By the exact same logic, the set $\Gamma_i$ of all maps in our collection which determine the same (up to conjugacy) maximal cyclic subgroup as $\mathcal{O}(w_i)$ is finite, and thus we can pass to a further subsequence all of whose terms determine pseudo-Anosov mapping classes which are distinct, up to conjugacy and powers, from those already obtained.

## 5. Ratio optimizers in the Johnson filtration and point pushing subgroups

Fix $S = S_{g,p}$ with $g \geq 2$ and $p \in \{0, 1\}$ and let $\alpha$ and $\beta$ be separating curves of $S$ which fill and intersect minimally, i.e. $i(\alpha, \beta) = i_{g,p}^{sep}$. Recall that by Lemma 2.2, there is a constant $C \geq 0$, independent of $S$, such that $i(\alpha_g, \beta_g) \leq C \cdot \omega(g, p)$. Let $a = T^D(\alpha)$ and $b = T^D(\beta)$.

THEOREM 5.1. *There is a constant $C_J \geq 0$ satisfying the following. Let $S = S_{g,0}$ or $S_{g,1}$ with $g \geq 2$ and denote by $J_k(S)$ the $k$th term of the Johnson filtration of* $\mathrm{Mod}(S)$. *Then there exist $f_k \in J_k(S)$ with*

$$\tau(f_k) = \frac{\ell_{\mathcal{T}}(f_k)}{\ell_{\mathcal{C}}(f_k)} \leq C_J \log \omega(S).$$

*In other words, there are ratio optimizers arbitrarily deep into the Johnson filtration.*

PROOF. Set $w_1 = aba$ and $w_2 = bab$ for $a$ and $b$ as defined at the beginning of this section. Set

$$f_k = [\ldots[[w_1, w_2], w_1] \ldots w_*]$$

which is $k$ iterated commutators alternating between $w_1$ and $w_2$. Note that by construction $|f_k| = |f_k|_s$, i.e. each syllable of $f_k$ has length 1. Since by definition $a, b \in J_1$ the same is true for $w_1, w_2$. Further, as the Johnson filtration $\{J_k\}$ is a central series, see [**BL94**] and [**Mor91**], we have $f_k \in J_k$.

By Corollary 3.2, $\ell_{\mathcal{C}}(f_k) \geq \frac{1}{3}|f_k|_s = \frac{1}{3}|f_k|$. Moreover, using Lemma 3.3 we can directly compute an upper bound for the dilatation, which (up to a uniform constant) is a product of $|f_k|$ with $\log(\omega(S))$. This completes the proof. □

We now construct ratio optimizers in the point-pushing subgroup $PP_g < \mathrm{Mod}(S_{g,1})$ of the mapping class group of a once-punctured surface. To achieve this, it suffices to construct a pair $(\alpha, \beta)$ of curves on $S_{g,1}$ which (1) fill the surface, (2) have geometric intersection number at most some fixed polynomial function of $g$, and (3) such that $\alpha$ and $\beta$ are isotopic after forgetting the puncture. Assuming the existence of such a pair, note that the pseudo-Anosov $T_\alpha^D T_\beta^{-D}$ lies in $PP_g$, and by Theorem 3.4, it will be a ratio optimizer. From this, we will obtain:

THEOREM 5.2. *There exists a uniform constant $C_P \geq 0$ satisfying the following. Let $S = S_{g,1}$ with $g \geq 2$ and let $PP_g \leq$* $\mathrm{Mod}(S)$ *be the point pushing subgroup of* *its mapping class group. Then there is $\phi \in PP_g$ with*

$$\tau(\phi) = \frac{\ell_{\mathcal{T}}(\phi)}{\ell_{\mathcal{C}}(\phi)} \leq C_P \log \omega(S).$$

To construct the desired filling pair, begin with a filling pair $(\rho, \delta)$ of non-separating curves on a closed surface $S_{g,0}$ with $i(\rho, \delta)$ bounded above by some fixed linear function of $g$. For example, $(\rho, \delta)$ could be a minimally intersecting filling pair on $S_{g,0}$.

Let $\delta_1, \delta_2$ be two parallel copies of $\delta$, and puncture the surface $S_{g,0}$ on the interior of the annulus bounded by $\delta_1$ and $\delta_2$ to form the surface $S = S_{g,1}$. Note that $f_\delta := T_{\delta_1}^3 \circ T_{\delta_2}^{-3}$ is a point-pushing map in $\mathrm{Mod}(S_{g,1})$. We claim that $\rho$ fills with $f_\delta(\rho)$, and that $i(\rho, f_\delta(\rho))$ is bounded above by a quadratic function of $g$.

We first show that these two curves jointly fill $S_{g,1}$; that is we must show that if $\gamma$ is any essential simple closed curve, $\gamma$ must intersect either $\rho$ or $f_\delta(\rho)$. We use the following inequality as seen in [**Iva92**]:

LEMMA 5.3. *Let $c_1, \ldots, c_m$ be a collection of pairwise disjoint, pairwise non-homotopic simple closed curves on a surface $S$ with negative Euler characteristic, let $\boldsymbol{S} := (s_1, \ldots s_m) \in \mathbb{Z}^m$, and let $T^{\mathbf{S}}$ denote the composition of Dehn twists $c_1^{s_1} \circ \cdots \circ c_m^{s_m}$. Then for any simple closed curves $\gamma, \rho$,*

$$\sum_{i=1}^{m} (|s_i - 2|) i(\rho, c_i) i(c_i, \gamma) - i(\rho, \gamma) \leq i(T^{\boldsymbol{S}}(\rho), \gamma)$$
$$\leq \sum_{i=1}^{m} |s_i| i(\rho, c_i) i(c_i, \gamma) + i(\rho, \gamma).$$

Now suppose $i(\gamma, \rho) = 0$. Then since $\rho$ fills with $\delta$ on the original closed surface, it follows that $i(\delta_j, \gamma) \neq 0$ for $j = 1, 2$. Thus the left hand side of the inequality of Lemma 5.3 is non-zero, so $\gamma$ must intersect $f_\delta(\rho)$.

The quadratic bound on $i(\rho, f_\delta(\rho))$ follows from the linear bound on $i(\rho, \delta)$, and another application of Lemma 5.3. This completes the proof of Theorem 5.2.

## References


[AH15] T. Aougab and S. Huang, *Minimally intersecting filling pairs on surfaces*, Algebr. Geom. Topol. **15** (2015), no. 2, 903–932, DOI 10.2140/agt.2015.15.903. MR3342680

[AT14] T. Aougab and S. J. Taylor, *Small intersection numbers in the curve graph*, Bull. Lond. Math. Soc. **46** (2014), no. 5, 989–1002, DOI 10.1112/blms/bdu057. MR3262200

[BL94] H. Bass and A. Lubotzky, *Linear-central filtrations on groups*, The mathematical legacy of Wilhelm Magnus: groups, geometry and special functions (Brooklyn, NY, 1992), Contemp. Math., vol. 169, Amer. Math. Soc., Providence, RI, 1994, pp. 45–98, DOI 10.1090/conm/169/01651. MR1292897

[Bow08] B. H. Bowditch, *Tight geodesics in the curve complex*, Invent. Math. **171** (2008), no. 2, 281–300, DOI 10.1007/s00222-007-0081-y. MR2367021

[Dow11] S. Dowdall, *Dilatation versus self-intersection number for point-pushing pseudo-Anosov homeomorphisms*, J. Topol. **4** (2011), no. 4, 942–984, DOI 10.1112/jtopol/jtr023. MR2860347

[FLM08] B. Farb, C. J. Leininger, and D. Margalit, *The lower central series and pseudo-Anosov dilatations*, Amer. J. Math. **130** (2008), no. 3, 799–827, DOI 10.1353/ajm.0.0005. MR2418928

[FM11] B. Farb and D. Margalit, *A primer on mapping class groups*, Princeton Mathematical Series, vol. 49, Princeton University Press, Princeton, NJ, 2012. MR2850125

[GHKL13] V. Gadre, E. Hironaka, R. P. Kent IV, and C. J. Leininger, *Lipschitz constants to curve complexes*, Math. Res. Lett. **20** (2013), no. 4, 647–656, DOI 10.4310/MRL.2013.v20.n4.a4. MR3188024

[Iva92] N. V. Ivanov, *Subgroups of Teichmüller modular groups*, Translations of Mathematical Monographs, vol. 115, American Mathematical Society, Providence, RI, 1992. Translated from the Russian by E. J. F. Primrose and revised by the author. MR1195787

[Joh83] D. Johnson, *A survey of the Torelli group*, Low-dimensional topology (San Francisco, Calif., 1981), Contemp. Math., vol. 20, Amer. Math. Soc., Providence, RI, 1983, pp. 165–179, DOI 10.1090/conm/020/718141. MR718141

[Lei04] C. J. Leininger, *On groups generated by two positive multi-twists: Teichmüller curves and Lehmer's number*, Geom. Topol. **8** (2004), 1301–1359, DOI 10.2140/gt.2004.8.1301. MR2119298

[Man13] J. Mangahas, *A recipe for short-word pseudo-Anosovs*, Amer. J. Math. **135** (2013), no. 4, 1087–1116, DOI 10.1353/ajm.2013.0037. MR3086070

[MM99] H. A. Masur and Y. N. Minsky, *Geometry of the complex of curves. I. Hyperbolicity*, Invent. Math. **138** (1999), no. 1, 103–149, DOI 10.1007/s002220050343. MR1714338

[MM00] H. A. Masur and Y. N. Minsky, *Geometry of the complex of curves. II. Hierarchical structure*, Geom. Funct. Anal. **10** (2000), no. 4, 902–974, DOI 10.1007/PL00001643. MR1791145

[Mor91] S. Morita, *On the structure of the Torelli group and the Casson invariant*, Topology **30** (1991), no. 4, 603–621, DOI 10.1016/0040-9383(91)90042-3. MR1133875

[Pen91] R. C. Penner, *Bounds on least dilatations*, Proc. Amer. Math. Soc. **113** (1991), no. 2, 443–450, DOI 10.2307/2048530. MR1068128

[Thu88] W. P. Thurston, *On the geometry and dynamics of diffeomorphisms of surfaces*, Bull. Amer. Math. Soc. (N.S.) **19** (1988), no. 2, 417–431, DOI 10.1090/S0273-0979-1988-15685-6. MR956596

[Val14] A. D. Valdivia, *Lipschitz constants to curve complexes for punctured surfaces*, Topology Appl. **216** (2017), 137–145, DOI 10.1016/j.topol.2016.11.016. MR3584129

[Web13] R. C. H. Webb, *Uniform bounds for bounded geodesic image theorems*, J. Reine Angew. Math. **709** (2015), 219–228, DOI 10.1515/crelle-2013-0109. MR3430880

Department of Mathematics, Brown University, 151 Thayer Street, Providence, Rhode Island 02912

*E-mail address*: tarik_aougab@brown.edu

Department of Mathematics, Yale University, 10 Hillhouse Avenue, New Haven, Connecticut 06510

*E-mail address*: s.taylor@yale.edu

Contemporary Mathematics
Volume **696**, 2017
http://dx.doi.org/10.1090/conm/696/14015

# Handlebody subgroups in a mapping class group

Mladen Bestvina and Koji Fujiwara

ABSTRACT. Suppose subgroups $A, B < MCG(S)$ in the mapping class group of a closed orientable surface $S$ are given and let $\langle A, B\rangle$ be the subgroup they generate. We discuss a question by Minsky asking when $\langle A, B\rangle \simeq A *_{A\cap B} B$ for handlebody subgroups $A, B$.

We also prove that acylindrically hyperbolic groups are not "invariably generated".

## 1. Introduction

Let $V$ be a handlebody and $S = \partial V$ the boundary surface. We have an inclusion of mapping class groups, $MCG(V) < MCG(S)$. This subgroup is called a *handlebody subgroup* of $MCG(S)$. The kernel of the map $MCG(V) \to Out(\pi_1(V))$ is denoted by $MCG^0(V)$.

If $M = V_+ \cup_S V_-$ is a Heegaard splitting of a closed orientable 3-manifold, we have two handlebody subgroups $\Gamma_\pm = MCG(V_\pm) < MCG(S)$ with $S = \partial V_\pm$. Minsky [**8**, Question 5.1] asked

QUESTION 1.1. *When is $\langle \Gamma_+, \Gamma_-\rangle < MCG(S)$ equal to the amalgamation $\Gamma_+ *_{\Gamma_+\cap\Gamma_-} \Gamma_-$?*

Let $\mathcal{C}(S)$ be the curve graph of $S$ and $D_\pm \subset \mathcal{C}(S)$ the set of isotopy classes of simple curves in $S$ which bound disks in $V_\pm$. The *Hempel/Heegaard distance* of the splitting is defined to be equal to $d(D_+, D_-) = \min\{d_{\mathcal{C}(S)}(x, y) \mid x \in D_+, y \in D_-\}$. The group $MCG(S)$ acts on $\mathcal{C}(S)$ by isometries. The stabilizer subgroups of $D_\pm$ in $MCG(S)$ are $\Gamma_\pm$. If the Hempel distance is sufficiently large, depending only on $S$ ($> 3$ suffices if the genus of $S$ is at least two [**10**]), then $\Gamma_+ \cap \Gamma_-$ is finite [**16**].

The following is the main result. It gives a partial answer to Question 1.1.

**Theorem 5.1.** *For the closed surface $S$ of genus $4g+1, g \geq 1$ and for any $N > 0$ there exists a Heegaard splitting $M = V_+ \cup_S V_-$ so that $\Gamma_+ \cap \Gamma_-$ is trivial, $\langle \Gamma_+, \Gamma_-\rangle$ is not equal to $\Gamma_+ * \Gamma_-$, and $d(D_+, D_-) \geq N$.*

---

2010 *Mathematics Subject Classification.* Primary 20F65; Secondary 57M50, 20F05.

*Key words and phrases.* Handlebody subgroups, acylindrical actions, invariable generation, acylindrically hyperbolic groups.

The first author was supported by the National Science Foundation grant DMS 1308178.

The second author is supported in part by Grant-in-Aid for Scientific Research (No. 23244005, 15H05739).

We will prove Theorem 5.1 by constructing an example. To explain the idea we first construct a similar example in an "acylindrical" group action on a simplicial tree (Theorem 4.1), then imitate it for the action of $MCG(S)$ on $\mathcal{C}(S)$.

By contrast, for the subgroups $MCG^0(V_\pm) = \Gamma^0_\pm < \Gamma_\pm$, Ohshika-Sakuma [**17**] showed the following.

THEOREM 1.2. *If $d(D_+, D_-)$ is sufficiently large (depending on $S$), $\Gamma^0_+ \cap \Gamma^0_-$ is trivial and $\langle \Gamma^0_+, \Gamma^0_- \rangle = \Gamma^0_+ * \Gamma^0_-$.*

We will review their argument from our point of view later. One key ingredient in their argument is the fact that $\Gamma^0_+, \Gamma^0_-$ are torsion free (attributed to [**19**, proof of Prop 1.7] in [**17**]).

Here is an alternative proof, suggested by Minsky, to show that $MCG^0(V) < MCG(S)$ is torsion free for a handlebody $V$. Let $f \in \Gamma^0$ be a torsion element. Since $f$ has finite order, we have a conformal structure on $S$ invariant by $f$. Moreover, since $f$ extends to $V$, by the classical deformation theory of Kleinian groups developed by Ahlfors, Bers, Kra, Marden, Maskit, and Sullivan, we have a unique hyperbolic structure on $V$ whose conformal structure at infinity is the prescribed one. Since the conformal structure is $f$-invariant, so is the hyperbolic structure. Moreover, since $f$ acts trivially on $\pi_1(V)$, each geodesic in $V$ is invariant by $f$. This implies that $f$ is identity on $V$, hence $f$ is trivial.

In the last section, we discuss "invariable generation" using "acylindrical" actions, which applies to the action of a mapping class group to its curve graph.

**Theorem 7.1.** *If $G$ is an acylindrically hyperbolic group, then $G$ is not IG (invariably generated).*

**Acknowledgements.** We would like to thank Yair Minsky for useful comments. We are also grateful to the referee whose comments improved the presentation of the paper.

## 2. Preliminaries

Let $S$ be a closed orientable surface. The mapping class group $MCG(S)$ of $S$ is the group of orientation preserving homeomorphisms modulo isotopy. The curve graph $\mathcal{C}(S)$ has a vertex for every isotopy class of essential simple closed curves in $S$, and an edge corresponding to pairs of simple closed curves that intersect minimally.

It is a fundamental theorem of Masur and Minsky [**13**] that the curve graph is $\delta$-hyperbolic. Moreover, they show that an element $F$ acts hyperbolically if and only if $F$ is pseudo-Anosov, and that the translation length

$$trans(F) = \lim_{n\to\infty} \frac{d(x_0, F^n(x_0))}{n}, \quad x_0 \in \mathcal{C}(S)$$

of $F$ is uniformly bounded below by a positive constant that depends only on $S$. It follows that $F$ has an invariant quasi-geodesic, called an *axis* denoted by $axis(F)$, whose quasi-geodesic constants depend only on $S$.

A subset $A \subset X$ in a geodesic space is *$Q$-quasi-convex* if any geodesic in $X$ joining two points of $A$ is contained in the $Q$-neighborhood of $A$. If $S$ is the boundary of a handlebody $V$, then the set $D \subset \mathcal{C}(S)$ of curves that bound disks in $V$ is quasi-convex (i.e. $Q$-quasi-convex for some $Q$), [**14**].

The stabilizer of the set $D$ in $MCG(S)$ is $MCG(V)$, the mapping class group of the handlebody $V$, i.e., the group of isotopy classes of diffeomorphisms of $V$.

Given a $Q$-quasi-convex subset $X$ in $\mathcal{C}(S)$, we define the nearest point projection $\mathcal{C}(S) \to X$. The nearest point projection is not exactly a map, but a *coarse map*, since for a given point maybe there is more than one nearest point, but the set of such points is bounded in diameter, and the bound depends only on $\delta$ and $Q$, but not on $X$.

In this paper we often take $axis(F)$ of a pseudo-Anosov element $F$ as $X$. We may take any $F$-orbit instead of $axis(F)$. Two pseudo-Anosov elements $F, G$ are independent (i.e., $\langle F, G\rangle$ is not virtually cyclic) if and only if the nearest point projection of $axis(E)$ to $axis(F)$ has a bounded image (cf. [**13**], [**2**]).

## 3. Coarse fixed sets and free products

The goal of this section is to prove Theorem 3.5. It is used to prove Theorem 1.2 (see the discussion at the end of the section), and its proof explains the background of our conterexamples in Theorem 5.1, where the assumption of Theorem 3.5 fails to hold. Theorem 3.5 is also used to prove Theorem 7.1.

**3.1. Acylindrical actions.** Acylindricity was introduced by Sela for group actions on trees and extended by Bowditch [**3**]. Suppose $G$ acts on a metric space $X$. The action is *acylindrical* if for given $R > 0$ there exist $L(R)$ and $N(R)$ such that for any points $v, w \in X$ with $|v - w| \geq L$, there are at most $N$ elements $g \in G$ with $|v - g(v)|, |w - g(w)| \leq R$. (Here $|x - y|$ denotes the distance $d(x, y)$.) Bowditch [**3**] showed that the action of $MCG(S)$ on $\mathcal{C}(S)$ is acylindrical.

The following criterion will be useful.

LEMMA 3.1. *Suppose $G$ acts on a simplicial tree $X$. If the cardinality of the edge stabilizers is uniformly bounded then the action is acylindrical.*

PROOF. Assume that every edge stabilizer contains at most $K$ elements. Suppose an integer $R > 0$ is given. Take $L >> R$. We will show that if $|v - w| \geq L$ then there are at most $(2R + 1)K$ elements $g$ with $|v - gv|, |w - gw| \leq R$.

Indeed, let $[v, w]$ be the geodesic from $v$ to $w$, $[v, w]'$ its subsegment after removing the $R$-neighborhood of $v, w$, and $[v, w]''$ the one after removing the $2R$-neighborhood of $v, w$. Then by the assumption, $g([v, w]'') \subset [v, w]'$. Moreover, for an edge $E \subset [v, w]''$ near the midpoint, $g(E)$ is contained in $[v, w]''$ and the distance between $E$ and $g(E)$ is at most $R$. Now fix such $E$. Then there are elements $h_1, \cdots, h_n \in G'$ with $n \leq 2R + 1$, where $h_1 = 1$, such that for any concerned element $g$, there exists $h_i$ with $h_i g(E) = E$. But since the stabilizer of $E$ contains at most $K$ elements, there are at most $nK \leq (2R + 1)K$ distinct choices for $g$. □

**3.2. Coarse fixed sets.**

LEMMA 3.2. *Let $X$ be a $\delta$-hyperbolic space and $A, B \subset X$ be $Q$-quasi-convex subsets. Let $\gamma$ be a shortest geodesic between $A$ and $B$. Then,*

*(1) for any $x \in \gamma$ with both $d(x, A), d(x, B) > Q + 2\delta$, and for any geodesic $\tau$ between $A$ and $B$, we have $d(x, \tau) \leq 2\delta$.*

*(2) Suppose $f$ is an isometry of $X$ with $f(A) = A, f(B) = B$. Then for any $x \in \gamma$ with both $d(x, A), d(x, B) > Q + 2\delta$, we have $d(x, f(x)) \leq 4\delta$. Hence for any $x \in \gamma$, we have $d(x, f(x)) \leq 2Q + 8\delta$.*

(3) *Suppose $f$ is an isometry of $X$ with $f(A) = A$. For $x \in X \backslash A$ let $\sigma$ be a shortest geodesic from $x$ to $A$. For $D > 0$ and an integer $N > 0$ assume $d(x, A) \geq Q + DN$ and $d(x, f(x)) \leq D$. Then for any point $y \in \sigma$ with $Q + 2\delta < d(y, A) < d(x, A) - DN - 2\delta$, we have $d(y, f^i(y)) \leq 4\delta$ for $1 \leq i \leq N$.*

PROOF. (1) Draw a geodesic quadrilateral with $\gamma, \tau$ a pair of opposite sides. By $\delta$-hyperbolicity, $x$ must be in the $2\delta$-neighborhood of one of the three sides not equal to $\gamma$, which must be $\tau$, for otherwise, $d(x, A) \leq Q + 2\delta$ or $d(x, B) \leq Q + 2\delta$, impossible.

(2) Put $f(\gamma) = \tau$. Then for a point $x \in \gamma$ satisfying the assumption, by (1) there is a point $p \in f(\gamma)$ with $d(x, p) \leq 2\delta$. But $d(p, f(x)) \leq 2\delta$ since $d(x, A) = d(f(x), A)$ and $|d(x, A) - d(p, A)| \leq 2\delta$. By triangle inequality $d(x, f(x)) \leq 4\delta$. It then implies $d(x, f(x)) \leq 4\delta + 2(Q + 2\delta)$ for $x \in \gamma$ in general.

(3) By triangle inequality, for each $1 \leq i \leq N$, we have $d(x, f^i(x)) \leq Di$. Let $q = \sigma \cap A$, and draw a geodesic quadrilateral with the corners $x, q, f^i(q), f^i(x)$. Then by $\delta$-hyperbolicity, a concerned point $y \in \sigma$ is in the $2\delta$-neighborhood of the side $f^i(\sigma)$, hence, as before $d(y, f^i(y)) \leq 4\delta$. □

LEMMA 3.3. *Let a group $G$ act acylindrically on a $\delta$-hyperbolic space $X$. Let $A$ be a $Q$-quasi-convex subset in $X$. Set $L_0 = L(4\delta) + 2Q + 4\delta + 10\delta N(4\delta)$. Suppose $f \in G$ is an element of infinite order such that $f(A) = A$. If $d(x, A) \geq L_0$, then $d(x, f(x)) > 10\delta$.*

PROOF. To argue by contradiction assume $d(x, f(x)) \leq 10\delta$. Let $\sigma$ be a shortest geodesic from $x$ to $A$. Apply Lemma 3.2 (3) with $N = N(4\delta)$ and $D = 10\delta$. Then for each point $y \in \sigma$ with $Q+2\delta < d(y, A) < d(x, A) - 10\delta N(4\delta) - 2\delta$ and each $1 \leq i \leq N$, we have $d(y, f^i(y)) \leq 4\delta$. Now the subsegment of $\sigma$, except for the end points, that those $y$ can belong to has length at least $L_0 - 10\delta N(4\delta) - 2\delta - (Q+2\delta) = L(4\delta) + Q$. Taking two points near each end of the subsegment, they are moved at most $4\delta$ by $1, f, \cdots, f^N$. But by acylindricity there are at most $N(4\delta)$ such elements. Since $N = N(4\delta)$, the order of $f$ must be at most $N(4\delta)$, hence trivial, contradiction. □

The above argument applies to $f$ whose order is $> N(4\delta)$. In view of Lemma 3.3 we introduce the following definition. For an isometry $f : X \to X$ define the *coarse fixed set* as

$$CFix(f) = \{x \in X \mid d(x, f(x)) \leq 10\delta\}.$$

Using this notion, we rephrase Lemma 3.3 as follows. We use the same constant $L_0$, which does not depend on $A$ and $f$.

LEMMA 3.4. *Suppose $G$ acts on $X$ acylindrically such that $A$ is a $Q$-quasi-convex set and $f \in G$ is an element of infinite order with $f(A) = A$. Then, $CFix(f) \subset N_{L_0}(A)$.*

**3.3. Ping-Pong theorem.** We state a version of a "Ping-Pong" theorem. This is a well-known result, but we make the role of a coarse fixed set clear. For that we introduce one more definition. Let $B_1, B_2, \cdots, B_n$ be subsets in a $\delta$-hyperbolic space $X$. For a given constant $K > 0$, we say that they are *K-terminal* if

- $d_X(B_i, B_j) \geq K$ for all pairs $i \neq j$, and
- For each $B_i$ and any other $B_j, B_k$, the distance between $B_i$ and any geodesic $\gamma$ joining two points in $B_j, B_k$ is at least $K$.

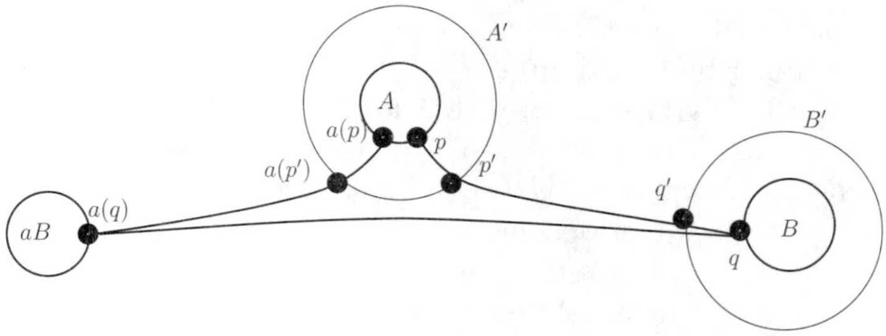


FIGURE 1. By $\delta$-hyperbolicity, $p' \in N_{2\delta}([p, a(p)] \cup [a(p), a(q)] \cup [a(q), q])$. By our assumption, in fact, $p' \in N_{2\delta}([a(q), q])$.

THEOREM 3.5. *Let a group $G$ act on a $\delta$-hyperbolic space $X$. Let $A_1, \cdots, A_n \subset X$ be $Q$-quasi-convex subsets, and $G_i < G$ subgroups with $G_i(A_i) = A_i$ for all $i$. Put $K = 110\delta + 2Q$.*

*Assume that there exists a subset $A'_i$ containing $A_i$ for each $i$ so that:*

*(a) for every element $1 \neq a \in G_i$ we have $CFix(a) \subset A'_i$ for each $i$;*
*(b) $A'_1, \cdots, A'_n$ are $K$-terminal.*

*Then $\langle G_1, \cdots, G_n \rangle = G_1 * \cdots * G_n$. In particular, $G_i \cap G_j$ is trivial for all $i \neq j$.*

PROOF. The case $n = 1$ is trivial, so we discuss the case $n = 2$ in detail. Set $A_1 = A, A'_1 = A', A_2 = B, A'_2 = B', G_1 = G_A, G_2 = G_B$. By assumption $d(A', B') \geq 110\delta + 2Q$. Taking $A', B'$ larger if necessary, we assume $N_{Q+5\delta}(A) \subset A', N_{Q+5\delta}(B) \subset B'$, so that a geodesic joining two points in $A$ (or $B$) is contained in $A'$ (or $B'$). We still have $d(A', B') \geq 100\delta$.

Let $\gamma = [p, q]$ be a shortest geodesic from $A$ to $B$, and let $\gamma' = [p', q'] \subset \gamma$ the closure of the longest connected component disjoint from $A'$ and $B'$. By our assumption $|\gamma'| \geq 100\delta$. Now, we apply $1 \neq a \in G_A$ to $A \cup \gamma \cup B$, and obtain $a(B) \cup a(\gamma) \cup A \cup \gamma \cup B$. We will give a lower bound of $|q - a(q)|$. Since $CFix(a) \subset A'$, we have $|p' - a(p')| \geq 10\delta$. Because of this, $p', a(p') \in N_{2\delta}([q, a(q)])$. To see this, draw a geodesic rectangle joining $p, q, a(q), a(p)$ and argue using $d(p', A) = d(a(p'), A)$. See Figure 1. Since $|p' - a(p')| \geq 10\delta$, (it makes sense to say that) $a(p')$ and $p'$ appears in this order along the segment from $a(q)$ to $q$ (as in Figure 1).

But this implies

$$|a(q) - q| \geq (|a(q) - a(p')| - 2\delta) + (|p' - q| - 2\delta) \geq 2 \cdot 90\delta.$$

In fact, we also have $d(a(B), B)) \geq 2 \cdot 90\delta$. Similarly, $d(A, b(A)) \geq 2 \cdot 90\delta$ for any non-trivial element $b \in G_B$.

We do one more example. Take a non-trivial element $a' \in G_A$, and consider the word $a'ba$. Letting $a, b, a'$ act on the union $A \cup \gamma \cup B$, we obtain the union of the sets $B \cup \gamma \cup A \cup a'(\gamma) \cup a'(B) \cup a'b(\gamma) \cup a'b(A) \cup a'ba(\gamma) \cup a'ba(B)$. Then as before, it follows both $d(B, a'ba(B))$ and $d(q, a'ba(q)) \geq |\gamma'| - 10\delta + |a'(\gamma')| - 20\delta + |a'b(\gamma')| - 20\delta + |a'ba(\gamma')| - 10\delta \geq 4 \cdot 80\delta$. (We subtract $20\delta$ for an orbit of $\gamma'$ which is not first or last.) For this estimate, we first prove, by drawing $\delta$-thin triangles, the points $p', a'(p'), a'(q'), a'b(q'), a'b(p'), a'ba(p')$ are in $N_{4\delta}([q, a'ba(q)])$, and those points appear in this order along the segment. So, the segments $\gamma', a'(\gamma'), a'b(\gamma'), a'ba(\gamma')$ are "almost contained" in the segment $[q, a'ba(q)]$, and they do not overlap. This implies the above estimate. It then follows that $a'ba$ is non-trivial in $G$.

We note that in this case, the points in question are only in the $4\delta$-neighborhood, and not in the $2\delta$-neighborhood, of $[q, a'ba(q)]$. But an important point in the following discussion is that if we deal with a longer word than $a'ba$, the size of the neighborhood we need is always $4\delta$. We remark that this point is related to a fundamental fact in $\delta$-hyperbolic geometry that "a local quasi-geodesic is in fact a (global) quasi-geodesic", and we could deduce our estimate from this fact. But we give an elementary argument in this paper.

The general case is similar. We consider the free product $G_A * G_B$, and suppose a non-trivial reduced word, $w$, is given. It will define an element in $G$, and we show the action by $w$ is non-trivial. To that end, suppose the reduced word is in the form $w = a_1 b_1 \cdots a_n b_n$, and we let the elements $b_n, a_n, \cdots, b_1, a_1$ successively act on $A \cup \gamma \cup B$. We then obtain the $(n+1)$ translates of $\gamma'$, and argue that the endpoins of those translates are in the $4\delta$-neighborhood of $[q, w(q)]$, and they do not overlap. We conclude, as before, $d(B, w(B)) \geq (|w| + 1)80\delta$ if $a_1$ is non-empty. If $a_1$ is empty, we replace $d(B, w(B))$ with $d(A, w(A))$. In either case, $w$ is non-trivial in $G$, and we showed $\langle G_A, G_B \rangle = G_A * G_B$.

The case $n > 2$ is similar. Suppose a non-trivial reduced word $w = p_1 \cdots p_n$ is given in the free product of $G_1 * \cdots * G_n$. Suppose $p_1 \in G_N$. Then $d(A_N, wA_N) > 0$. To see that we join each pair of $A_i, A_j$ by a shortest geodesic, then as before let the word $w$ act on $A_i$'s and the shortest geodesics, and argue. We omit details. It implies that the element $w$ is non-trivial in $G$. □

We quote a main technical result from [**17**]. This is an immediate consequence of Theorem 3.5 since the action is acylindrical, by Lemma 3.4, $CFix(a) \subset N_{L_0}(A_i)$ for any non-trivial element $a \in A_i$, where $a$ has infinite order. Now put $A_i' = N_{L_0}(A_i)$ and $M = K + 2L_0$. Then the distance between $A_1'$ and $A_2'$ is at least $K$, so that they are $K$-terminal, and that Theorem 3.5 applies.

THEOREM 3.6 ([**17**]). *Let a group $G$ act acylindrically on a $\delta$-hyperbolic space $X$. Then for a given $Q > 0$, there exists $M > 0$ with the following property. Let $A_1, A_2 \subset X$ be $Q$-quasi-convex subsets, and $G_1, G_2 < G$ torsion-free subgroups with $G_1(A_i) = A_i$. If $d_X(A_1, A_2) \geq M$ then $\langle G_1, G_2 \rangle = G_1 * G_2$. In particular, $G_1 \cap G_2$ is trivial.*

Applying the theorem to the action of $MCG(S)$ on $\mathcal{C}(S)$ with $A_1 = D_+, A_2 = D_-$, and $G_1 = \Gamma_+^0 < \Gamma_+ = stab(A_1)$, $G_2 = \Gamma_-^0 < \Gamma_- = stab(A_2)$, we obtain Theorem 1.2: if $d(D_+, D_-)$ is sufficiently large, depending only on $S$, then $\Gamma_+^0 \cap \Gamma_-^0$ is trivial and $\langle \Gamma_+^0, \Gamma_-^0 \rangle = \Gamma_+^0 * \Gamma_-^0$.

Elements of finite order play an essential role in our counterexamples. They will have the property that $A_1' \cap A_2' \neq \emptyset$ (namely, the two coarse fixed sets intersect although $A_1$ and $A_2$ are far away. Lemma 3.4 applies only to elements of infinite order, where $CFix(f)$ is contained in the $L_0$-neighborhood of $A_i$ for $f \in G_i$).

## 4. Example on a tree

We will show that Theorem 3.6 does not hold if we do not assume that $G_1$ and $G_2$ are torsion-free. We construct a counterexample in the action of $MCG(S)$ on $\mathcal{C}(S)$ (Theorem 5.1).

To explain the idea we start with a counterexample when $X$ is a simplicial tree. The key geometric feature is that, if we keep the previous notations, $\gamma$ and $a(\gamma)$

may stay close along an arbitrarily long segment if $a$ has finite order (each point on that segment does not move very much by $a$).

THEOREM 4.1. *There exists an acylindrical group action on a simplicial tree $X$ by a group $G$ such that for any number $N > 0$ there exist vertices $v, w \in X$ with $|v - w| \geq N$ such that $stab_G(v) \cap stab_G(w)$ is trivial and $\langle stab_G(v), stab_G(w)\rangle$ is not equal to the free product $stab_G(v) * stab_G(w)$.*

PROOF. We first construct an example with $N = 2$. Start with abelian groups $A, B$ with non-trivial torsion elements $a \in A$ and $b \in B$, for example, $A, B \simeq \mathbb{Z}/2\mathbb{Z}$.

Define the group

$$G = A *_{\langle a\rangle} (\langle a\rangle \times \langle b\rangle) *_{\langle b\rangle} B$$

and let $T$ be the Bass-Serre tree of this graph of groups decomposition.

There are two vertices $v, w$ in $T$ at distance two whose stabilizers are $A$ and $B$. The intersection $A \cap B$ is trivial in $G$ since $\langle a\rangle \cap \langle b\rangle$ is trivial in $\langle a\rangle \times \langle b\rangle$. On the other hand, $\langle A, B\rangle = G$ is not equal to $A * B$ since $G$ is the quotient of the free product $A * B$ by the relation $ab = ba$. The geometric reason for why $\langle a, b\rangle$ is not equal to $\langle a\rangle * \langle b\rangle$ is that $Fix(a)$ and $Fix(b)$ intersect non-trivially in $T$.

The action on $T$ is acylindrical by Lemma 3.1 since the edge stabilizer is a conjugate of $\langle a\rangle$ or $\langle b\rangle$.

To produce an acylindrical action that works for all $N > 1$ we modify the previous example. Take the direct product of $\mathbb{Z} = \langle t\rangle$ and the subgroup $\langle a\rangle \times \langle b\rangle$ in $G$. One can write the new group as

$$A *_{\langle a\rangle} \{\langle a\rangle \times \langle b\rangle \times \langle t\rangle\} *_{\langle b\rangle} B$$

Further, add a new element $s$ to $A$ with a relation $sa = as$ to get $A' = A \times \langle s\rangle$ and set $C = \langle a\rangle \times \langle b\rangle \times \langle t\rangle$ and

$$G' = A' *_{\langle a\rangle} C *_{\langle b\rangle} B$$

This is a two edge decomposition.

In the Bass-Serre tree of this decomposition, consider the "fundamental domain", i.e. the subtree spanned by two vertices $v, w$ at distance two with stabilizers $A'$ and $B$ respectively. Let $x$ be the vertex between them with stabilizer $C$. See Figure 2.

Now consider the ray based at $x$ that contains the vertices $x$, $t(v)$, $ts(x)$, $tst(v)$, $tsts(x), \cdots$. The stabilizer of every edge on this ray is $\langle a\rangle$ since both $t$ and $s$ commute with $a$.

So, the intersection of $B$, the vertex group of $w$, and any of the vertex groups along the ray except for $C$ is $\langle a\rangle \cap \langle b\rangle = 1$.

But for each $n > 0$ the subgroup $\langle C^{(ts)^n}, B\rangle < G'$ is not equal to $C^{(ts)^n} * B$ since $a \in C^{(ts)^n}$ and $b \in B$ generate $\langle a\rangle \times \langle b\rangle$ and not $\langle a\rangle * \langle b\rangle$.

The action of $G'$ is acylindrical by Lemma 3.1 since any edge stabilizer is a conjugate of $\langle a\rangle$ or $\langle b\rangle$. □

## 5. Example on $\mathcal{C}(S)$ and proof of theorem

We will prove the main theorem.

THEOREM 5.1. *For the closed surface $S$ of genus $4g+1$ with $g \geq 1$ and for any $N > 0$ there exists a Heegaard splitting $M = V_+ \cup_S V_-$ so that $\Gamma_+ \cap \Gamma_-$ is trivial, $\langle \Gamma_+, \Gamma_-\rangle$ is not equal to $\Gamma_+ * \Gamma_-$, and $d(D_+, D_-) \geq N$.*

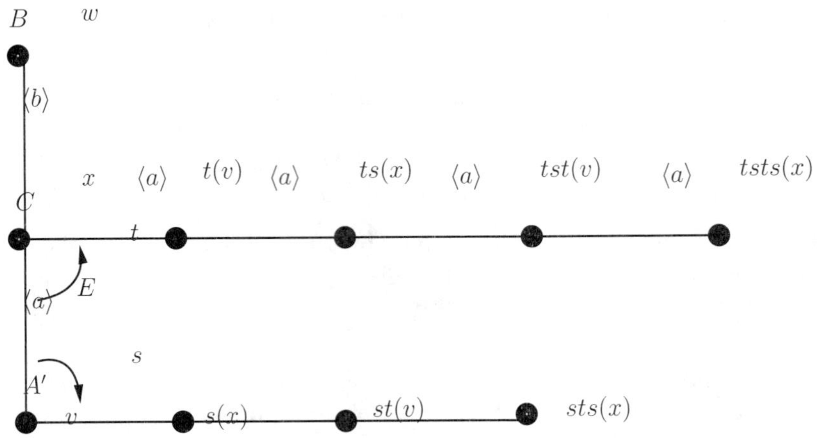


FIGURE 2

We will need two properties of pseudo-Anosov elements to prove the theorem (Lemma 5.2, Lemma 5.4).

**5.1. Pseudo-Anosov elements by Masur-Smillie.** Let $S$ be a closed surface and $F$ a pseudo-Anosov mapping class on $S$. The *elementary closure* of $F$ is the subgroup $E(F)$ of $MCG(S)$ that consists of mapping classes preserving the stable and unstable foliations of $F$. Equivalently, $E(F)$ is the centralizer of $F$ in $MCG(S)$. The group $E(F)$ contains a unique finite normal subgroup $N(F)$ such that $E(F)/N(F)$ is infinite cyclic. Note that $E(F^k) = E(F)$ and $N(F^k) = N(F)$ for every $k \neq 0$.

If $S' \to S$ is a regular cover with deck group $\Delta$ and if $F : S \to S$ is a pseudo-Anosov mapping class with $N(F) = 1$, then we certainly have $N(F') \supseteq \Delta$ for any lift $F' : S' \to S'$ of any power of $F$, but strict inclusion may hold. It is an interesting question whether one can construct $F$ so that equality holds for all regular covers. We call such $F$ *prime* and we discuss a construction of prime pseudo-Anosov mapping classes in Section 6. For our purposes we need quite a bit less.

LEMMA 5.2. *Suppose $F : S \to S$ is a pseudo-Anosov mapping class whose stable and unstable foliations have two singular points, one of order $p$, the other of order $q$, with both $p, q$ odd and relatively prime. Let $S' \to S$ be a double cover with deck group $\langle a\rangle$ and $F'$ a lift of a power of $F$ to $S'$. Then $N(F) = 1$ and $N(F') = \langle a\rangle$.*

PROOF. We first argue that $N(F) = 1$. Suppose $g \in N(F)$. Then $g$ can be represented by a homeomorphism, also denoted $g : S \to S$, that preserves both measured foliations. In particular, $g$ is an isometry in the associated flat metric on $S$ with cone type singularities. The homeomorphism $g$ fixes both singular points and satisfies both $g^p = 1$ and $g^q = 1$, since an isometry that fixes a nonempty open set is necessarily the identity. Since $p, q$ are relatively prime it follows that $g = 1$.

We now argue that $N(F') = \langle a\rangle$. We have $\langle a\rangle < N(F')$. Let $g \in N(F')$, then $g : S' \to S'$ is a finite order homeomorphism that preserves the lift of stable and unstable foliations of $F$. Composing with $a$ if necessary we may assume that $g$ fixes both $p$-prong singularities. Arguing as above, we see that $g^p = 1$. Since $g^2$ fixes both $q$-prong singularities, similarly we have $g^{2q} = 1$ and since $(p, 2q) = 1$ we have $g = 1$. We showed $N(F') = \langle a \rangle$. $\square$

COROLLARY 5.3. *Pseudo-Anosov mapping classes as in Lemma 5.2 exist in every genus* $\geq 3$.

PROOF. Write $4g = p + q$ where $p, q$ are relatively prime odd numbers. For example, we can take $p = 2g - 1$ and $q = 2g + 1$. By the work of Masur-Smillie [**15**] $F$ as above exists. $\square$

**5.2. Masur domain and Hempel elements.** Suppose $V$ is a handlebody and $S$ its boundary. Let $D \subset \mathcal{C}(S)$ be the set of curves that bound disks in $V$. Denote by $L \subset \mathcal{PML}(S)$ the closure of $D$, viewed as a subset of $\mathcal{PML}(S)$. Then $L$ is nowhere dense in $\mathcal{PML}(S)$ [**12**], and its complement $\Omega$ is called the *Masur domain.*

Hempel [**9**] found that if the stable lamination of a pseudo-Anosov element $F$ is in $\Omega$ then $\lim_{n\to\infty} d_{\mathcal{C}(S)}(D, F^n(D)) = \infty$. We say a pseudo-Anosov element $F : S \to S$ is *Hempel* for $D$ if the nearest point projection of $D$ to $axis(F)$ is a bounded set.

$F$ is Hempel if and only if the end points of $axis(F)$ are in $\Omega$, [**1**]. On the other hand, both endpoints of $axis(F)$ are in $L$ if and only if $axis(F)$ is contained in a $K$-neighborhood of $D$ for some $K > 0$ since both $D$ and $axis(F)$ are quasi-convex subsets in the $\delta$-hyperbolic space $\mathcal{C}(S)$ (cf. [**1**]). Since $L$ is nowhere dense in $\mathcal{PML}(S)$ and the set of pairs of endpoints $(\lambda^+, \lambda^-)$ of pseudo-Anosov mapping classes is dense in $\mathcal{PML}(S) \times \mathcal{PML}(S)$, there is a pseudo-Anosov element $F$ whose stable and unstable laminations are not in $L$, so that $F$ is Hempel.

Masur found a condition in terms of the intersection number for a curve to be in $D$ [**12**, Lemma 1.1] and used it to prove $L$ is nowhere dense [**12**, Theorem 1.2]. The following lemma is proved using his ideas.

LEMMA 5.4. *Let* $V' \to V$ *be a double cover between handlebodies with the deck group* $\langle a \rangle$, $S' = \partial V', S = \partial V$, *and* $D' \subset \mathcal{C}(S'), D \subset \mathcal{C}(S)$ *the set of curves that bound disks in* $V', V$, *respectively.*

*If the genus of* $S$ *is* $\geq 3$, *then* $MCG(S)$ *contains a pseudo-Anosov element* $F$ *such that:*

(i) $F$ *is Hempel for* $D$,
(ii) $F$ *lifts to* $F' : S' \to S'$ *and* $F'$ *is Hempel for* $D'$,
(iii) $N(F') = \langle a \rangle$.

PROOF. Let $\Omega$ be the Masur domain for $V$ and $\Omega'$ for $V'$. We first find a lamination $\Lambda$ on $S$ that is in $\Omega$ such that its lift $\Lambda'$ on $S'$ is also in $\Omega'$. Choose a pants decomposition of $S$ using curves in $D$ and a lamination $\Lambda \in \mathcal{PML}(S)$ whose support intersects each pair of pants in this decomposition in 3 (non-empty) families of arcs connecting distinct boundary components (so there are no arcs connecting a boundary component to itself). In the proof of [**12**, Theorem 1.2] Masur shows that $\Lambda \in \Omega$ (for example, take the curve $\beta$ in his proof as $\Lambda$). This is done by verifying the conditions in Lemma 1.1 for $\beta$ with respect to the pants decomposition in the last two paragraphs of the proof of Theorem 1.2. Now the lift $\Lambda'$ of $\Lambda$ to $S'$ satisfies the same condition with respect to the lifted pants decomposition (it lifts since our covering is between handlebodies and the boundary curves bound disks), so we have $\Lambda' \in \Omega'$.

Choose a pseudo-Anosov homeomorphism $G : S \to S$ both of whose fixed points in $\mathcal{PML}(S)$ are close to $\Lambda$ and in particular they are in $\Omega$ since $\Omega$ is open. The lift $G'$ of $G$ (or its power) to $S'$ similarly has endpoints close to $\Lambda'$ and in particular in $\Omega'$. It follows that both $G$ and $G'$ are Hempel.

To finish the proof we need to arrange that $G$ has the extra property (iii). Let $H : S \to S$ be an arbitrary pseudo-Anosov mapping class that satisfies the assumption of Lemma 5.2. Such $H$ exists by Corollary 5.3. Then $F = G^n H G^{-n}$ also satisfies the assumptions, and hence also conclusion of Lemma 5.2 for any $n > 0$ and has an axis whose endpoints are close to $\Lambda$ if $n > 0$ is sufficiently large. Therefore $F$ is Hempel, and similarly, the lift $F'$ has an axis whose endpoints close to $\Lambda'$, therefore $F'$ is Hempel. □

**5.3. Proof of Theorem 5.1.** We prove Theorem 5.1 by constructing an example.

PROOF OF THEOREM 5.1. Let $H \simeq \mathbb{Z}/2\mathbb{Z} + \mathbb{Z}/2\mathbb{Z}$ with generators $a_1, a_2$, and let $V' \to V$ be a normal cover between handlebodies with the deck group $H$. If $g \geq 2$ is the genus of $V$, then the genus of $V'$ is $4g-3$. Let $D' \subset \mathcal{C}(S')$ be the set of curves in $S' = \partial V'$ that bound disks in $V'$. We have two double covers $S' \to S'/a_i$. Let $D_i \subset \mathcal{C}(S'/a_i)$ be the set of curves in $S'/a_i$ that bound disks in $V'/a_i$. Put $S_i = S'/a_i$. The genus of $S_i$ is $2g-1$. Let $Q$ be a common quasi-convexity constant for $D', D_1, D_2$, and $\delta$ the hyperbolicity constant of $\mathcal{C}(S')$.

Using Lemma 5.4, take a pseudo-Anosov element $F_i$ on $S_i$ that is Hempel for $D_i$ such that the lift $F_i'$ of $F_i$ to $S'$ is also Hempel for $D'$, and that $N(F_i') = \langle a_i \rangle$. Note that $F_1', F_2'$ are independent pseudo-Anosov elements on $S'$ since their elementary closures are different. In particular, the projection of $axis(F_1')$ to $axis(F_2')$ is bounded, and vice versa.

Note that $a_i \in stab(D')$ since $a_i \in H$. Set $D_i' = F_i'^N(D')$ for $N > 0$. Then, $a_i \in stab(D_i')$ since $F_i'$ centralizes $a_i$.

Form the Heegaard splitting $V_+' \cup_{S'} V_-'$ such that $D_+ = D_1', D_- = D_2' \subset \mathcal{C}(S')$. The surface $S'$ is fixed but the splitting depends on $N$. We will argue this is a desired splitting.

Set $\Gamma_i = stab(D_i') < MCG(S')$. In other words, $\Gamma_1 = \Gamma_+, \Gamma_2 = \Gamma_-$ in the Heegaard splitting convention. Since $a_i \in \Gamma_i$ and $a_1 a_2 = a_2 a_1$, $\langle \Gamma_1, \Gamma_2 \rangle$ is not the free product of $\Gamma_1, \Gamma_2$.

To prove the theorem we are left to verify $d(D_1', D_2') \to \infty$ as $N \to \infty$ and $\Gamma_1 \cap \Gamma_2 = 1$ for any large $N > 0$.

LEMMA 5.5. *$d(D_1', D_2') \to \infty$ as $N \to \infty$.*

PROOF. We claim that there is a constant $A$ such that for any $N > 0$,

$$d(D_1', D_2') \geq (trans(F_1') + trans(F_2'))N - A.$$

Let $\pi_1$ denote the projection to $axis(F_1')$, and $\pi_2$ the projection to $axis(F_2')$. As we said they are coarse maps but we pretend they are maps for simplicity. Also, we pretend that both $axis(F_1'), axis(F_2')$ are geodesics.

Let $L$ be a common bound of the diameter of the sets $\pi_1(D')$, $\pi_1(axis(F_2'))$, $\pi_2(D')$ and $\pi_2(axis(F_1'))$. Then $L$ is a bound of $\pi_1(D_1')$ and $\pi_2(D_2')$ for all $N > 0$. Choose points $q_1 \in \pi_1(D')$, $q_2 \in \pi_2(D')$, $r_1 \in \pi_1(axis(F_2'))$ and $r_2 \in \pi_2(axis(F_1'))$.

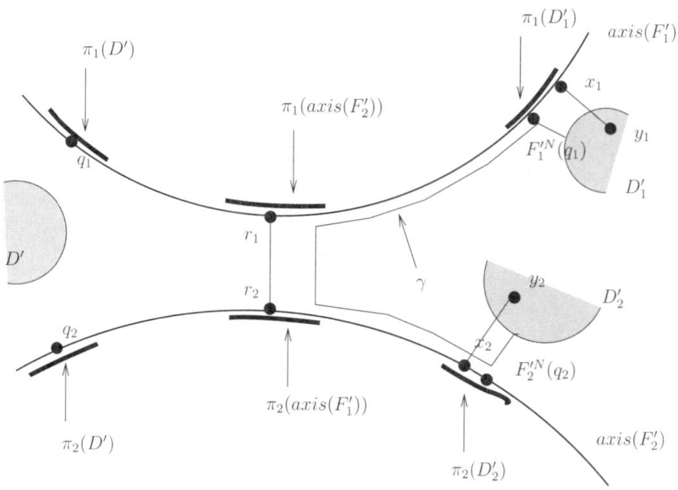


FIGURE 3. If $N > 0$ is large, the projection $\pi_1(axis(F'_2))$ and $\pi_1(D'_1)$ are far apart on $axis(F'_1)$, Also, $\pi_2(axis(F'_1))$ and $\pi_2(D'_2)$ are far apart on $axis(F'_2)$. As a consequence any shortest geodesic $\gamma$ between $D'_1$ and $D'_2$ must enter a bounded neighborhood of each of those four projection sets, and the segment in $\gamma$ near $axis(F'_i)$ is almost fixed by $a_i$ pointwise.

Now assume $N > 0$ is so large that $\pi_1(axis(F'_2))$ and $\pi_1(D'_1)$ are far apart, and also $\pi_2(axis(F'_1))$ and $\pi_2(D'_2)$ are far apart (compared to $L$ and $\delta$). It suffices to show the above inequality under this assumption.

Let $y_1 \in D'_1$ and $y_2 \in D'_2$ be any points, and put $x_1 = \pi_1(y_1), x_2 = \pi_2(y_2)$. Then, by a standard argument using $\delta$-hyperbolicity, the piecewise geodesic $[y_1, x_1] \cup [x_1, r_1] \cup [r_1, r_2] \cup [r_2, x_2] \cup [x_2, y_2]$ is a quasi-geodesic with uniform quasi-geodesic constants that depends only on $L$ and $\delta$. See Figure 3. Hence the Hausdorff distance between the piecewise geodesic and the geodesic $[y_1, y_2]$ is bounded (the bound depends only on $L$ and $\delta$).

It follows that there is a constant $C$ such that $d(D'_1, D'_2) \geq d(y_1, x_1) + d(x_1, r_1) + d(r_1, r_2) + d(r_2, x_2) + d(x_2, y_2) - C \geq d(x_1, r_1) + d(r_2, x_2) - C \geq d(F_1'^N(q_1), r_1) - L + d(F_2'^N(q_2), r_2) - L - C$. On the other hand since $q_1, r_1 \in axis(F'_1)$ and $q_2, r_2 \in axis(F'_2)$ there is a constant $B$ such that for all $N > 0$ we have $d(F_1'^N(q_1), r_1) + d(F_2'^N(q_2), r_2) \geq (trans(F'_1) + trans(F'_2))N - B$. Combining them we get a desired estimate with $A = 2L + B + C$. □

We note that any geodesic joining a point in $D'_1$ and a point in $D'_2$ passes through a bounded neighborhood of each of $F_1'^N(q_1), r_1, r_2, F_2'^N(q_2)$ provided that $N > 0$ is large enough. The bound depends only on $L$ and $\delta$. See Figure 3. In the argument we did not use that $D'$ (as well as $D'_1, D'_2$) are quasi-convex.

To argue $\Gamma_1 \cap \Gamma_2 = 1$, we will need the following lemma from [**2**, Proposition 6]. This is a consequence of the fact that $F$ is a "WPD element".

LEMMA 5.6. *Let $F$ be a pseudo-Anosov mapping class on a hyperbolic surface $S$. There is a constant $M > 0$ such that for any $g \in MCG(S)$ the diameter of the projection of $g(axis(F))$ to $axis(F)$ in $\mathcal{C}(S)$ is larger than $L$, then $g \in E(F)$.*

LEMMA 5.7. *$\Gamma_1 \cap \Gamma_2 = 1$ for any large $N > 0$.*

PROOF. By Lemma 5.5 choose $N$ large such that $d(D_1', D_2')$ is very large compared to $\delta$ and $L$. Let $\gamma$ be a shortest geodesic from $D_1'$ to $D_2'$. Then as we noted after the proof of Lemma 5.5, $\gamma$ passes through the bounded neighborhood of each of $F_1'^N(q_1), r_1, r_2, F_2'^N(q_2)$.

Now let $f \in \Gamma_1 \cap \Gamma_2$. Then we have $d(x, f(x)) \leq 2Q + 8\delta$ for any $x \in \gamma$ by Lemma 3.2 (2). Since all of $F_1'^N(q_1), r_1, r_2, F_2'^N(q_2)$ are in bounded distance from $\gamma$ we conclude each of those four points is moved by $f$ a bounded amount (the bound depends only on $\delta, L, Q$).

But since $F_1'^N(q_1), r_1$ are contained in $axis(F_1')$ and are far apart from each other for any large $N > 0$, we find $f(axis(F_1'))$ has a long ($> M$) projection to $axis(F_1')$, hence $f \in E(F_1')$ by Lemma 5.6. By the same reason $f \in E(F_2')$. We conclude $f \in E(F_1') \cap E(F_2')$. But since $F_1'$ and $F_2'$ are independent, $f$ must be a torsion element, so $f \in N(F_1') \cap N(F_2')$. By Lemma 5.2, $f \in \langle a_1 \rangle \cap \langle a_2 \rangle = 1$. We showed the lemma. □

We proved the theorem. □

## 6. Prime pseudo-Anosov elements

In view of Lemma 5.2 we introduce a property that looks interesting for its own sake. We say a pseudo-Anosov mapping class $F$ is *prime* if its stable/unstable foliations are not lifts of any foliations of a (possibly orbifold) quotient of $S$.

If $F$ is prime then $E(F)$ is cyclic and $N(F) = 1$. Indeed, if $N(F) \neq 1$ then the two foliations lift from $S/N(F)$, with $N(F)$ realized as a group of isometries of $S$ using Nielsen realization. Moreover,

LEMMA 6.1. *(cf. Lemma 5.2) Suppose $S' \to S$ is a finite cover with the Deck group $\Delta$. Let $F$ be a prime pseudo-Anosov element on $S$ and $F'$ a lift of a power of $F$ to $S'$. Then $N(F') = \Delta$.*

PROOF. It is clear that $\Delta < N(F')$. If the inclusion is strict, then $S = S'/\Delta$ covers $S'/N(F')$ (maybe a branched cover). Then the stable and unstable foliations of $F$ on $S$ can be obtained by pulling back the quotient foliation of $F$ by $N(F')$ on $S'/N(F')$. So we have a contradiction. □

Note that if we have a prime pseudo-Anosov element on $S$, we can use Lemma 6.1 instead of Lemma 5.2 in the proof of Lemma 5.4 and Theorem 5.1. We will give a construction of prime pseudo-Anosov elements when the genus of $S$ is 3, so this will also prove the theorem for the genus 5 case.

Recall that if $a, b, c, d$ are 4 vectors in $\mathbb{R}^2$ then the cross ratio is

$$[a, b; c, d] = \frac{[a, c][b, d]}{[a, d][b, c]}$$

where $[x, y] = x_1 y_2 - x_2 y_1$ for $x = (x_1, x_2), y = (y_1, y_2)$. We do not define it when one of $[a, c], [b, d], [a, d], [b, c]$ is 0. The cross ratio is invariant under changing signs and scaling individual vectors and applying matrices in $SL_2(\mathbb{R})$. It follows that for any flat structure on the torus the cross ratio for the vectors in the directions of four distinct closed geodesics is (well-defined and) rational.

A singular Euclidean structure (or just a *flat* structure) on a surface $S$ is *good* if the cone angle is a multiple of $\pi$ at each singularity. A geodesic segment connecting two singular points, or a closed geodesic is *good* if the angle along the geodesic at each singular point is a multiple of $\pi$.

The developing map $\widetilde{S - \Sigma} \to \mathbb{R}^2$ defined on the universal cover of the complement of the cone points will take a good geodesic to a straight line, or a line segment. So, for any four good geodesics, the cross ratio for the four directions, if they are distinct, is well-defined.

Next, if $S' \to S$ is a branched cover between good flat structures, then the cross ratio of four good geodesics in $S'$ is equal to the cross ratio of their images in $S$, simply because $S, S'$ have the "same" developing map. In particular, all cross ratios between good geodesics on a torus or a sphere with 4 cone points are rational, and to prove that a particular good flat surface is not commensurable with a torus it suffices to produce four good geodesics whose cross ratio is irrational.

LEMMA 6.2. *Suppose $F : S \to S$ is a pseudo-Anosov homeomorphism such that:*

*(1) the stable foliation of $F$ has two singular points $x$ and $y$, with $p$ and $q$ prongs respectively, and with $p$ and $q$ distinct odd primes, and*
*(2) a flat structure on $S$ determined by the stable and unstable foliations has four good closed geodesics with the cross ratio of their (distinct) direction vectors $a, b, c, d \in \mathbb{R}^2$ irrational.*

*Then $F$ is prime.*

We note that there is a 2-parameter family of flat structures determined by the two foliations; they depend on the choice of the transverse measure in a projective class on each foliation. However, since scaling and linear transformations do not change the cross ratio, the assumption is independent of these choices.

PROOF. Let $p, q, F$ be as in the statement. Now suppose $\pi : S \to S'$ is a branched cover of degree $d > 1$ and $\mathcal{F} = \pi^{-1}\mathcal{F}'$. The local degree of $\pi$ at $x$ is either 1 or $p$. It cannot be 1, since at any other preimage of $\pi(x)$ the singularity would have to have $kp$ prongs, and there aren't any. Thus at $x$ the map is modeled on $z \mapsto z^p$, and similarly at $y$ it looks like $z \mapsto z^q$. There are now two cases.

*Case 1.* $\pi(x) \neq \pi(y)$.

It follows that the other points that map to $\pi(x)$ have 2 prongs and so the map there has local degree 2. Thus $d$ is odd and away from the images of singular points the foliation $\mathcal{F}'$ is regular (since otherwise $d$ would have to be even). Thus there are $(d-p)/2$ other preimages of $\pi(x)$, and deleting these and the same for the $q$-prong singularity we get that the Euler characteristic of $S - \pi^{-1}(\{\pi(x), \pi(y)\})$ is

$$(2 - 2g) - (d - p)/2 - (d - q)/2 - 2 = -d$$

So the Euler characteristic of the quotient $S'$ minus 2 singular points is $-1$, i.e. the quotient is the twice punctured $\mathbb{R}P^2$, which does not support any pseudo-Anosov homeomorphisms (e.g. the curve complex is finite, see [**20**]). On the other hand, let $\tilde{S}$ be a finite cover of the punctured $S$ so that the induced cover to $\mathbb{R}P^2$ minus two points is regular. Some power of $F$ lifts to $\tilde{F}$ on $\tilde{S}$, and since the cover is regular, a further power of $\tilde{F}$ descends to $\mathbb{R}P^2$ (since each element, $a$, of the Deck group leaves the stable and unstable foliations invariant, so that $(a\tilde{F}a^{-1})^N = \tilde{F}^N$ for some $N > 0$, so $a$ and $\tilde{F}^N$ commute), contradiction.

*Case 2.* $\pi(x) = \pi(y)$.

Again the other points that map to $\pi(x) = \pi(y)$ are regular and the map has local degree 2, so there are $\frac{d-(p+q)}{2}$ such points. Here $d$ is even and we may have some number, say $k \geq 0$, of 1-prong singularities $z_1, \cdots, z_k$ in the quotient, with each singularity having $\frac{d}{2}$ preimages where local degree is 2. Now we have that the Euler characteristic of $S - \pi^{-1}(\{\pi(x), z_1, \cdots, z_k\})$ is

$$(2-2g) - \frac{d-(p+q)}{2} - 2 - \frac{kd}{2} = -\frac{(k+1)d}{2}$$

So $k$ is odd and the Euler characteristic of the quotient $S'$ minus the singular points is $-\frac{k+1}{2}$. So, the Euler characteristic of $S'$ is $-\frac{k+1}{2} + (k+1) = \frac{k+1}{2}$. The only possibilities are $k = 1$ and $k = 3$, and the quotient is twice punctured $\mathbb{R}P^2$ or 4 times punctured $S^2$. The first possibility is ruled out as in Case 1.

In the second case the good flat structure on $S$ descends to a good flat structure on $S^2$ with 4 singular points, then lifts to a flat structure on the branch double cover $T^2$, with four closed geodesics such that the cross ratio of the four direction vectors is $[a, b; c, d]$ that is not rational, contradiction.

Indeed, using the same notation as in Case 1, a power of $F$ lifts to $\tilde{F}$ on $\tilde{S}$ that regularly covers $S^2$ minus 4 points. We lift the flat structure on $S$ and the stable and unstable foliations of $F$ to $\tilde{S}$. Then their regular leaves are straight lines. Each deck transformation preserves the foliations, so that it is an isometry of $\tilde{S}$, and that the good flat structure on $\tilde{S}$, with cone angle at each singular point at least $2\pi$, descends to a good flat structure of $S^2$ minus 4 points (and the angle at each puncture is $\pi$). We obtain a good flat structure on $S^2$ with four good closed geodesics and the cross ratio is $[a, b; c, d]$. Also, the cross ratio will not change when we take a double cover that is a flat torus, contradiction. □

REMARK 6.3. *Regarding the assumption (1), if $g$ is the genus of $S$, by an Euler characteristic count we must have $p + q = 4g$. Conversely, the Goldbach conjecture predicts that every even integer $> 2$ can be written as a sum of two primes. When the integer is $\geq 8$ and divisible by 4, the two primes are necessarily distinct and odd. For example, $12 = 5 + 7$ satisfies the Goldbach conjecture. The work of Masur-Smillie* [**15**] *shows that if $g \geq 3$ and $4g = p + q$ then the surface $S$ admits a pseudo-Anosov homeomorphism whose stable and unstable foliations have two singular points, one of order $p$, the other of order $q$.*

EXAMPLE 6.4. *We now construct an explicit example in genus 3 satisfying the assumption of Lemma 6.2. We take $p = 5$, $q = 7$. Consider the flat square tiled surface $S$ pictured below. Edges labeled by the same letter are to be identified. If the edges are on opposite sides of the parallelogram they are identified by a translation, and otherwise by a rotation by $\pi$. The square tiling of $\mathbb{R}^2$ induces one on the surface $S$. There are two cone points, with cone angles $5\pi$ and $7\pi$ respectively. So, $S$ has a good flat structure.*

*We will use Thurston's construction of pseudo-Anosov homeomorphisms* [**6**, Theorem 14.1] *to construct $F$. The lines bisecting the squares form three horizontal and three vertical geodesics. The matrix $N$ of intersection numbers, where the $jk$ entry is the intersection number $i(\alpha_j, \alpha'_k)$, is*

$$N = \begin{pmatrix} 1 & 1 & 0 \\ 1 & 1 & 1 \\ 1 & 0 & 0 \end{pmatrix}$$

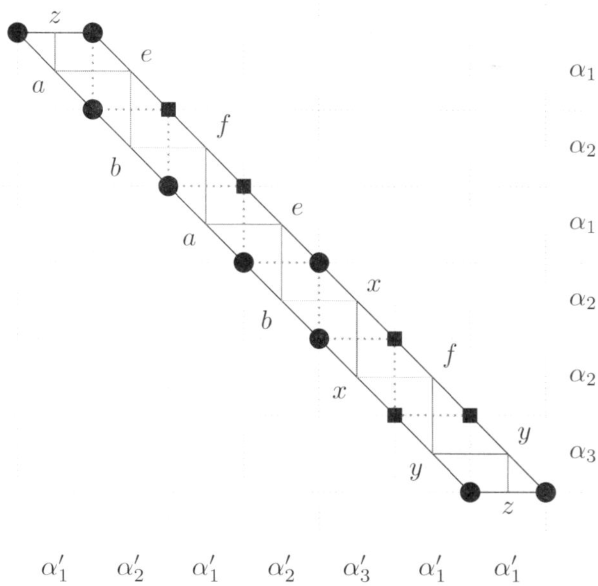


FIGURE 4. The square tiled surface. The round vertex has 7 prongs and the square vertex has 5. The surface has a good flat structure.

*The largest eigenvalue of* $NN^t$ *is* $\mu = 5.0489\ldots$ *satisfying the minimal polynomial* $\mu^3 - 6\mu^2 + 5\mu - 1 = 0$. *Then one can choose the lengths and heights of the squares (making them into rectangles, which gives a tiling of* $S$*) so that the twist in the horizontal multicurve is given by the matrix*

$$\begin{pmatrix} 1 & \mu^{1/2} \\ 0 & 1 \end{pmatrix}$$

*and the twist in the vertical multicurve by the matrix*

$$\begin{pmatrix} 1 & 0 \\ -\mu^{1/2} & 1 \end{pmatrix}$$

*This means that the product of the first and the inverse of the second is*

$$\begin{pmatrix} 1 & \mu^{1/2} \\ 0 & 1 \end{pmatrix} \begin{pmatrix} 1 & 0 \\ \mu^{1/2} & 1 \end{pmatrix} = \begin{pmatrix} 1+\mu & \mu^{1/2} \\ \mu^{1/2} & 1 \end{pmatrix} = A$$

*whose trace is* $2+\mu$. *So this product is pseudo-Anosov and its dilatation is the larger,* $\lambda$*, of the eigenvalues of* $A$ *and the eigenvector is* ${}^t(1,\sigma)$ *with* $\mu^{1/2} = (1-\sigma^2)/\sigma$ *and* $\lambda = 1/\sigma^2$.

*The heights and widths of the rectangles are coordinates of the* $\mu$*-eigenvectors* $V$ *of* $NN^t$ *and* $V'$ *of* $N^tN$. *We compute*

$$V = \begin{pmatrix} 1 \\ \mu^2 - 5\mu + 1 \\ -2\mu^2 + 11\mu - 4 \end{pmatrix} = \begin{pmatrix} v_1 \\ v_2 \\ v_3 \end{pmatrix}$$

*and*

$$V' = \mu^{-1/2} N^t V = \mu^{-1/2} \begin{pmatrix} -\mu^2 + 6\mu - 2 \\ \mu^2 - 5\mu + 2 \\ \mu^2 - 5\mu + 1 \end{pmatrix} = \begin{pmatrix} v_1' \\ v_2' \\ v_3' \end{pmatrix}$$

*To show that $F$ is prime it suffices to find 4 closed geodesics whose slopes have irrational cross ratio. We take $a = (1,0)$, $b = (0,1)$, $c = (-v_2', v_1 + v_2)$, $d = (-v_1' - v_2', 2v_1 + v_2)$, where $c$ and $d$ connect second, respectively third, vertex on the lower left side in the figure with the upper right vertex. They are good closed geodesics on $S$ based at the round vertex (to compute the angle at the round vertex, it helps first to identify the two edges labeled by $z$). The cross ratio is*

$$\frac{[a,c][b,d]}{[a,d][b,c]} = \frac{(v_1+v_2)(v_1'+v_2')}{(2v_1+v_2)v_2'} = \frac{(\mu^2-5\mu+2)\mu}{(\mu^2-5\mu+3)(\mu^2-5\mu+2)} = \frac{\mu}{\mu^2-5\mu+3}$$
$$= \frac{1}{3\mu^2 - 17\mu + 10},$$

*which is irrational (for the last equality use $\mu^3 - 6\mu^2 + 5\mu - 1 = 0$).*

## 7. Invariable generation

In this section we discuss another application of Theorem 3.6 on "invariable generation" of a group. For this we need the version stated in Theorem 3.5. We would like to thank A. Lubotzky for introducing the subject to us.

**7.1. Definitions and result.** Following Dixon [**5**] a group $G$ is *invariably generated* by a subset $S$ of $G$ if $G = \langle s^{g(s)} | s \in S \rangle$ for any choice of $g(s) \in G, s \in S$. The group $G$ is *IG* if it is invariably generated by some subset $S$ in $G$, or equivalently, if $G$ is invariably generated by $G$. Abelian groups are IG, and also it is easy to see that finite groups are IG.

$G$ is *FIG* if is is invariably generated by some finite subset of $G$. FIG implies IG. Finite groups are FIG, and for infinite groups, Kantor-Lubotzky-Shalev [**11**] prove that a linear group is FIG if and only if it is finitely generated and virtually solvable.

In this section, we are interested in groups $G$ that are not IG. In other words, $G$ contains a proper subgroup such that each element in $G$ is conjugate to some element in the subgroup. Here are some examples. $SL(n, \mathbb{C}), n > 1$ are not IG since each element is conjugate to an upper triangular matrix using Jordan normal forms. By a similar reason, $SL(n, \mathbb{R}), n > 2$ are not IG. Also, $SO(3, \mathbb{R})$ is not IG (consider its action on $S^2$, then each element fixes a point). It is unknown if $SL(n, \mathbb{Z}), n > 2$ is IG or not.

Gelander proves that non-elementary hyperbolic groups (and also relatively hyperbolic groups as well) are not IG [**7**]. We generalize this result to acylindrically hyperbolic groups. A group $G$ is *acylindrically hyperbolic* if it admits an acylindrical action on a hyperbolic space and $G$ is not virtually cyclic [**18**]. Examples are non-elementary hyperbolic groups, $MCG(S_{g,p})$ except for the genus $g = 0$ and the number of punctures $p \leq 3$, and $Out(F_n)$ with $n \geq 2$ (cf. [**18**]).

We prove

THEOREM 7.1. *If $G$ is an acylindrically hyperbolic group, then $G$ is not IG.*

REMARK 7.2. *In fact our argument shows that $G$ is IG if $G$ acts on a hyperbolic space $X$ with at least one hyperbolic isometry $f$ that is "WPD" ([**2**]), and $G$ is not* *virtually cyclic. It it trivial that an acylindrically hyperbolic group satisfies this assumption, and it may seem that the class of groups with this assumption is larger than the class of acylindrically hyperbolic groups. It turns out they are equal ([**19**]), but we do not need this fact to prove the theorem.*

**7.2. Invariant quasi-convex subsets.** Suppose $G$ acts on a $\delta$-hyperbolic space $X$. This is the setting through this section unless otherwise said.

For $g \in G$, define its *minimal translation length* by

$$min(g) = \inf_{p \in X} |p - g(p)|.$$

It is a well-known fact that if $min(g) \geq 10\delta$ then $g$ is hyperbolic. (To be precise, we assume $\delta > 0$ here.)

For $L > 0$ define a subset

$$X(g, L) = \{x \in X | |gx - x| \leq L\},$$

and put

$$M(g) = X(g, min(g) + 1000\delta).$$

$M(g)$ is a $g$-invariant non-empty set. If $g$ is hyperbolic, then $M(g)$ is contained in a Hausdorff neighborhood of an axis of $g, axis(g)$. This is an easy exercise and we leave it to the reader (If $g$ has an unbounded orbit in $X$ then $|x - g(x)|$ grows almost linearly on the distance $|x - M(g)|$).

LEMMA 7.3. *If $g$ has a bounded orbit, then $min(g) \leq 6\delta$.*

PROOF. This is well known too. Let $Z$ be the set of centers of the orbit of a point $x$ by $g$. $Z$ is invariant by $g$ and its diameter is at most $6\delta$, therefore is contained in $X(g, 6\delta)$. □

LEMMA 7.4. *If $L \geq min(g) + 1000\delta$, then $X(g, L)$ is $100\delta$-quasi-convex.*

PROOF. We define a function on $X$ as follows: $t_g(x) = |x - g(x)|$.
*Case 1. $min(g) \leq 10\delta$.*
Fix $p \in X(g, 10\delta) \subset X(g, L)$. Suppose $x \in X(g, L)$ is given. We will show that $[p, x]$ is contained in the $20\delta$-neighborhood of $X(g, L)$. If $|x - p| \leq 100\delta$, then by triangle inequality, $t_g(z) \leq 210\delta$ for every $z \in [p, x]$, so that $[p, x] \subset X(g, L)$. So assume $|x - p| > 100\delta$. To compute $t_g(z)$ for $z \in [p, x]$, draw a triangle $\Delta$ for $p, x, g(x)$, and let $c \in [p, x]$ be a branch point of this triangle, i.e., the distance to each side of $\Delta$ from $c$ is at most $\delta$. Since $|p - g(p)| \leq 10\delta$, $t_g(z) \leq 20\delta$ for any point $z \in [p, c]$. For $z \in [c, x]$, $t_g(z)$ is roughly equal to $2d(c, z)$, with an additive error at most $20\delta$. Also it is roughly maximal at $x$ on $[p, x]$. (Imagine the case that $X$ is a tree and $p = g(p)$.) It follows that $[p, x]$ is contained in the $20\delta$-neighborhood of $X(g, L)$.

Now suppose another point $y \in X(g, L)$ is given. Then $[x, y]$ is contained in the $\delta$-neighborhood of $[p, x] \cup [p, y]$, so that $[x, y]$ is contained in the $21\delta$-neighborhood of $X(g, L)$.
*Case 2. $min(g) \geq 10\delta$.*
Then $g$ is hyperbolic. To simplify the argument, let's assume that there is a geodesic axis for $g$. Then for any $x \in X$,

$$|t_g(x) - \{trans(g) + 2d(x, axis(g))\}| \leq 20\delta$$

To see this let $x' \in axis(g)$ be a nearest point from $x$. Then the Hausdorff distance between $[x, g(x)]$ and $[x, x'] \cup [x', g(x')] \cup [g(x'), g(x)]$ is at most $5\delta$, and the estimate follows.

It follows from the above estimate that if $x \in X(g, L)$, then $[x, x']$ is contained in the $20\delta$-neighborhood of $X(g, L)$. For $y \in X(g, L)$, let $y' \in axis(g)$ be a nearest point from another point $y$ to $axis(g)$. Then $[x, x'] \cup [x', y'] \cup [y', y]$ is contained in the $20\delta$-neighborhood of $X(g, L)$. But since $[x, y]$ is contained in the $5\delta$-neighborhood of $[x, x'] \cup [x', y'] \cup [y', y]$, $[x, y]$ is contained in the $25\delta$-neighborhood of $X(g, L)$.

The argument for the case that $g$ has only a quasi-geodesic (with uniform quasi-geodesic constants depending only on $\delta$) as an axis is similar and we only need to modify the constants in the argument. We omit the details. □

Lemma 7.4 implies

LEMMA 7.5. *$M(g)$ is $100\delta$-quasi-convex for every $g \in G$.*

**7.3. Projection to axes and elementary closures.** Let $\partial X$ denote the Gromov boundary at infinity of $X$. For a quasi-convex subset $Y \subset X$, let $\partial Y \subset \partial X$ be the boundary at infinity of $Y$.

LEMMA 7.6. *If $p \in \partial M(g) \subset \partial X$, then $g(p) = p$.*

PROOF. Let $\gamma$ be a geodesic ray from a point in $M(g)$ that tends to $p$. Then the ray is contained in the $100\delta$-neighborhood of $M(g)$. So every point of the ray is moved by $g$ by a bounded amount, therefore $p$ is fixed by $g$. □

When $f$ is hyperbolic, $\partial(axis(f))$ consists of two points. The subgroup of elements in $G$ that fix each point of $\partial(axis(f))$ is called the *elementary closure* of $f$, denoted by $E(f)$.

LEMMA 7.7. *Assume the action of $G$ is acylindrical on $X$. If $g \in G$ fixes one point in $\partial(axis(f))$, then $g \in E(f)$.*

PROOF. Let $\gamma$ be a ray in $axis(f)$ that tends to the point fixed by $g$. Then $|x - gx|$ is bounded for $x \in \gamma$. Assume that $\gamma$ tends to the direction that $f$ translates $axis(f)$ (otherwise, we let $N < 0$ below). Then for any $N > 0$ and $x \in \gamma$, $|f^{-N} g f^N(x) - x|$ is bounded. Now by acylindricity (apply it to $x, y \in \gamma$ that are far from each other), there are only finitely many possibilities for $f^{-N} g f^N$, so $g$ commutes with a non-trivial power of $f$. So $g$ moves each point in $axis(f)$ by a bounded amount, therefore $g \in E(f)$. □

$M(g)$ is a $100\delta$-quasi-convex subset that is invariant by $< g >$. In view of Theorem 3.5 we define a larger set that contains $M(g)$ as follows. If the order of $g \in G$ is infinite, define

$$M'(g) = \cup_{n \neq 0} M(g^n),$$

and if the order is $P < \infty$, then define

$$M'(g) = \cup_{0 < n < P} M(g^n).$$

Clearly $M'(g)$ contains $M(g)$ and invariant by $g$, also the following is obvious from the definition.

LEMMA 7.8. *For any $x \in X - M'(g)$ and any non-trivial $h \in < g >$, we have $|h(x) - x| > 1000\delta$. In particular, $CFix(h) \subset M'(g)$.*

LEMMA 7.9. *Suppose the action of $G$ is acylindrical on $X$, and $f \in G$ is hyperbolic on $X$.*

*For some non-trivial $g \in G$, assume $< g > \cap E(f)$ is trivial. Then for each $g^n \neq 1$, the projection of $M(g^n)$ to $axis(f)$ is bounded. Moreover, the projection of $M'(g)$ is bounded.*

PROOF. By assumption $g \notin E(f)$. We first show the projection of $M(g)$ is bounded. Suppose not, and let $p$ be a point in $\partial(axis(f))$ that the projection of $M(g)$ tends to. Then $p \in \partial M(g)$. This is because since both $axis(f)$ and $M(g)$ are quasi-convex, a half of $axis(f)$ to the direction of $p$ is contained in a bounded neighborhood of $M(g)$. By Lemma 7.6 $g(p) = p$. Then by Lemma 7.7, $g$ is in $E(f)$, a contradiction.

The same argument applies to all $M(g^n)$ with $g^n \neq 1$.

Now assume the projection of $M'(g)$ is not bounded. In particular $g$ has infinite order. Choose an element $g^n \neq 1$ such that the projection of $M(g)$ and $M(g^n)$ are very far (we specify later) from each other on $axis(f)$. Let $B$ be a bound of the projection of $M(g)$. Notice both $M(g)$ and $M(g^n)$ are invariant by $g^n$ and quasi-convex, so that a shortest geodesic between the two sets is almost invariant by $g^n$. It implies that the end points of the (long, compared to $B$) subsegment of $axis(f)$ joining the projections of $M(g)$ and $M(g^n)$ are moved by $g^n$ by a bounded amount, say, $R$ (the bound does not depend on $g^n$, but only on $\delta$).

Moreover, this holds for any power of $g^{nI}, I = 1, 2, ....$ But this is impossible by the acylindricity of the action (arrange the distance between the projections of $M(g)$ and $M(g^n)$ is much larger than $L(R)$, then there are at most $N(R)$ possibilities for $g^{nI}$). □

REMARK 7.10. *(1) Although we will not use this fact, we observe that the set $M'(g)$ is $Q$-quasi-convex, and $Q$ does not depend on $g$. For example, if $g$ has finite order $P$, then $M'(g)$ is $101\delta$-quasi-convex. This is because $g$ has an orbit whose diameter is at most $6\delta$ (see the proof of Lemma 7.3), and $M'(g) = \cup_{0<n<P} M(g^n)$ is a union of $100\delta$- quasi-convex sets all of which contain the bounded orbit. Thus $M'(g)$ is $101\delta$-quasi-convex as desired. (Fix a point $x$ from the bounded orbit. Then for any points $y, z \in M'(g)$, draw a $\delta$-thin triangle for $x, y, z$. $[y, z]$ is in the $\delta$-neighborhood of $[x, y] \cup [x, z]$.)*

*If $g$ has infinite order, then we may assume that for each $N \neq 0$, the distance between $M(g)$ and $M(g^N)$ is uniformly bounded, since otherwise, take a shortest geodesic $[p, q]$ from $M(g)$ to $M(g^N)$, which is almost invariant by $g$, so that the orbit of $p$ by $< g >$ is bounded. Now we can argue as in the previous paragraph. But if the distance between $M(g)$ and $M(g^N)$ is uniformly bounded for all $N \neq 0$, then $M'(g)$ is uniformly quasi-convex.*

*(2) In Lemma 7.9, we can show that a bound on the projection of $M'(g)$ does not depend on $g$.*

**7.4. Proof of Theorem 7.1.** We quote the following result from [**4**]. First, if the action is acylindrical, $G$ contains a "hyperbolically embedded subgroup" that is non-degenerate, i.e., proper and infinite, see Theorem 1.2 in [**18**]. Then [**4**, Theorem 6.14 (c)] applies to $G$. It is not explicitly stated that $E(f)$ is cyclic, but it easily follows from their proof.

By a *Schottky subgroup* $F < G$ we mean a non-abelian, free subgroup such that an orbit map $F \to X$ is a quasi-isometric embedding.

PROPOSITION 7.11 ([**4**, Theorem 6.14]). *Suppose the action of $G$ on $X$ is acylindrical and $G$ is not virtually cyclic. Assume that $G$ does not contain any non-trivial finite normal subgroup.*

*Then $G$ contains a Schottky subgroup $F$ so that $E(f)$ is infinite cyclic for every non-trivial $f \in F$.*

We will use elements from a Schottky group $F$ to move $M(g)$ and $M'(g)$ in $X$. Note that $M(hgh^{-1}) = hM(g), M'(hgh^{-1}) = hM'(g)$.

LEMMA 7.12. *Suppose there is a Schottky free subgroup $F < G$ such that any non-trivial $f \in F$ is hyperbolic and $E(f)$ is cyclic.*

*Let $S = \{g_1, g_2, g_3, \cdots\}$ be a (finite or infinite) set of non-trivial elements in $G$.*

*Then for any given constant $K > 0$ there is a set $S' = \{g'_i\}$ in $G$ such that*

- *$g_i$ and $g'_i$ are conjugate in $G$ for each $i$.*
- *For any $N > 0$, the sets $M'(g'_i), \cdots, M'(g'_N)$ are $K$-terminal.*

PROOF. We first prepare a sequence of elements in $F$ that we will use to conjugate $g_i$ to $g'_i$ to satisfy the second bullet. Take two elements $a, b \in F$ that produce a free subgroup of rank two. Put $f_i = ab^i, i \geq 1$. Note that $E(f_i) \cap E(f_j)$ is trivial if $i \neq j$.

We describe a geometric property we use about the sequence of elements. Fix a bounded set $Y \subset X$. Given a constant $K > 0$, if we choose $L$ sufficiently large, then for any $N > 0$, and any $L_i \geq L$, the following sets are $K$-terminal.

$$Y, f_1^{L_1}(Y), f_2^{L_2}(Y), \cdots, f_N^{L_N}(Y).$$

This is an easy consequence of the property such that the embedding of the Cayley graph of $F$ in $X$ using the orbit of a point in $X$ is quasi-isometric to the image.

To define $g'_1, g'_2, g'_3, \cdots$, choose a sequence

$$1 \neq n_1 < n_2 < n_3 < \cdots$$

such that for each $i$, $< g_i > \cap E(f_{n_i})$ is trivial. This is possible since if the order of $g_i$ is finite then this holds for every $E(f_j)$; if the order of $g_i$ is infinite, then the intersection is non-trivial for at most one $E(f_j)$.

By Lemma 7.9, the projection of $M'(g_i)$ to $axis(f_{n_i})$ is bounded. Now take a sequence of sufficiently large constants $L_i > 0$, depending on the given constant $K$ such that for the elements $g'_i = f_{n_i}^{L_i} g_i f_{n_i}^{-L_i}$, the projections of $M'(g'_1), \cdots, M'(g'_N)$ to the axes of $f_{n_1}, \cdots, f_{n_N}$, respectively, are $K$-terminal. We used that $M'(g'_i) = f_{n_i}^{L_i}(M'(g_i))$ and each of the projections of $M'(g_i)$ is bounded. It then follows (for maybe slightly larger constants $L_i$) that the sets $M'(g'_1), \cdots, M'(g'_N)$ themselves are $K$-terminal. □

*Proof of Theorem* 7.1.
*Case 1.* Assume $G$ does not contain any non-trivial finite normal subgroup.

By Proposition 7.11, there is a Schottky subgroup $F < G$ such that any non-trivial element $f \in F$ is hyperbolic and $E(f)$ is cyclic.

Let $K > 0$ be a constant from Theorem 3.5 for $Q = 100\delta$. Let $\mathcal{C} = \{g_1, g_2, \cdots\}$ be a set of all conjugacy classes of $G$ except for the class for the trivial element. We assume $\mathcal{C}$ contains at least two elements, otherwise the argument is straightforward. Apply lemma 7.12 to the set $\mathcal{C}$ and the constant $K$, and obtain a new set $\mathcal{C}' = \{g'_i\}$.

For each $i > 0$, $M(g_i')$ is a non-empty, $g_i'$-invariant, $100\delta$-quasi-convex subset. Also, for each $N$, the sets $M'(g_1'), \cdots, M'(g_N')$ are $K$-terminal.

For each $i$, since $CFix(h) \subset M'(g_i')$ for any non-trivial $h \in < g_i' >$ by Lemma 7.8, Theorem 3.5 applies to $< g_1' >, \cdots, < g_N' >$ with respect to the sets $M(g_i')$ and $M'(g_i')$, $1 \le i \le N$, so the subgroup generated by $g_1', \cdots, g_N'$ is the free product $< g_1' > * \cdots * < g_N' >$.

If the subgroup generated by the set $\mathcal{C}'$ is a proper subgroup, then we are done since $\mathcal{C}'$ meets all the conjugacy classes. So suppose that $G$ is generated by $\mathcal{C}'$.

Then we replace $g_2'$ by $g_2'' = (g_2'g_1')g_2'(g_2'g_1')^{-1}$, and obtain a new set $\mathcal{C}''$. $\mathcal{C}''$ meets all conjugacy classes of $G$. Notice that the subgroup generated by $g_1'$ and $g_2''$ is a free product $< g_1' > * < g_2'' >$, which is a proper subgroup in $< g_1' > * < g_2' >$.

It follows that the subgroup, $G_1$, generated by $\mathcal{C}''$ is a free product of cyclic subgroups generated by the elements in $\mathcal{C}''$, and this subgroup is proper in $G$. To see that, define a quotient homomorphism from $G$ to the group $H =< g_1' > * < g_2' >$ by sending all $g_i', i > 2$ to 1. Then the image of $G_1$ is a proper subgroup in $H$, so $G_1$ is a proper subgroup in $G$. We showed $G$ is not IG.

In conclusion we always find a proper subgroup that meets all conjugacy classes, which is a free product of cyclic subgroups.

*Case 2.* Assume that $G$ contains a non-trivial finite normal subgroup.

Let $K$ be the maximal finite normal subgroup in $G$. Then $G' = G/K$ does not contain any non-trivial finite normal subgroup. Moreover $G'$ is acylindrically hyperbolic group. Probably this fact is well known to specialists, and we postpone giving an argument till the end (Proposition 7.13).

By Case 1, $G'$ contains a proper subgroup $H'$ that contains all conjugacy classes of $G'$. Let $H < G$ be the pull-back of $H'$ by the quotient map $G \to G'$. Then $H$ is a proper subgroup that contains all conjugacy classes of $G$, therefore $G$ is not IG. □

PROPOSITION 7.13. *Let $G$ be an acylindrically hyperbolic group and $N < G$ a finite normal subgroup. Then $G' = G/N$ is an acylindrically hyperbolic group.*

PROOF. By assumption $G$ acts on a hyperbolic space $X$ such that the action is acylindrical and $G$ is not virtually cyclic. Moreover we may assume that the action is co-compact. In fact, we may assume that $X$ is a Cayley graph with a certain generating set, which is maybe infinite, [**18**, Theorem 1.2].

We will produce a new $G$-graph $Y$ from $X$ such that the kernel of the action contains $N$ and that $Y$ and $X$ are quasi-isometric. This is a desired action for $G'$.

For each $N$-orbit of a vertex of $X$, we assign a vertex of $Y$. Note that $G$ is transitive on the set of $N$-orbits of vertexes of $X$, so $G$ acts transitively on the vertex set of $Y$. Now join two vertices of $Y$ if the distance of the corresponding $N$-orbits in $X$ is 1. $Y$ is a connected $G$-graph and it is easy to check that $Y$ and $X$ are quasi-isometric (by the obvious map sending a vertex $x$ of $X$ to the vertex of $Y$ corresponding to the orbit of $x$), so that $Y$ is hyperbolic, and that the $G$-action on $Y$ is acylindrical. By construction, $N$ acts trivially on $Y$. □

## References


[1] A. Abrams and S. Schleimer, *Distances of Heegaard splittings*, Geom. Topol. **9** (2005), 95–119, DOI 10.2140/gt.2005.9.95. MR2115669

[2] M. Bestvina and K. Fujiwara, *Bounded cohomology of subgroups of mapping class groups*, Geom. Topol. **6** (2002), 69–89, DOI 10.2140/gt.2002.6.69. MR1914565

[3] B. H. Bowditch, *Tight geodesics in the curve complex*, Invent. Math. **171** (2008), no. 2, 281–300, DOI 10.1007/s00222-007-0081-y. MR2367021

[4] F. Dahmani, V. Guirardel, and D. Osin, *Hyperbolically embedded subgroups and rotating families in groups acting on hyperbolic spaces*, Mem. Amer. Math. Soc. **245** (2017), no. 1156, v+152, DOI 10.1090/memo/1156. MR3589159

[5] J. D. Dixon, *Random sets which invariably generate the symmetric group*, Discrete Math. **105** (1992), no. 1-3, 25–39, DOI 10.1016/0012-365X(92)90129-4. MR1180190

[6] B. Farb and D. Margalit, *A primer on mapping class groups*, Princeton Mathematical Series, vol. 49, Princeton University Press, Princeton, NJ, 2012. MR2850125

[7] T. Gelander, *Convergence groups are not invariably generated*, Int. Math. Res. Not. IMRN **19** (2015), 9806–9814, DOI 10.1093/imrn/rnu248. MR3431613

[8] C. McA. Gordon, *Problems*, Workshop on Heegaard Splittings, Geom. Topol. Monogr., vol. 12, Geom. Topol. Publ., Coventry, 2007, pp. 401–411, DOI 10.2140/gtm.2007.12.401. MR2408256

[9] J. Hempel, *3-manifolds as viewed from the curve complex*, Topology **40** (2001), no. 3, 631–657, DOI 10.1016/S0040-9383(00)00033-1. MR1838999

[10] J. Johnson, *Mapping class groups of medium distance Heegaard splittings*, Proc. Amer. Math. Soc. **138** (2010), no. 12, 4529–4535, DOI 10.1090/S0002-9939-2010-10545-2. MR2680077

[11] W. M. Kantor, A. Lubotzky, and A. Shalev, *Invariable generation of infinite groups*, J. Algebra **421** (2015), 296–310, DOI 10.1016/j.jalgebra.2014.08.030. MR3272383

[12] H. Masur, *Measured foliations and handlebodies*, Ergodic Theory Dynam. Systems **6** (1986), no. 1, 99–116, DOI 10.1017/S014338570000331X. MR837978

[13] H. A. Masur and Y. N. Minsky, *Geometry of the complex of curves. I. Hyperbolicity*, Invent. Math. **138** (1999), no. 1, 103–149, DOI 10.1007/s002220050343. MR1714338

[14] H. A. Masur and Y. N. Minsky, *Quasiconvexity in the curve complex*, In the tradition of Ahlfors and Bers, III, Contemp. Math., vol. 355, Amer. Math. Soc., Providence, RI, 2004, pp. 309–320, DOI 10.1090/conm/355/06460. MR2145071

[15] H. Masur and J. Smillie, *Quadratic differentials with prescribed singularities and pseudo-Anosov diffeomorphisms*, Comment. Math. Helv. **68** (1993), no. 2, 289–307, DOI 10.1007/BF02565820. MR1214233

[16] H. Namazi, *Big Heegaard distance implies finite mapping class group*, Topology Appl. **154** (2007), no. 16, 2939–2949, DOI 10.1016/j.topol.2007.05.011. MR2355879

[17] K. Ohshika and M. Sakuma, *Subgroups of mapping class groups related to Heegaard splittings and bridge decompositions*, Geom. Dedicata **180** (2016), 117–134, DOI 10.1007/s10711-015-0094-4. MR3451460

[18] D. Osin, *Acylindrically hyperbolic groups*, Trans. Amer. Math. Soc. **368** (2016), no. 2, 851–888, DOI 10.1090/tran/6343. MR3430352

[19] J-P. Otal, Courants géodésiques et produits libres, Thèse d'état, Université de Paris- Sud, Orsay (1988).

[20] M. Scharlemann, *The complex of curves on nonorientable surfaces*, J. London Math. Soc. (2) **25** (1982), no. 1, 171–184, DOI 10.1112/jlms/s2-25.1.171. MR645874



Department of Mathematics, University of Utah, Salt Lake City, Utah 84112
*E-mail address*: bestvina@math.utah.edu

Department of Mathematics, Kyoto University, Kyoto, 606-8502, Japan
*E-mail address*: kfujiwara@math.kyoto-u.ac.jp

Contemporary Mathematics
Volume **696**, 2017
http://dx.doi.org/10.1090/conm/696/14016

# The convex real projective orbifolds with radial or totally geodesic ends: A survey of some partial results

Suhyoung Choi

ABSTRACT. A real projective orbifold has a *radial end* if a neighborhood of the end is foliated by projective geodesics that develop into geodesics ending at a common point. It has a *totally geodesic end* if the end can be completed to have a totally geodesic boundary.

We will prove a homeomorphism between the deformation space of convex real projective structures on an orbifold $\mathcal{O}$ with radial or totally geodesic ends with various conditions, and the union of open subspaces of strata of a subset

$$\mathrm{Hom}_{\mathcal{E}}(\pi_1(\mathcal{O}), \mathrm{PGL})/\,\mathrm{PGL}$$

of the PGL-character variety for $\pi_1(\mathcal{O})$ given by corresponding end conditions for holonomy representations.

Lastly, we will talk about the openness and closedness of the properly (resp. strictly) convex real projective structures on a class of orbifold with generalized admissible ends.

CONTENTS



2010 *Mathematics Subject Classification.* Primary 57M50; Secondary 53A20, 53C10, 20C99, 20F65.

*Key words and phrases.* Geometric structures, real projective structures, relatively hyperbolic groups, $\mathrm{PGL}(n, \mathbb{R})$, $\mathrm{SL}(n, \mathbb{R})$, character of groups.

This work was supported by the National Research Foundation of Korea(NRF) grant funded by the Korea government(MEST) (No. 2013R1A1A2056698).

## 1. Introduction

**1.1. Preliminary.** Let $H$ be a closed upper-half space $\{x \in \mathbb{R}^n | x_n \geq 0\}$ with boundary $\partial H = \{x \in \mathbb{R}^n | x_n = 0\}$. An *orbifold* $\mathcal{O}$ is a second countable Hausdorff space where each point $x$ has a neighborhood with a chart $(U, G, \phi)$ consisting of

- a finite group $G$ acting on $U$ an open subset of $H$,
- $\phi : U \to \phi(U)$ inducing a homeomorphism $U/G \to \phi(U)$ to a neighborhood $\phi(U)$ of $x$.

Also, these charts are compatible in some obvious sense as explained by Satake and Thurston. Such a triple $(U, G, \phi)$ is called a model of a neighborhood of $\mathcal{O}$. The *orbifold boundary* $\partial\mathcal{O}$ is the set of points with only models of form $(U, G, \phi)$ where $U$ meets $\partial H$. A *closed orbifold* is a compact orbifold with empty boundary. The orbifolds in this paper are the quotient space of a manifold under the action of a finite group. (See Chapter 13 of Thurston [**57**] or more modern Moerdijk [**53**].)

We will study properly convex real projective structures on such orbifolds. A properly convex real projective orbifold is the quotient $\Omega/\Gamma$ of a properly convex domain $\Omega$ in an affine space in $\mathbb{RP}^n$ by a group $\Gamma$, $\Gamma \subset \mathrm{PGL}(n+1, \mathbb{R})$, of projective automorphisms acting on $\Omega$ properly but maybe not freely. Let $\pi_1(\mathcal{O})$ denote the orbifold-fundamental group of $\mathcal{O} = \Omega/\Gamma$, which is isomorphic to $\Gamma$. Given an orbifold $\mathcal{O}$, a *properly convex real projective structure* on $\mathcal{O}$ is a diffeomorphism $f : \mathcal{O} \to \Omega/\Gamma$ for a properly convex real projective orbifold of form $\Omega/\Gamma$ as above. Finite volume complete hyperbolic manifolds are examples since we can use the

Klein model and identify $\Omega$ as the model space where $\Gamma$ is the hyperbolic isometry group, which acts projectively on $\Omega$. (See Example 2.5.)

For closed $n$-dimensional orbifolds, these structures are somewhat well studied by Benoist [**7**] generalizing the previous work for surfaces by Goldman [**39**] and Choi-Goldman [**26**]. The work of Cooper, Long, and Thistlethwaite [**31**] and [**30**] showed the existence of deformations for some hyperbolic 3-orbifolds that are explicitly computable. Currently, there seems to be more interest in this field due to these and other developments. (For a recent survey, see [**29**].)

A *strongly tame* orbifold is an orbifold that has a compact suborbifold whose complement is homeomorphic to a disjoint union of closed $(n-1)$-orbifolds times intervals. The theory is mostly applicable to strongly tame orbifolds that are not manifolds, and is most natural in this setting. In fact, the theory is mostly adopted for Coxeter orbifolds, i.e., orbifolds based on convex polytopes with faces silvered, and also, orbifolds that are doubles of these. (See Section 1.3.)

One central example to keep in mind is the tetrahedron with silvered sides and edge orders equal to 3. This orbifold admits a complete hyperbolic structure. Also, it admits deformations to convex real projective orbifolds. The deformation space of real projective structures is homeomorphic to a four cell. (See Choi [**24**] and Marquis [**52**].) We can also take the double of this orbifold. The deformation space is 5-dimensional and can be explicitly computed in [**21**]. (See Chapter 7 of [**19**].) Except for ones based on tetrahedra, complete hyperbolic Coxeter 3-orbifolds with all edge orders 3 have at least six dimensional deformation spaces by Theorem 1 of Choi-Hodgson-Lee [**28**].

Another well-known prior example is due to Tillmann: This is a complete hyperbolic orbifold on a complement of two-points $p, q$ in the 3-sphere where the singularities are two simple arcs connecting $p$ to itself and $q$ to itself forming a link of index 1 and another simple arc connecting $p$ and $q$. These arcs have $\mathbb{Z}_3$ as the local group. Heard, Hodgson, Martelli, and Petronio [**44**] labelled this orbifold $2h\_1\_1$. The dimension is computed to be 2 by Porti and Tillman [**56**]. (See Figure 1.)

For all these examples, we know that some horospherical ends deform to lens-type radial ones and vice versa. We can also obtain totally geodesic ends by "cutting off" some radial ends. We call the phenomenon "cusp opening". Benoist [**10**] first found such phenomena for a Coxeter 3-orbifold. We also had the many numerical and theoretical results for above Coxeter 3-orbifolds which we plan to write more explicitly in a later paper [**27**]. Also, Greene [**41**] found many such examples using explicit computations. Recently, Ballas, Danciger, and Lee [**4**] found these phenomena using cohomology arguments for complete finite-volume hyperbolic 3-manifolds as explained in their MSRI talks in 2015. Some of the computations are available from the authors.

### 1.2. Main results.

We concentrate on studying the ends that are well-behaved, i.e., ones that are foliated by lines or are totally geodesic. In this setting we wish to study the deformation spaces of the convex real projective structures on orbifolds with some boundary conditions using the character varieties. Our main aim is

- to identify the deformation space of convex real projective structures on an orbifold $\mathcal{O}$ with certain boundary conditions with the union of some open subsets of strata of $\mathrm{Hom}(\pi_1(\mathcal{O}), \mathrm{PGL}(n+1, \mathbb{R}))/\mathrm{PGL}(n+1, \mathbb{R})$ defined by conditions corresponding to the boundary conditions.

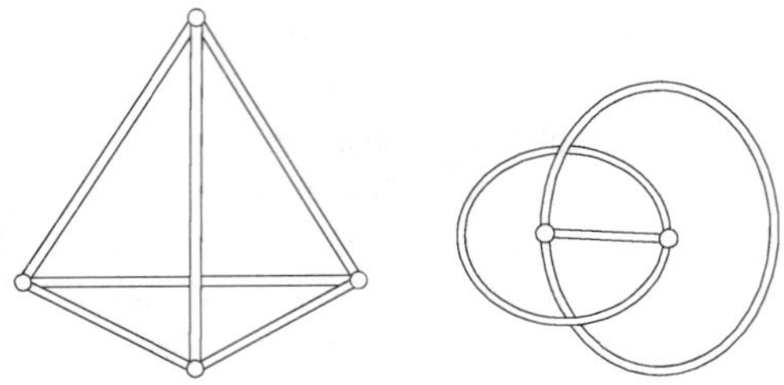

FIGURE 1. The singularities of the our orbifolds, the double of a tetrahedral reflection orbifold with orders 3 and Tillmann's orbifold in the 3-spheres. The white dots indicate the points removed. The edges are all of order 3.

This is an example of the so-called Ehresmann-Thurston-Weil principle [**63**]. (See Canary-Epstein-Green [**14**], Goldman [**38**], Lok [**51**], Bergeron-Gelander [**11**], and Choi [**23**].) The precise statements are given in Theorems 6.7 and 6.8.

See Definition 5.1 for the condition (IE) and (NA). We use the notion of *strict convexity with respect to ends* as defined in Definition 5.3. Our main result is the following as a corollary of Theorem 6.12:

COROLLARY 1.1. *Let $\mathcal{O}$ be a noncompact strongly tame SPC $n$-orbifold with generalized admissible ends and satisfies* (IE) *and* (NA). *Assume $\partial\mathcal{O} = \emptyset$, and that the nilpotent normal subgroups of every finite-index subgroup of $\pi_1(\mathcal{O})$ are trivial. Then* hol *maps the deformation space* $\mathrm{CDef}_{\mathcal{E},\mathrm{u},\mathrm{ce}}(\mathcal{O})$ *of SPC-structures on $\mathcal{O}$ homeomorphically to a union of components of*

$$\mathrm{rep}^s_{\mathcal{E},\mathrm{u},\mathrm{ce}}(\pi_1(\mathcal{O}), \mathrm{PGL}(n+1, \mathbb{R})).$$

*The same can be said for* $\mathrm{SDef}_{\mathcal{E},\mathrm{u},\mathrm{ce}}(\mathcal{O})$.

These terms will be defined more precisely later on in Sections 2.1.8 and 6.1.1. Roughly speaking, $\mathrm{CDef}_{\mathcal{E}}(\mathcal{O})$ (resp. $\mathrm{SDef}_{\mathcal{E}}(\mathcal{O})$) is the deformation spaces of properly convex (resp. strictly properly convex) real projective structures with conditions on ends that each end holonomy group fixes a point.

$$\mathrm{rep}^s_{\mathcal{E}}(\pi_1(\mathcal{O}), \mathrm{PGL}(n+1, \mathbb{R}))$$

is the space of characters each of whose end holonomy group fixes a point.

$$\mathrm{CDef}_{\mathcal{E},\mathrm{u},\mathrm{ce}}(\mathcal{O}) \text{ (resp. } \mathrm{SDef}_{\mathcal{E},\mathrm{u},\mathrm{ce}}(\mathcal{O}))$$

is the deformation space of properly convex (resp. strictly properly convex) real projective structures with conditions on ends that each end has a lens-cone neighborhood or a horospherical one, and each end holonomy group fixes a unique point.

$$\mathrm{rep}^s_{\mathcal{E},\mathrm{u},\mathrm{ce}}(\pi_1(\mathcal{O}), \mathrm{PGL}(n+1, \mathbb{R}))$$

is the space of characters each of whose end holonomy group fixes a unique point and acts on a lens-cone or a horosphere.

Our main examples satisfy this condition: Suppose that a strongly tame properly convex 3-orbifold $\mathcal{O}$ with radial ends admits a finite volume complete hyperbolic structure and has radial ends only and any end neighborhood contains singularities of dimension 1 of order 3 or 6, and end orbifolds have base spaces homeomorphic to disks or spheres. The theory simplifies by Corollary 6.6, i.e., each end is always of lens-type or horospherical, so that

$$\mathrm{SDef}_{\mathcal{E},\mathrm{u},\mathrm{ce}}(\mathcal{O}) = \mathrm{SDef}_{\mathcal{E}}(\mathcal{O}).$$

Corollary 6.13 applies to these cases, and the space under hol maps homeomorphically to a union of components of

$$\mathrm{rep}^{s}_{\mathcal{E}}(\pi_1(\mathcal{O}), \mathrm{PGL}(4, \mathbb{R})).$$

For a strongly tame Coxeter orbifold $\mathcal{O}$ of dimension $n \geq 3$ admitting a complete hyperbolic structure, hol is a homeomorphism from

$$\mathrm{SDef}_{\mathcal{E},\mathrm{u},\mathrm{ce}}(\mathcal{O}) = \mathrm{SDef}_{\mathcal{E}}(\mathcal{O})$$

to a union of components of

$$\mathrm{rep}^{s}_{\mathcal{E}}(\pi_1(\mathcal{O}), \mathrm{PGL}(n+1, \mathbb{R}))$$

by Corollary 6.14. For this theory, we can consider a Coxeter orbifold based on a convex polytope admitting a complete hyperbolic structure with all edge orders equal to 3. More specifically, we can consider a hyperbolic ideal simplex or a hyperbolic ideal cube with such structures. (See Choi-Hodgson-Lee [**28**] for examples of 6-dimensional deformations.)

One question is whether we can remove the stability condition on the target character varieties. (For closed orbifolds, see Theorem 4.1 of [**29**] essentially following from Benoist [**9**].). We plan to prove this for many interesting orbifolds, such as strongly tame orbifolds.

**1.3. Remarks.** We give some remarks on our results here: The theory here is by no means exhaustive final words. We have a somewhat complicated theory of ends [**15**], [**16**], [**17**], [**18**], which is used in this paper. Instead of publishing some of these and [**19**]. We will try to refine and generalize the theory and put into a monograph [**20**] in a near future. Our boundary condition is very restrictive in some sense. With this it could be said that the above theory is not so surprising.

Ballas, Cooper, Leitner, Long, and Tillman have different restrictions on ends and they are working with manifolds. The associated end neighborhoods have nilpotent holonomy groups. (See [**30**], [**33**], [**49**], [**50**], [**48**], and [**3**]). They are currently developing the theory of ends and the deformation theory based on this assumption. Of course, we expect to benefit and thrive from many interactions between the theories as they happen in multitudes of fields.

Originally, we developed the theory for orbifolds as given in papers of Choi [**24**], Choi, Hodgson, and Lee [**28**], and [**19**]. However, the recent examples of Ballas [**1**], [**2**], and Ballas, Danciger, and Lee [**4**] can be covered using fixing sections. Also, differently from the above work, we can allow ends with hyperbolic holonomy groups.

**1.4. Outline.** In Section 2, we will go over real projective structures. We will define end structures of convex real projective orbifolds. We first discuss totally geodesic ends and define lens condition for these totally geodesic ends. Then we define radial ends and radial foliation marking for radial ends. We define end orbifolds, and horospherical ends. We define the space of characters and the deformation spaces of convex real projective structures on orbifolds with radial or totally geodesic ends. Finally, we discuss the local homeomorphisms between the subsets of deformation spaces and those of character varieties. Here, we are not yet concerned with convexity.

In Section 3, we will discuss the known facts about the convex real projective orbifolds including the Vinberg duality result.

In Section 4, we will discuss the end theory. We introduce pseudo-ends and pseudo-end neighborhoods, and pseudo-end fundamental groups. We introduce admissible groups and admissible ends. We introduce the lens-conditions for radial ends and totally geodesic ends.

In Section 5, we will relate the relative hyperbolicity of the fundamental groups of strongly tame properly convex real projective orbifolds with the "relative" strict convexity of the real projective structures. (See Section 5.2.) In Section 5.1, we define the stable properly convex real projective orbifolds relative to ends. We will show that under mild conditions properly convex strongly tame real projective orbifolds with generalized admissible ends have stable holonomy groups.

In Section 6, we will state our main results. In Section 6.2, we will state that a holonomy homomorphism map from a deformation space with end conditions to the character variety is a homeomorphism to a union of open subsets of strata of the character variety. This map is injective. In Section 6.3, we will say about the closedness.

**1.5. Acknowledgement.** We thank many discussions with Samuel Ballas, Daryl Cooper, Jeffrey Danciger, William Goldman, Craig Hodgson, Gye-Seon Lee, Darren Long, Joan Porti, and Stephan Tillmann. This research are supported and inspired by many difficult computer computations done by them. We benefited from the helpful discussions with Stephan Tillman in May of 2008, and ones with Cooper at ICERM, Brown University, in September 2013, and the ones with Samuel Ballas, Daryl Cooper, and Darren Long at the UC Santa Barbara in June of 2015. We appreciate their generosity as well as their deep insights and clarifications. We also thank the hospitality of MSRI during Spring of 2015 where some parts of this research was carried out.

## 2. Preliminary

### 2.1. Basic definitions.

2.1.1. *Topological notation.* Define bd$A$ for a subset $A$ of $\mathbb{RP}^n$, $n \geq 2$, (resp. in $\mathbb{S}^n$) to be the *topological boundary* in $\mathbb{RP}^n$ (resp. in $\mathbb{S}^n$) and define $\partial A$ for a manifold or orbifold $A$ to be the *manifold or orbifold boundary* and $A^o$ denote the manifold interior. The closure $\mathrm{Cl}(A)$ of a subset $A$ of $\mathbb{RP}^n$ (resp. of $\mathbb{S}^n$) is the topological closure in $\mathbb{RP}^n$ (resp. in $\mathbb{S}^n$).

2.1.2. *The Hausdorff metric.* Recall the standard elliptic metric $\mathbf{d}$ on $\mathbb{RP}^n$ (resp. in $\mathbb{S}^n$). Given two sets $A$ and $B$ of $\mathbb{RP}^n$ (resp of $\mathbb{S}^n$),

$$\mathbf{d}(A, B) := \inf\{\mathbf{d}(x, y) | x \in A, y \in B\}.$$

We can let $A$ or $B$ be points as well obviously.

The *Hausdorff distance* between two convex subsets $K_1, K_2$ of $\mathbb{RP}^n$ (resp. of $\mathbb{S}^n$) is defined by

$$\mathbf{d}^H(K_1, K_2) = \inf\left\{\epsilon \geq 0 \,\middle|\, \mathrm{Cl}(K_1) \subset N_\epsilon(\mathrm{Cl}(K_2)), \mathrm{Cl}(K_2) \subset N_\epsilon(\mathrm{Cl}(K_1))\right\}$$

where $N_\epsilon(A)$ is the $\epsilon$-$\mathbf{d}$-neighborhood of $A$ under the standard metric $\mathbf{d}$ of $\mathbb{RP}^n$ (resp. of $\mathbb{S}^n$) for $\epsilon > 0$. $\mathbf{d}^H$ gives a compact Hausdorff topology on the set of all compact subsets of $\mathbb{RP}^n$ (resp. of $\mathbb{S}^n$). (See p. 281 of [**55**].)

We say that a sequence of sets $\{K_i\}$ *geometrically converges* to a set $K$ if $\mathbf{d}^H(K_i, K) \to 0$. If $K$ is assumed to be closed, then the geometric limit is unique.

LEMMA 2.1. *Suppose that a sequence $\{K_i\}$ of compact convex domains geometrically converges to a compact convex domain $K$ in $\mathbb{RP}^n$ (resp. in $\mathbb{S}^n$). Then*

$$\mathbf{d}^H(\partial K_i, \partial K) \to 0. \tag{2.1}$$

2.1.3. *Real projective structures.* Given a vector space $V$, we let $\mathbb{P}(V)$ denote the space obtained by taking the quotient space of $V - \{O\}$ under the equivalence relation

$$v \sim w \text{ for } v, w \in \mathbb{R}^{n+1} - \{O\} \text{ iff } v = sw, \text{ for } s \in \mathbb{R} - \{0\}.$$

We let $[v]$ denote the equivalence class of $v \in V - \{O\}$. Recall that the projective linear group $\mathrm{PGL}(n+1, \mathbb{R})$ acts on $\mathbb{RP}^n$, i.e., $\mathbb{P}(\mathbb{R}^{n+1})$, in a standard manner.

Let $\mathcal{O}$ be a noncompact strongly tame $n$-orbifold where the orbifold boundary is not necessarily empty.

- A *real projective orbifold* is an orbifold with a geometric structure modelled on $(\mathbb{RP}^n, \mathrm{PGL}(n+1, \mathbb{R}))$. (See Thurston [**58**], [**23**] and Chapter 6 of [**25**].)
- $\mathcal{O}$ has a universal cover $\tilde{\mathcal{O}}$ where the deck transformation group $\pi_1(\mathcal{O})$ acts on.
- The underlying space of $\mathcal{O}$ is homeomorphic to the quotient space $\tilde{\mathcal{O}}/\pi_1(\mathcal{O})$.
- A real projective structure on $\mathcal{O}$ gives us a so-called development pair $(\mathbf{dev}, h)$ where
  - $\mathbf{dev} : \tilde{\mathcal{O}} \to \mathbb{RP}^n$ is an immersion, called the *developing map*,
  - and $h : \pi_1(\mathcal{O}) \to \mathrm{PGL}(n+1, \mathbb{R})$ is a homomorphism, called a *holonomy homomorphism*, satisfying

$$\mathbf{dev} \circ \gamma = h(\gamma) \circ \mathbf{dev} \text{ for } \gamma \in \pi_1(\mathcal{O}).$$

Let $\mathbb{R}^{n+1*}$ denote the dual of $\mathbb{R}^{n+1}$. Let $\mathbb{RP}^{n*}$ denote the dual projective space $\mathbb{P}(\mathbb{R}^{n+1*})$. $\mathrm{PGL}(n+1, \mathbb{R})$ acts on $\mathbb{RP}^{n*}$ by taking the inverse of the dual transformation. Then a representation $h : \pi_1(\mathcal{O}) \to \mathrm{PGL}(n+1, \mathbb{R})$ has a dual representation $h^* : \pi_1(\mathcal{O}) \to \mathrm{PGL}(n+1, \mathbb{R})$ sending elements of $\pi_1(\mathcal{O})$ to the inverse of the dual transformation of $\mathbb{R}^{n+1*}$.

The complement of a codimension-one subspace of $\mathbb{RP}^n$ can be identified with an affine space $\mathbb{R}^n$ where the geodesics are preserved. We call the complement an *affine subspace*. A *convex domain* in $\mathbb{RP}^n$ is a convex subset of an affine subspace. A *properly convex domain* in $\mathbb{RP}^n$ is a convex domain contained in a precompact subset of an affine subspace.

A *convex real projective orbifold* is a real projective orbifold projectively diffeomorphic to the quotient $\Omega/\Gamma$ where $\Omega$ is a convex domain in an affine subspace of $\mathbb{RP}^n$ and $\Gamma$ is a discrete group of projective automorphisms of $\Omega$ acting properly. If an open orbifold has a convex real projective structure, it is covered by a convex domain $\Omega$ in $\mathbb{RP}^n$. Equivalently, this means that the image of the developing map $\mathbf{dev}(\tilde{\mathcal{O}})$ for the universal cover $\tilde{\mathcal{O}}$ of $\mathcal{O}$ is a convex domain for the developing map $\mathbf{dev}$ with associated holonomy homomorphism $h$. Here we may assume $\mathbf{dev}(\tilde{\mathcal{O}}) = \Omega$, and $\mathcal{O}$ is projectively diffeomorphic to $\mathbf{dev}(\tilde{\mathcal{O}})/h(\pi_1(\mathcal{O}))$. In our discussions, since **dev** is an imbedding and so is $h$, $\tilde{\mathcal{O}}$ will be regarded as an open domain in $\mathbb{RP}^n$ and $\pi_1(\mathcal{O})$ as a subgroup of $\mathrm{PGL}(n+1,\mathbb{R})$ in such cases.

REMARK 2.2. Given a vector space $V$, we denote by $\mathbb{S}(V)$ the quotient space of

$$(V - \{O\})/\sim \text{ where } v \sim w \text{ iff } v = sw \text{ for } s > 0.$$

We will represent each element of $\mathrm{PGL}(n+1,\mathbb{R})$ by a matrix of determinant $\pm 1$; i.e., $\mathrm{PGL}(n+1,\mathbb{R}) = \mathrm{SL}_{\pm}(n+1,\mathbb{R})/\langle \pm \mathrm{I} \rangle$. Recall the covering map $\mathbb{S}^n = \mathbb{S}(\mathbb{R}^{n+1}) \to \mathbb{RP}^n$. For each $g \in \mathrm{PGL}(n+1,\mathbb{R})$, there is a unique lift in $\mathrm{SL}_{\pm}(n+1,\mathbb{R})$ preserving each component of the inverse image of $\mathbf{dev}(\tilde{\mathcal{O}})$ under $\mathbb{S}^n \to \mathbb{RP}^n$. We will use this representative.

2.1.4. *End structures.* A *strongly tame* $n$-orbifold is one where the complement of a compact set is diffeomorphic to a union of $(n-1)$-dimensional orbifolds times intervals. Of course it can be compact.

Let $\mathcal{O}$ be a strongly tame $n$-orbifold. Each end has a neighborhood diffeomorphic to a closed $(n-1)$-orbifold times an interval. The *fundamental group* of an end is the fundamental group of such an end neighborhood. It is independent of the choice of the end neighborhood by Proposition 4.1.

An end $E$ of a real projective orbifold $\mathcal{O}$ is *totally geodesic* or *of type* $T$ if the following hold:

- The end has an end neighborhood homeomorphic to a closed connected $(n-1)$-dimensional orbifold $B$ times a half-open interval $(0,1]$.
- $B$ completes to a compact orbifold $U$ diffeomorphic to $B \times [0,1]$ in an ambient real projective orbifold.
- The subset of $U$ corresponding to $B \times \{0\}$ is the added boundary component.
- Each point of the added boundary component has a neighborhood projectively diffeomorphic to the quotient orbifold of an open set $V$ in an affine half-space $P$ so that $V \cap \partial P \neq \emptyset$ by a projective action of a finite group.

The completion is called a *compactified end neighborhood* of the end $E$. The boundary component is called the *ideal boundary component* of the end. Such ideal boundary components may not be uniquely determined as there are two projectively nonequivalent ways to add boundary components of elementary annuli (see Section 1.4 of [**22**]). Two compactified end neighborhoods of an end are *equivalent* if they contain a common compactified end neighborhood.

We also define as follows:

- The equivalence class of the chosen compactified end neighborhood is called a *marking* of the totally geodesic end.
- We will also call the ideal boundary the *end orbifold* of the end.

DEFINITION 2.3. A *lens* is a properly convex domain $L$ so that $\partial L$ is a union of two smooth strictly convex open disks. A properly convex domain $K$ is a *generalized lens* if $\partial K$ is a union of two open disks one of which is strictly convex and smooth and the other is allowed to be just a topological disk. A *lens-orbifold* is a compact quotient orbifold of a lens by a properly discontinuous action of a projective group $\Gamma$. Thus, for two boundary components $A$ and $B$, $A/\Gamma$ and $B/\Gamma$ are homotopy equivalent to $L/\Gamma$ by the obvious inclusion maps.

**(Lens condition):** The ideal boundary is realized as a totally geodesic suborbifold in the interior of a lens-orbifold in some ambient real projective orbifold of $\mathcal{O}$.

If the lens condition is satisfied for an T-end, we will call it the *T-end of lens-type*.

Let $\tilde{\mathcal{O}}$ denote the universal cover of $\mathcal{O}$ with the developing map **dev**. An end $E$ of a real projective orbifold is *radial* or *of type R* if the following hold:

- The end has an end neighborhood $U$ foliated by properly imbedded projective geodesics.
- Choose a map $f : \mathbb{R} \times [0,1] \to \mathcal{O}$ so that $f|\mathbb{R} \times \{t\}$ for each $t$ is a geodesic leaf of such a foliation of $U$. Then $f$ lifts to $\tilde{f} : \mathbb{R} \times [0,1] \to \tilde{\mathcal{O}}$ where $\mathbf{dev} \circ \tilde{f}|\mathbb{R} \times \{t\}$ for each $t, t \in [0,1]$, maps to a geodesic in $\mathbb{RP}^n$ ending at a point of concurrency common for $t$.

The foliation is called a *radial foliation* and leaves *radial lines* of $E$. Two such radial foliations $\mathcal{F}_1$ and $\mathcal{F}_2$ of radial end neighborhoods of an end are *equivalent* if the restrictions of $\mathcal{F}_1$ and $\mathcal{F}_2$ in an end neighborhood agree. A *radial foliation marking* is an equivalence class of radial foliations. The marking will give us an ideal boundary component on the end.

DEFINITION 2.4. A *real projective orbifold with radial or totally geodesic ends* is a strongly tame orbifold with a real projective structure where each end is an R-end or a T-end with a marking given for each.

Let $\mathbb{RP}^{n-1}_x$ denote the space of concurrent lines to a point $x$ where $\mathbb{RP}^{n-1}_x$ is projectively diffeomorphic to $\mathbb{RP}^{n-1}$. The real projective transformations fixing $x$ induce real projective transformations of $\mathbb{RP}^{n-1}_x$. Radial lines in an R-end neighborhood are *equivalent* if they agree outside a compact subset. The space of equivalent classes of radial lines in an R-end neighborhood is an $(n-1)$-orbifold by the properness of the radial lines. The *end orbifold* associated with an R-end is defined as the equivalence space of radial lines in $\mathcal{O}$. The equivalence space of radial lines in an R-end has the local structure of $\mathbb{RP}^{n-1}$ since we can lift a local neighborhood to $\tilde{\mathcal{O}}$, and these radial lines lift to lines developing into concurrent lines. The end orbifold has a unique induced real projective structure of one dimension lower.

EXAMPLE 2.5. Let $\mathbb{R}^{n+1}$ have standard coordinates $x_0, x_1, \ldots, x_n$, and let $B$ be the subset in $\mathbb{RP}^n$ corresponding to the cone given by

$$x_0 > \sqrt{x_1^2 + \cdots + x_n^2}.$$

The Klein model gives a hyperbolic space as $B \subset \mathbb{RP}^n$ with the isometry group $\mathrm{PO}(1,n)$, a subgroup of $\mathrm{PGL}(n+1,\mathbb{R})$ acting on $B$. Thus, a complete hyperbolic orbifold is projectively diffeomorphic to a real projective orbifold of $B/\Gamma$ for $\Gamma$ in $\mathrm{PO}(1,n)$. The interior of a finite-volume hyperbolic $n$-orbifold with rank $(n-1)$ horospherical ends and totally geodesic boundary forms an example of a properly convex strongly tame real projective orbifold with radial or totally geodesic ends. For horospherical ends, the end orbifolds have Euclidean structures. (Also, we could allow hyperideal ends by attaching radial ends. Section 3.1.1 in [**15**].)

EXAMPLE 2.6. For examples, if the end orbifold of an R-end $E$ is a 2-orbifold based on a sphere with three singularities of order 3, then a line of singularity is a leaf of a radial foliation. End orbifolds of Tillman's orbifold and the the double of a tetrahedral reflection orbifold are examples. A double orbifold of a cube with edges having orders 3 only has eight such end orbifolds. (See Proposition 4.6 of [**16**] and their deformations are computed in [**28**]. Also, see Ryan Greene [**41**] for the theory.)

2.1.5. *Horospherical ends.* An *ellipsoid* in $\mathbb{RP}^n = \mathbb{P}(\mathbb{R}^{n+1})$ (resp. in $\mathbb{S}^n = \mathbb{S}(\mathbb{R}^{n+1})$) is the projection $C - \{O\}$ of the null cone

$$C := \{x \in \mathbb{R}^{n+1} | B(x, x) = 0\}$$

for a nondegenerate symmetric bilinear form $B : \mathbb{R}^{n+1} \times \mathbb{R}^{n+1} \to \mathbb{R}$. Ellipsoids are always equivalent by projective automorphisms of $\mathbb{RP}^n$. An *ellipsoid ball* is the closed contractible domain in an affine subspace $A$ of $\mathbb{RP}^n$ (resp. $\mathbb{S}^n$) bounded by an ellipsoid contained in $A$. A *horoball* is an ellipsoid ball with a point $p$ of the boundary removed. An ellipsoid with a point $p$ on it removed is called a *horosphere.* The *vertex* of the horosphere or the horoball is defined as $p$.

Let $U$ be a horoball with a vertex $p$ in the boundary of $B$. A real projective orbifold that is real projectively diffeomorphic to an orbifold $U/\Gamma_p$ for a discrete subgroup $\Gamma_p \subset \mathrm{PO}(1, n)$ fixing a point $p \in \mathrm{bd}B$ is called a *horoball orbifold.* A *horospherical end* is an end with an end neighborhood that is such an orbifold.

2.1.6. *Deformation spaces and the space of holonomy homomorphisms.* An *isotopy* $i : \mathcal{O} \to \mathcal{O}$ is a self-diffeomorphism so that there exists a smooth orbifold map $J : \mathcal{O} \times [0, 1] \to \mathcal{O}$, so that

$$i_t : \mathcal{O} \to \mathcal{O} \text{ given by } i_t(x) = J(x, t)$$

are self-diffeomorphisms for $t \in [0, 1]$ and $i = i_1, i_0 = \mathrm{I}_{\mathcal{O}}$. We will extend this notion strongly.

- Two real projective structures $\mu_0$ and $\mu_1$ on $\mathcal{O}$ with R-ends or T-ends with end markings are *isotopic* if there is an isotopy $i$ on $\mathcal{O}$ so that $i^*(\mu_0) = \mu_1$ where $i^*(\mu_0)$ is the induced structure from $\mu_0$ by $i$ where we require for each $t$
  - $i_{t*}(\mu_0)$ has a radial end structure for each radial end,
  - $i_t$ sends the radial end foliation for $\mu_0$ from an R-end neighborhood to the radial end foliation for real projective structure $\mu_t = i_{t*}(\mu_0)$ with corresponding R-end neighborhoods,
  - $i_t$ extends to diffeomorphisms of the compactifications of $\mathcal{O}$ using the radial foliations and the totally geodesic ideal boundary components for $\mu_0$ and $\mu_t$.

We define $\mathrm{Def}_{\mathcal{E}}(\mathcal{O})$ as the deformation space of real projective structures on $\mathcal{O}$ with end marks; more precisely, this is the quotient space of the real projective structures on $\mathcal{O}$ satisfying the above conditions for ends of type R and T under the isotopy equivalence relations. We put on $\mathcal{O}$ a radial foliation on each end neighborhood of type R and attach an ideal boundary component for each end neighborhood of type T to obtain a new compactified orbifold $\overline{\mathcal{O}}$. We introduce the equivalence relation based on isotopies and end neighborhood structures and ideal boundary components. We may assume that the developing maps extend to the smooth maps of the universal cover $\widehat{\mathcal{O}}$ of $\overline{\mathcal{O}}$. The topology of such a space is defined by the compact open $C^2$-topology for the space of developing maps $\mathbf{dev}|\widehat{\mathcal{O}}$. (See [**23**], [**14**] and [**38**] for more details. )

2.1.7. *The end restrictions.* To discuss the deformation spaces, we introduce the following notions. The end will be either assigned an $\mathcal{R}$-*type* or a $\mathcal{T}$-*type*.

- An $\mathcal{R}$-type end is required to be radial.
- A $\mathcal{T}$-type end is required to have totally geodesic properly convex ideal boundary components of lens-type or be horospherical.

A strongly tame orbifold will always have such an assignment in this paper, and finite-covering maps will always respect the types. We will fix the types for ends of our orbifolds in consideration.

2.1.8. *Character spaces of relevance.* Since $\mathcal{O}$ is strongly tame, the fundamental group $\pi_1(\mathcal{O})$ is finitely generated. Let $\{g_1, \dots, g_m\}$ be a set of generators of $\pi_1(\mathcal{O})$. As usual $\mathrm{Hom}(\pi_1(\mathcal{O}), G)$ for a Lie group $G$ has an *algebraic topology* as a subspace of $G^m$. This topology is given by the notion of *algebraic convergence*

$$\{h_i\} \to h \text{ if } h_i(g_j) \to h(g_j) \in G \text{ for each } j, j = 1, \dots, m.$$

A conjugacy class of representation is called a *character* in this paper.

The $\mathrm{PGL}(n+1,\mathbb{R})$-*character variety* $\mathrm{rep}(\pi_1(\mathcal{O}), \mathrm{PGL}(n+1,\mathbb{R}))$ is the quotient space of the homomorphism space

$$\mathrm{Hom}(\pi_1(\mathcal{O}), \mathrm{PGL}(n+1,\mathbb{R}))$$

where $\mathrm{PGL}(n+1,\mathbb{R})$ acts by conjugation

$$h(\cdot) \mapsto gh(\cdot)g^{-1} \text{ for } g \in \mathrm{PGL}(n+1,\mathbb{R}).$$

Similarly, we define

$$\mathrm{rep}(\pi_1(\mathcal{O}), \mathrm{SL}_\pm(n+1,\mathbb{R})) := \mathrm{Hom}(\pi_1(\mathcal{O}), \mathrm{SL}_\pm(n+1,\mathbb{R}))/\mathrm{SL}_\pm(n+1,\mathbb{R})$$

as the $\mathrm{SL}_\pm(n+1,\mathbb{R})$-character variety.

A representation or a character is *stable* if the orbit of it or its representative is closed and the stabilizer is finite under the conjugation action in

$$\mathrm{Hom}(\pi_1(\mathcal{O}), \mathrm{PGL}(n+1,\mathbb{R})) \text{ (resp. } \mathrm{Hom}(\pi_1(\mathcal{O}), \mathrm{SL}_\pm(n+1,\mathbb{R}))).$$

By Theorem 1.1 of [**45**], a representation $\rho$ is stable if and only if it is irreducible and no proper parabolic subgroup contains the image of $\rho$. The stability and the irreducibility are open conditions in the Zariski topology. Also, if the image of $\rho$ is Zariski dense, then $\rho$ is stable. $\mathrm{PGL}(n+1,\mathbb{R})$ acts properly on the open set of stable representations in $\mathrm{Hom}(\pi_1(\mathcal{O}), \mathrm{PGL}(n+1,\mathbb{R}))$. Similarly, $\mathrm{SL}_\pm(n+1,\mathbb{R})$ acts so on $\mathrm{Hom}(\pi_1(\mathcal{O}), \mathrm{SL}_\pm(n+1,\mathbb{R}))$. (See [**45**] for more details.)

A representation of a group $G$ into $\mathrm{PGL}(n+1,\mathbb{R})$ or $\mathrm{SL}_\pm(n+1,\mathbb{R})$ is *strongly irreducible* if the image of every finite index subgroup of $G$ is irreducible. Actually, many of the orbifolds have strongly irreducible and stable holonomy homomorphisms by Theorem 5.4.

An *eigen-1-form* of a linear transformation $\gamma$ is a linear functional $\alpha$ in $\mathbb{R}^{n+1}$ so that $\alpha \circ \gamma = \lambda\alpha$ for some $\lambda \in \mathbb{R}$. We recall the lifting of Remark 2.2.

- $$\mathrm{Hom}_{\mathcal{E}}(\pi_1(\mathcal{O}), \mathrm{PGL}(n+1,\mathbb{R}))$$

  to be the subspace of representations $h$ satisfying

  **The vertex condition for $\mathcal{R}$-ends:** $h|\pi_1(\tilde{E})$ has a nonzero common eigenvector of positive eigenvalues for the lift of $h(\pi_1(\tilde{E}))$ in $\mathrm{SL}_\pm(n+1,\mathbb{R})$ for each $\mathcal{R}$-type p-end fundamental group $\pi_1(\tilde{E})$, and

**The hyperplane condition for $\mathcal{T}$-ends :** $h|\pi_1(\tilde E)$ acts on a hyperplane $P$ for each $\mathcal{T}$-type p-end fundamental group $\pi_1(\tilde E)$ discontinuously and cocompactly on a lens $L$, a properly convex domain with $L^o \cap P = L \cap P \neq \emptyset$ or a horoball tangent to $P$.

- We denote by
$$\mathrm{Hom}^s(\pi_1(\mathcal{O}), \mathrm{PGL}(n+1,\mathbb{R}))$$
the subspace of stable and irreducible representations, and define
$$\mathrm{Hom}^s_{\mathcal{E}}(\pi_1(\mathcal{O}), \mathrm{PGL}(n+1,\mathbb{R}))$$
to be
$$\mathrm{Hom}_{\mathcal{E}}(\pi_1(\mathcal{O}), \mathrm{PGL}(n+1,\mathbb{R})) \cap \mathrm{Hom}^s(\pi_1(\mathcal{O}), \mathrm{PGL}(n+1,\mathbb{R})).$$
- We define
$$\mathrm{Hom}_{\mathcal{E},\mathrm{u}}(\pi_1(\mathcal{O}), \mathrm{PGL}(n+1,\mathbb{R}))$$
to be the subspace of representations $h$ where
  - $h|\pi_1(\tilde E)$ has a unique common eigenspace of dimension 1 in $\mathbb{R}^{n+1}$ with positive eigenvalues for its lift in $\mathrm{SL}_\pm(n+1,\mathbb{R})$ for each p-end fundamental group $\pi_1(\tilde E)$ of $\mathcal{R}$-type and
  - $h|\pi_1(\tilde E)$ has a common null-space $P$ of eigen-1-forms uniquely satisfying the following:
    * $\pi_1(\tilde E)$ acts properly on a lens $L$ with $L \cap P$ with nonempty interior in $P$ or
    * $H - \{p\}$ for a horosphere $H$ tangent to $P$ at $p$

  for each p-end fundamental group $\pi_1(\tilde E)$ of the end of $\mathcal{T}$-type.

REMARK 2.7. The above condition for type $\mathcal{T}$ generalizes the principal boundary condition for real projective surfaces.

Suppose that there are no ends of $\mathcal{T}$-type. Since each $\pi_1(\tilde E)$ is finitely generated and there is only finitely many conjugacy classes of $\pi_1(\tilde E)$,
$$\mathrm{Hom}_{\mathcal{E}}(\pi_1(\mathcal{O}), \mathrm{PGL}(n+1,\mathbb{R}))$$
is a closed semi-algebraic subset.
$$\mathrm{Hom}_{\mathcal{E},\mathrm{u}}(\pi_1(\mathcal{O}), \mathrm{PGL}(n+1,\mathbb{R}))$$
is an open subset of this closed semi-algebraic subset by Lemma 2.8.

LEMMA 2.8. *Let $V$ be a semialgebraic subset of $\mathrm{PGL}(n+1,\mathbb{R})^n$. For each $(g_1,\dots,g_n) \in V$, we arbitrarily choose a maximal eigenspace $E_i(g_i) \subset \mathbb{R}^{n+1}$ corresponding to the eigenvalue $\lambda(g_i)$ where $\bigcap_{i=1}^n E_i(g_i) \neq \{0\}$ on every point of semialgebraic subset $V$. We assume that for each $i$, $(g_1,\dots,g_n) \in V \mapsto E_i(g_i) \subset \mathbb{R}^{n+1}$ has a nonzero continuous section on $V$. Then the dimension function of the intersection $\bigcap_{i=1}^n E_i(g_i)$ is upper semi-continuous in $V$.*

PROOF. Since the limit subspace of $E_i(g)$ is contained in an eigenspace of $g$, this follows. ☐

When there are ends of $\mathcal{T}$-type,
$$\mathrm{Hom}_{\mathcal{E}}(\pi_1(\mathcal{O}), \mathrm{PGL}(n+1,\mathbb{R})) \text{ and } \mathrm{Hom}_{\mathcal{E},\mathrm{u}}(\pi_1(\mathcal{O}), \mathrm{PGL}(n+1,\mathbb{R}))$$

are the unions of open subsets of this closed semi-algebraic subsets since we have to consider the lens condition.

We define

- $$\mathrm{rep}_{\mathcal{E}}(\pi_1(\mathcal{O}), \mathrm{PGL}(n+1, \mathbb{R}))$$
  to be
  $$\mathrm{Hom}_{\mathcal{E}}(\pi_1(\mathcal{O}), \mathrm{PGL}(n+1, \mathbb{R}))/\mathrm{PGL}(n+1, \mathbb{R}).$$
- We denote by
  $$\mathrm{rep}^s_{\mathcal{E}}(\pi_1(\mathcal{O}), \mathrm{PGL}(n+1, \mathbb{R}))$$
  the subspace of
  $$\mathrm{rep}_{\mathcal{E}}(\pi_1(\mathcal{O}), \mathrm{PGL}(n+1, \mathbb{R}))$$
  of stable and irreducible characters.
- We define
  $$\mathrm{rep}_{\mathcal{E},\mathrm{u}}(\pi_1(\mathcal{O}), \mathrm{PGL}(n+1, \mathbb{R}))$$
  to be
  $$\mathrm{Hom}_{\mathcal{E},\mathrm{u}}(\pi_1(\mathcal{O}), \mathrm{PGL}(n+1, \mathbb{R}))/\mathrm{PGL}(n+1, \mathbb{R}).$$
- We define

$$\begin{aligned} &\mathrm{rep}^s_{\mathcal{E},\mathrm{u}}(\pi_1(\mathcal{O}), \mathrm{PGL}(n+1, \mathbb{R})) \\ &:= \mathrm{rep}^s(\pi_1(\mathcal{O}), \mathrm{PGL}(n+1, \mathbb{R})) \cap \mathrm{rep}_{\mathcal{E},\mathrm{u}}(\pi_1(\mathcal{O}), \mathrm{PGL}(n+1, \mathbb{R})). \end{aligned} \tag{2.2}$$

Note that when there are no $\mathcal{T}$-type ends,

$$\mathrm{rep}^s_{\mathcal{E}}(\pi_1(\mathcal{O}), \mathrm{PGL}(n+1, \mathbb{R}))$$

is a closed subset of

$$\mathrm{rep}^s(\pi_1(\mathcal{O}), \mathrm{PGL}(n+1, \mathbb{R})), \text{ and}$$

$$\mathrm{rep}^s_{\mathcal{E},\mathrm{u}}(\pi_1(\mathcal{O}), \mathrm{PGL}(n+1, \mathbb{R}))$$

is an open subset of

$$\mathrm{rep}^s_{\mathcal{E}}(\pi_1(\mathcal{O}), \mathrm{PGL}(n+1, \mathbb{R})).$$

Note that elements of $\mathrm{Def}_{\mathcal{E}}(\mathcal{O})$ have characters in

$$\mathrm{rep}_{\mathcal{E}}(\pi_1(\mathcal{O}), \mathrm{PGL}(n+1, \mathbb{R})).$$

Denote by $\mathrm{Def}_{\mathcal{E},\mathrm{u}}(\mathcal{O})$ the subspace of $\mathrm{Def}_{\mathcal{E}}(\mathcal{O})$ of equivalence classes of real projective structures with characters in

$$\mathrm{rep}_{\mathcal{E},\mathrm{u}}(\pi_1(\mathcal{O}), \mathrm{PGL}(n+1, \mathbb{R})).$$

Also, we denote by $\mathrm{Def}^s_{\mathcal{E}}(\mathcal{O}) \subset \mathrm{Def}_{\mathcal{E}}(\mathcal{O})$ and $\mathrm{Def}^s_{\mathcal{E},\mathrm{u}}(\mathcal{O}) \subset \mathrm{Def}_{\mathcal{E},\mathrm{u}}(\mathcal{O})$ the subspaces of equivalence classes of real projective structures with stable and irreducible characters.

**2.2. Oriented real projective structures.** Recall that $\mathrm{SL}_\pm(n+1,\mathbb{R})$ is isomorphic to $\mathrm{GL}(n+1,\mathbb{R})/\mathbb{R}^+$. Then this group acts on $\mathbb{S}^n = \mathbb{S}(\mathbb{R}^{n+1})$. We let $[v]$ denote the equivalence class of $v \in \mathbb{R}^{n+1} - \{O\}$. There is a double covering map $\mathbb{S}^n \to \mathbb{RP}^n$ with the deck transformation group generated by $\mathcal{A}$. This gives a projective structure on $\mathbb{S}^n$. The group of projective automorphisms is identified with $\mathrm{SL}_\pm(n+1,\mathbb{R})$.

An $(\mathbb{S}^n, \mathrm{SL}_\pm(n+1,\mathbb{R}))$-structure on $\mathcal{O}$ is said to be an *oriented real projective structure* on $\mathcal{O}$. We define $\mathrm{Def}_{\mathbb{S}^n}(\mathcal{O})$ as the deformation space of $(\mathbb{S}^n, \mathrm{SL}_\pm(n+1,\mathbb{R}))$-structures on $\mathcal{O}$.

Again, we can define the *radial end structures* and *totally geodesic ideal boundary* for oriented real projective structures and also horospherical end neighborhoods in obvious ways. They correspond in the direct way in the following theorem also.

THEOREM 2.9. *There is a one-to-one correspondence between the space of real projective structures on an orbifold $\mathcal{O}$ with the space of oriented real projective structures on $\mathcal{O}$. Moreover, a real projective diffeomorphism of real projective orbifolds is an $(\mathbb{S}^n, \mathrm{SL}_\pm(n+1,\mathbb{R}))$-diffeomorphism of oriented real projective orbifolds and vice versa.*

PROOF. Straightforward. See p. 143 of Thurston [**58**]. □

**2.3. The local homeomorphism theorems.** For technical reasons, we will be assuming $\partial\mathcal{O} = \emptyset$ in most cases. Here, we are not yet concerned with convexity of orbifolds. The following map hol, the so-called *Ehresmann-Thurston map*, is induced by sending $(\mathbf{dev}, h)$ to the conjugacy class of $h$ as isotopies preserve $h$:

THEOREM 2.10 ([**19**]). *Let $\mathcal{O}$ be a noncompact strongly tame real projective $n$-orbifold with radial ends or totally-geodesic ends of lens-type with markings and given types $\mathcal{R}$ or $\mathcal{T}$. Assume $\partial\mathcal{O} = \emptyset$. Then the following map is a local homeomorphism:*

$$\mathrm{hol}: \mathrm{Def}^s_{\mathcal{E},\mathrm{u}}(\mathcal{O}) \to \mathrm{rep}^s_{\mathcal{E},\mathrm{u}}(\pi_1(O), \mathrm{PGL}(n+1,\mathbb{R})).$$

Also, we define

$$\mathrm{rep}^s_{\mathcal{E}}(\pi_1(\mathcal{O}), \mathrm{SL}_\pm(n+1,\mathbb{R})), \mathrm{rep}^s_{\mathcal{E},\mathrm{u}}(\pi_1(\mathcal{O}), \mathrm{SL}_\pm(n+1,\mathbb{R}))$$

similarly to Section 2.1.6.

By lifting $(\mathbf{dev}, h)$ by the method of Section 2.2, we obtain that

$$\mathrm{hol}: \mathrm{Def}^s_{\mathcal{E},\mathrm{u}}(\mathcal{O}) \to \mathrm{rep}^s_{\mathcal{E},\mathrm{u}}(\pi_1(O), \mathrm{SL}_\pm(n+1,\mathbb{R}))$$

is a local homeomorphism.

REMARK 2.11. The restrictions of end types are necessary for this theorem to hold. (See Goldman [**38**], Canary-Epstein-Green [**14**], Choi [**23**], and Bergeron-Gelander [**11**] for many versions of similar results.)

## 3. Convex real projective structures

**3.1. Metrics.** Let $\Omega$ be a properly convex open domain. A line or a subspace of dimension-one in $\mathbb{RP}^n$ has a two-dimensional homogenous coordinate system. Define a metric by defining the distance for $p, q \in \Omega$,

$$d_\Omega(p,q) = \log |[o,s,q,p]|$$

where $o$ and $s$ are endpoints of the maximal segment in $\Omega$ containing $p, q$ where $o, q$ separates $p, s$ and $[o, s, q, p]$ denotes the cross ratio.

Given a properly convex real projective structure on $\mathcal{O}$, there is a Hilbert metric which we denote by $d_{\tilde{\mathcal{O}}}$ on $\tilde{\mathcal{O}}^o$. Since the metric $d_{\tilde{\mathcal{O}}}$ is invariant under the deck transformation group, we obtain a metric $d_{\mathcal{O}}$ on $\mathcal{O}$.

Assume that $K_i \to K$ geometrically for a sequence of properly convex domains $K_i$ and a properly convex domain $K$. Suppose that two sequences of points $\{x_i | x_i \in K_i^o\}$ and $\{y_i | y_i \in K_i^o\}$ converge to $x, y \in K^o$ respectively. Since the end of a maximal segments always are in $\partial K_i$ and $\partial K_i \to \partial K$ by Lemma 2.1, we obtain

$$d_{K_i^o}(x_i, y_i) \to d_{K^o}(x, y) \tag{3.1}$$

holds.

### 3.2. Convexity and convex domains.

PROPOSITION 3.1 (Kuiper [**47**], Koszul [**46**], Vey [**60**]).

- *A strongly tame real projective orbifold is properly convex if and only if each developing map sends the universal cover to a properly convex open domain bounded in an affine subspace of* $\mathbb{RP}^n$.
- *If a strongly tame convex real projective orbifold is not properly convex, then its holonomy homomorphism is virtually reducible.*

PROPOSITION 3.2 (Corollary 2.13 of Benoist [**9**]). *Suppose that a discrete subgroup* $\Gamma$ *of* $\mathrm{PGL}(n, \mathbb{R})$ *(resp.* $\mathrm{SL}_\pm(n, \mathbb{R})$*) acts on a properly convex* $(n-1)$*-dimensional open domain* $\Omega$ *in* $\mathbb{RP}^{n-1}$ *(resp.* $\mathbb{S}^{n-1}$*) so that* $\Omega/\Gamma$ *is compact. Then the following statements are equivalent.*

- *Every subgroup of finite index of* $\Gamma$ *has a finite center.*
- *Every subgroup of finite index of* $\Gamma$ *has a trivial center.*
- *Every subgroup of finite index of* $\Gamma$ *is irreducible in* $\mathrm{PGL}(n, \mathbb{R})$ *(resp.* $\mathrm{SL}_\pm(n, \mathbb{R})$*). That is,* $\Gamma$ *is strongly irreducible.*
- *The Zariski closure of* $\Gamma$ *is semisimple.*
- $\Gamma$ *does not contain a normal infinite nilpotent subgroup.*
- $\Gamma$ *does not contain a normal infinite abelian subgroup.*

THEOREM 3.3 (Theorem 1.1 of Benoist [**9**]). *Let* $\Gamma$ *be a discrete subgroup of* $\mathrm{PGL}(n, \mathbb{R})$ *(resp.* $\mathrm{SL}_\pm(n, \mathbb{R})$*) with a trivial virtual center. Suppose that a discrete subgroup* $\Gamma$ *of* $\mathrm{PGL}(n, \mathbb{R})$ *(resp.* $\mathrm{SL}_\pm(n, \mathbb{R})$*) acts on a properly convex* $(n-1)$*-dimensional open domain* $\Omega$ *so that* $\Omega/\Gamma$ *is a compact orbifold. Then every representation of a component of* $\mathrm{Hom}(\Gamma, \mathrm{PGL}(n, \mathbb{R}))$ *(resp.* $\mathrm{Hom}(\Gamma, \mathrm{SL}_\pm(n, \mathbb{R}))$*) containing the inclusion representation also acts on a properly convex* $(n-1)$*-dimensional open domain cocompactly and properly discontinuously.*

In general, a *join* of two convex sets $C_1$ and $C_2$ in an affine subspace $A$ of $\mathbb{RP}^n$ is defined as

$$\{[tv_1 + (1-t)v_2] | v_i \in C_{C_i}, i = 1, 2, 0 \leq t \leq 1\}$$

where $C_{C_i}$ is a cone in $\mathbb{R}^{n+1}$ corresponding to $C_i$, $i = 1, 2$. The join is denoted by $C_1 * C_2$ in this paper.

Given subspaces $V_1, \ldots, V_m \subset \mathbb{RP}^n$ (resp. $\subset \mathbb{S}^n$) that are from linear independent subspaces in $\mathbb{R}^{n+1}$ and a subset $C_i \subset V_i$ for each $i$, we define a *strict join* of

$m$ sets $C_1, \dots, C_m$

$$C_1 * \cdots * C_m := \left\{ \left[ \sum_{i=1}^m t_i v_i \right] \middle| \sum_{i=1}^m t_i = 1, 0 \leq t_i \leq 1, v_i \in C_{C_i} \right\},$$

where $C_{C_i}$ is a cone in $\mathbb{R}^{n+1}$ corresponding to $C_i$. (Of course, this depends on the choices of $C_{C_i}$ up to $\mathcal{A}$.)

A *cone-over* a strictly joined domain is one containing a strictly joined domain $A$ and is a union of segments from a cone-point $\notin A$ to points of $A$.

PROPOSITION 3.4 (Theorem 1.1 of Benoist [**9**]). *Assume $n \geq 2$. Let $\Sigma$ be a closed $(n-1)$-dimensional properly convex projective orbifold and let $\Omega$ denote its universal cover in $\mathbb{RP}^{n-1}$ (resp. in $\mathbb{S}^{n-1}$). Then*

(i) *$\Omega$ is projectively diffeomorphic to the interior of a strict join $K_1 * \cdots * K_{l_0}$ where $K_i$ is a properly convex open domain of dimension $n_i \geq 0$ corresponding to a convex open cone $C_i \subset \mathbb{R}^{n_i+1}$.*

(ii) *$\Omega$ is the image of the interior of $C_1 \oplus \cdots \oplus C_{l_0}$.*

(iii) *The fundamental group $\pi_1(\Sigma)$ is virtually isomorphic to a cocompact subgroup of $\mathbb{Z}^{l_0-1} \times \Gamma_1 \times \cdots \times \Gamma_{l_0}$ for $l_0 - 1 + \sum_{i=1}^{l_0} n_i = n$ with following properties:*
   - *Each $\Gamma_i$ has the property that each finite index subgroup has a trivial center.*
   - *Each $\Gamma_j$ acts on $K_j^o$ cocompactly and the Zariski closure is a semisimple Lie group in $\mathrm{PGL}(n_j+1, \mathbb{R})$ (resp. in $\mathrm{SL}_\pm(n_j+1, \mathbb{R})$), and acts trivially on $K_m$ for $m \neq j$.*
   - *The subgroup corresponding to $\mathbb{Z}^{l_0-1}$ acts trivially on each $K_j$.*

**3.3. The duality.** We starts from linear duality. Let us choose the origin $O$ in $\mathbb{R}^{n+1}$. Let $\Gamma$ be a group of linear transformations $\mathrm{GL}(n+1, \mathbb{R})$. Let $\Gamma^*$ be the *affine dual group* defined by $\{g^{*-1} | g \in \Gamma\}$ acting on the dual space $\mathbb{R}^{n+1*}$. Suppose that $\Gamma$ acts on a properly convex cone $C$ in $\mathbb{R}^{n+1}$ with the vertex $O$.

An open convex cone $C^*$ in $\mathbb{R}^{n+1,*}$ is *dual* to an open convex cone $C$ in $\mathbb{R}^{n+1}$ if $C^* \subset \mathbb{R}^{n+1*}$ equals

$$\{\phi \in \mathbb{R}^{n+1*} | \, \phi | \mathrm{Cl}(C) - \{O\} > 0\}.$$

$C^*$ is a cone with vertex as the origin again. Note $(C^*)^* = C$.

Let $\mathbb{R}_+$ denote the set of positive real numbers. Now $\Gamma^*$ acts on $C^*$. Also, if $\Gamma$ acts cocompactly on $C$ if and only if $\Gamma^*$ acts on $C^*$ cocompactly. A *central dilatation extension* $\Gamma'$ of $\Gamma$ is the subgroup of $\mathrm{GL}(n+1, \mathbb{R})$ generated by $\Gamma$ and a dilatation $s\mathrm{I}$ by a scalar $s \in \mathbb{R}_+ - \{1\}$ with the fixed $O$. The dual of $\Gamma'$ is a central dilatation extension of $\Gamma^*$.

Given a subgroup $\Gamma$ in $\mathrm{PGL}(n+1, \mathbb{R})$, an *affine lift* in $\mathrm{GL}(n+1, \mathbb{R})$ is any subgroup that maps to $\Gamma$ isomorphically under the projection. Given a subgroup $\Gamma$ in $\mathrm{PGL}(n+1, \mathbb{R})$, the dual group $\Gamma^*$ is the image in $\mathrm{PGL}(n+1, \mathbb{R})$ of the dual of any affine lift of $\Gamma$.

A properly convex open domain $\Omega$ in $\mathbb{P}(\mathbb{R}^{n+1})$ (resp. in $\mathbb{S}(\mathbb{R}^{n+1})$) is *dual* to a properly convex open domain $\Omega^*$ in $\mathbb{P}(\mathbb{R}^{n+1*})$ (resp. in $\mathbb{S}(\mathbb{R}^{n+1*})$) if $\Omega$ corresponds to an open convex cone $C$ and $\Omega^*$ to its dual $C^*$. We say that $\Omega^*$ is dual to $\Omega$. We also have $(\Omega^*)^* = \Omega$ and $\Omega$ is properly convex if and only if so is $\Omega^*$.

We call $\Gamma$ a *dividing group* if a central dilatational extension acts cocompactly on $C$. $\Gamma$ is dividing if and only if so is $\Gamma^*$. (See [**62**] and [**8**]).

THEOREM 3.5 (Vinberg [**61**]). *Let $\mathcal{O}$ be a properly convex real projective orbifold of form $\Omega/\Gamma$. Let $\mathcal{O}^* = \Omega^*/\Gamma^*$ be a properly convex real projective orbifold. Then $\mathcal{O}^*$ is diffeomorphic to $\mathcal{O}$.*

Given a convex domain $\Omega$ in an affine subspace $A \subset \mathbb{R}^n$, a *supporting hyperplane* $h$ at a point $x \in \mathrm{bd}\Omega$ is a hyperplane so that a component of $A - h$ contains the interior of $\Omega$.

A hyperspace is an element of $\mathbb{RP}^{n*}$ since it is represented as a 1-form, and an element of $\mathbb{RP}^n$ can be considered as a hyperspace in $\mathbb{RP}^{n*}$. The following definition applies to $\Omega \subset \mathbb{RP}^n$ (resp. $\mathbb{S}(\mathbb{R}^{n+1*})$) and $\Omega^* \subset \mathbb{RP}^{n*}$ (resp. $\mathbb{S}(\mathbb{R}^{n+1*})$). Given a properly convex domain $\Omega$, we define the *augmented boundary* of $\Omega$

$$\mathrm{bd}^{\mathrm{Ag}}\Omega := \big\{(x,h)\big|x \in \mathrm{bd}\Omega, h \text{ is a supporting hyperplane of } \Omega, h \ni x\big\}.$$

REMARK 3.6. For open properly convex domains $\Omega_1$ and $\Omega_2$, we have

$$\Omega_1 \subset \Omega_2 \text{ if and only if } \Omega_2^* \subset \Omega_1^*. \tag{3.2}$$

The following standard results are proved in Section 3 of [**16**]. We will call the homeomorphism below as the *duality map*.

PROPOSITION 3.7. *Suppose that $\Omega \subset \mathbb{RP}^n$ (resp. $\mathbb{S}(\mathbb{R}^{n+1*})$) and its dual $\Omega^* \subset \mathbb{RP}^{n*}$ (resp. $\mathbb{S}(\mathbb{R}^{n+1*})$) are properly convex domains.*

(i) *There is a proper quotient map $\Pi_{\mathrm{Ag}} : \mathrm{bd}^{\mathrm{Ag}}\Omega \to \mathrm{bd}\Omega$ given by sending $(x,h)$ to $x$.*
(ii) *Let $\Gamma$ act on properly discontinuously $\Omega$ if and only if so acts $\Gamma^*$ on $\Omega^*$.*
(iii) *There exists a homeomorphism*

$$\mathcal{D} : \mathrm{bd}^{\mathrm{Ag}}\Omega \leftrightarrow \mathrm{bd}^{\mathrm{Ag}}\Omega^*$$

*given by sending $(x,h)$ to $(h,x)$.*
(iv) *Let $A \subset \mathrm{bd}^{\mathrm{Ag}}\Omega$ be a subspace and $A^* \subset \mathrm{bd}^{\mathrm{Ag}}\Omega^*$ be the corresponding dual subspace $\mathcal{D}(A)$. If a group $\Gamma$ acts properly discontinuously on $A$ if and only if $\Gamma^*$ so acts on $A^*$.*

Given a convex domain $\Omega$, we denote by $R_p(\Omega)$ the space of directions of open rays in $\Omega$ from $p$ in $\mathbb{S}^{n-1}_p$.

PROPOSITION 3.8. *Let $\Omega^*$ be the dual of a properly convex domain $\Omega$. Then*

(i) $\mathrm{bd}\Omega$ *is $C^1$ and strictly convex if and only if* $\mathrm{bd}\Omega^*$ *is $C^1$ and strictly convex.*
(ii) *$\Omega$ is a horospherical orbifold if and only if so is $\Omega^*$.*
(iii) *Let $p \in \mathrm{bd}\Omega$. Then $\mathcal{D}$ sends in a one-to-one and onto manner*

$$\{(p,h)|h \text{ is a supporting hyperplane of } \Omega \text{ at } p\}$$

*to $\{(h^*,p^*)|h^* \in D = p^* \cap \mathrm{bd}\Omega^*\}$ where $D$ is a properly convex set in* $\mathrm{bd}\Omega$.
(iv) $\mathrm{bd}\Omega^*$ *contains a properly convex domain $D = P \cap \mathrm{bd}\Omega^*$ open in a totally geodesic hyperplane $P$ if and only if* $\mathrm{bd}\Omega$ *contains a vertex $p$ with $R_p(\Omega)$ a properly convex domain. Moreover, $D^o$ and $R_p(\Omega)$ are properly convex and are projectively diffeomorphic to dual domains in $\mathbb{RP}^{n-1}$.*

## 4. The end theory

We will now discuss in detail the end theory. The following is simply the notions useful in relative hyperbolic group theory as can be found in Bowditch [**12**].

**4.1. p-ends, p-end neighborhoods, and p-end fundamental groups.** Let $\mathcal{O}$ be a real projective orbifold with the universal cover $\tilde{\mathcal{O}}$ and the covering map $p_{\tilde{\mathcal{O}}}$. Each end neighborhood $U$, diffeomorphic to $S_E \times (0,1)$, of an end $E$ lifts to a connected open set $\tilde{U}$ in $\tilde{\mathcal{O}}$. A subgroup $\Gamma_{\tilde{U}}$ of $\Gamma$ acts on $\tilde{U}$ where

$$p_{\tilde{\mathcal{O}}}^{-1}(U) = \bigcup_{g \in \pi_1(\mathcal{O})} g(\tilde{U}).$$

Each component $\tilde{U}$ is said to be a *proper pseudo-end neighborhood*.

- An *exiting sequence* of sets $U_1, U_2, \cdots$ in $\tilde{\mathcal{O}}$ is a sequence so that for each compact subset $K$ of $\mathcal{O}$ there exists an integer $N$ satisfying $p_{\tilde{\mathcal{O}}}^{-1}(K) \cap U_i = \emptyset$ for $i > N$.
- A *pseudo-end sequence* is an exiting sequence of proper pseudo-end neighborhoods
$$\{U_i | i = 1, 2, 3, \dots\}, \text{ where } U_{i+1} \subset U_i \text{ for every } i.$$
- Two pseudo-end sequences $\{U_i\}$ and $\{V_j\}$ are *compatible* if for each $i$, there exists $J$ such that $V_j \subset U_i$ for every $j$, $j > J$ and conversely for each $j$, there exists $I$ such that $U_i \subset V_j$ for every $i$, $i > I$.
- A compatibility class of a proper pseudo-end sequence is called a *pseudo-end* of $\tilde{\mathcal{O}}$. Each of these corresponds to an end of $\mathcal{O}$ under the universal covering map $p_{\mathcal{O}}$.
- For a pseudo-end $\tilde{E}$ of $\tilde{\mathcal{O}}$, we denote by $\Gamma_{\tilde{E}}$ the subgroup $\Gamma_{\tilde{U}}$ where $U$ and $\tilde{U}$ is as above. We call $\Gamma_{\tilde{E}}$ a *pseudo-end fundamental group*. We will also denote it by $\pi_1(\tilde{E})$.
- A *pseudo-end neighborhood* $U$ of a pseudo-end $\tilde{E}$ is a $\Gamma_{\tilde{E}}$-invariant open set containing a proper pseudo-end neighborhood of $\tilde{E}$. A proper pseudo-end neighborhood is an example.

(From now on, we will replace "pseudo-end" with the abbreviation "p-end".)

PROPOSITION 4.1. *The p-end fundamental group $\Gamma_{\tilde{E}}$ is independent of the choice of $U$.*

PROOF. Given $U$ and $U'$ that are end-neighborhoods for an end $E$, let $\tilde{U}$ and $\tilde{U}'$ be p-end neighborhoods for a p-end $\tilde{E}$ that are components of $p^{-1}(U)$ and $p^{-1}(U')$ respectively. Let $\tilde{U}''$ be the component of $p^{-1}(U'')$ that is a p-end neighborhood of $\tilde{E}$. Then $\Gamma_{\tilde{U}''}$ injects into $\Gamma_{\tilde{U}}$ since both are subgroups of $\Gamma$. Any $\mathcal{G}$-path in $U$ in the sense of Bridson-Haefliger [**13**] is homotopic to a $\mathcal{G}$-path in $U''$ by a translation in the $I$-factor. Thus, $\pi_1(U') \to \pi_1(U)$ is surjective. Since $\tilde{U}$ is connected, any element $\gamma$ of $\Gamma_{\tilde{U}}$ is represented by a $\mathcal{G}$-path connecting $x_0$ to $\gamma(x_0)$. (See Example 3.7 in Chapter III.$\mathcal{G}$ of [**13**].) Thus, $\Gamma_{\tilde{U}}$ is isomorphic to the image of $\pi_1(U) \to \pi_1(\mathcal{O})$. Since $\Gamma_{\tilde{U}''}$ is surjective to the image of of $\pi_1(U'') \to \pi_1(\mathcal{O})$, it follows that $\Gamma_{\tilde{U}''}$ is isomorphic to $\Gamma_{\tilde{U}}$. □

Let $\mathcal{O}$ be a strongly tame real projective orbifold. We give each end of $\mathcal{O}$ a marking. We fix a developing map $\mathbf{dev} : \tilde{\mathcal{O}} \to \mathbb{RP}^n$ in this subsection.

A *ray* from a point $v$ of $\mathbb{RP}^n$ is a segment with endpoint equal to $v$ oriented away from $v$.

Let $E$ be an R-end of $\mathcal{O}$.

- Let $\tilde{E}$ denote a p-R-end corresponding to $E$ and $U$ denote a p-R-end neighborhood of $\tilde{E}$ with a radial foliation $\mathcal{F}$ induced from the end marking on an end neighborhood of $E$.
- Two radial leaves of equivalent radial foliations of proper p-end neighborhoods of $\tilde{E}$ are *equivalent* if they agree on a proper p-end neighborhood of $\tilde{E}$.
- Let $x$ be the common end point of the images under the developing map of leaves of $\mathcal{F}$. We call $x$ the *p-end vertex* of $\tilde{\mathcal{O}}$. $x$ will be denoted by $v_{\tilde{E}}$ if its associated p-end neighborhood corresponds to a p-end $\tilde{E}$.
- Let $\mathbb{S}^{n-1}_{v_{\tilde{E}}}$ denote the space of equivalence classes of rays from $v_{\tilde{E}}$ diffeomorphic to an $(n-1)$-sphere where $\pi_1(\tilde{E})$ acts as a group of projective automorphisms. Here, $\pi_1(\tilde{E})$ acts on $v_{\tilde{E}}$ and sends leaves to leaves in $U_1$.
- We denote by $R_{v_{\tilde{E}}}(\tilde{\mathcal{O}}) = \tilde{S}_{\tilde{E}}$ as the following space
$$\{[l] \,|\, l \subset \tilde{\mathcal{O}}, \mathbf{dev}(l) \text{ is a ray from } v_{\tilde{E}}\},$$
which is an $(n-1)$-dimensional open manifold.
- The map $\mathbf{dev}$ induces an immersion
$$\tilde{S}_{\tilde{E}} \to \mathbb{S}^{n-1}_{v_{\tilde{E}}}.$$
Also, $\Gamma_{\tilde{E}}$ projectively acts on $\tilde{S}_{\tilde{E}}$ by $g([l]) = [g(l)]$ for each leaf $l$ and $g \in \Gamma_{\tilde{E}}$.
- Recall that $\tilde{S}_{\tilde{E}}/\Gamma_{\tilde{E}}$ is diffeomorphic to the end orbifold denoted by $S_E$. Thus, $S_E$ has a convex real projective structure. (However, the projective structure and the differential topology on $S_E$ does depend on the end markings.)

Given a T-end of $\mathcal{O}$ and an end neighborhood $U$ of the product form $S_E \times [0,1)$ with a compactification by a totally geodesic orbifold $S_E$, we take a component $U_1$ of $p^{-1}(U)$ and a convex domain $\tilde{S}_{\tilde{E}}$ developing into a totally geodesic hypersurface $P$ under $\mathbf{dev}$. Here $\tilde{E}$ is the p-end corresponding to $E$ and $U_1$. $\Gamma_{\tilde{E}}$ acts on $U_1$ and hence on $\tilde{S}_{\tilde{E}}$. Again $\tilde{S}_{\tilde{E}}/\Gamma_{\tilde{E}}$ is projectively diffeomorphic to the *end orbifold* to be denote by $S_E$ again. We call $\tilde{S}_{\tilde{E}}$ the *p-ideal boundary component* of $\tilde{\mathcal{O}}$. Generalizing further an open subset $U$ of $\tilde{\mathcal{O}}$ containing a proper p-end-neighborhood of $\tilde{E}$, where $\pi_1(\tilde{E})$ acts on, is said to be a *p-end neighborhood.*

**4.2. The admissible groups.** If every subgroup of finite index of a group $\Gamma$ has a finite center, $\Gamma$ is said to be a *virtual center-free group.* An *admissible group* $G$ acting on projective $\mathbb{S}^{n-1}$ is a finite extension of the finite product $\mathbb{Z}^{l-1} \times \Gamma_1 \times \cdots \times \Gamma_l$ for infinite hyperbolic or trivial groups $\Gamma_i$ with following properties:

- $G$ acts on a properly convex domain of form $K_1 * \cdots * K_l$ for a strictly convex domain $K_j \subset \mathbb{S}^{n-1}$ for each $j$, $j = 1, \dots, l$, in an $n_j$-dimensional subspace of $\mathbb{S}^{n-1}$, $0 \leq n_j \leq n-1$,
- $\Gamma_j$ is the restriction of $G$ to each $K_j$ and extended on $\mathbb{S}^{n-1}$ to act trivially on $K_m$ for $m \neq j$ where $K_j^o/\Gamma_j$ is an orbifold of dimension $n_j$.
- $\mathbb{Z}^{l-1}$ acts trivially on each $K_j$ and is a virtual center of $G$.

This is strictly stronger than the conclusion of Proposition 3.4 of Benoist and is needed for now. Here, we conjecture that we do not need the stronger condition but the conclusion of Proposition 3.4 is enough assumption for everything in this paper. (See also Example 5.5.3 of [**54**] as pointed out by M. Kapovich. )

We have $l = 1$ if and only if the end fundamental group is hyperbolic or is trivial. If our orbifold has a complete hyperbolic structure, then end fundamental groups are virtually free abelian.

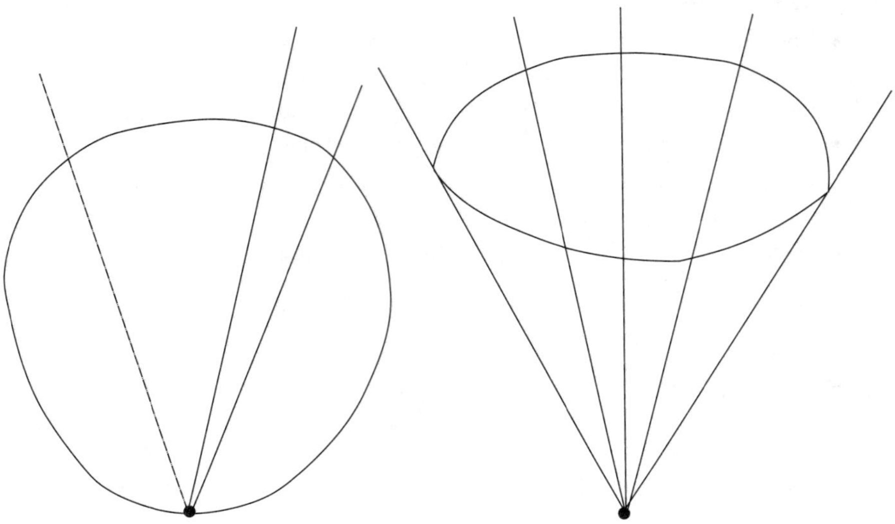

FIGURE 2. The universal covers of horospherical and lens shaped ends. The radial lines form cone-structures.

**4.3. The admissible ends.** Let $\mathcal{O}$ be a convex real projective orbifold with the universal cover $\tilde{\mathcal{O}}$.

- A *cone* over a point $x$ and a set $A$ in an affine subspace of $\mathbb{RP}^n$ (resp. in $\mathbb{S}^n$), $x \notin \mathrm{Cl}(A)$ is the set given by $x * A$ in $\mathbb{RP}^n$ (resp. in $\mathbb{S}^n$).
- Take a cone $C := \{x\} * L$ over a lens $L$ and a point $x$, $x \notin \mathrm{Cl}(L)$, so that every maximal segment $l$ from $x$ in $C$ ends in one component $\partial_1 L$ of $\partial L$ and meets $\partial_1 L$ and $\partial_2 L$ exactly once. A *lens-cone* is $C - \{x\}$.
- Take a cone $C := \{x\} * L$ over a generalized lens $L$ with the same properties as above where a nonsmooth component has to be in the boundary of the cone. A *generalized lens-cone* is $C - \{x\}$.
- For two components $A_1$ and $A_2$ of $\partial L$ for $L$ as above in the lens-cone, $A_1$ is called a *top hypersurface* if it is in $\mathrm{bd}(\{x\} * L)$ and $A_2$ is then called a *bottom hypersurface.*
- A (generalized) *lens* of a (generalized) lens-cone $C$ is the lens-shaped domain $A$ so that $C = \{x\} * A - \{x\}$ for a point $x \notin \mathrm{Cl}(A)$ and with the properties in the first item.
- A *totally-geodesic subdomain* is a convex domain in a hyperspace.
- A *cone-over* a totally-geodesic open domain $A$ is $\{x\} * A - \{x\}$ for the cone $\{x\} * A$ over a point $x$ not in the hyperspace.

(See Figure 2.) We will also call a real projective orbifold with boundary to be

- a *lens-cone* or
- a *lens*, provided it is compact,

if it is covered by such domains and is diffeomorphic to a closed $(n-1)$-orbifold times an interval.

We introduce some relevant adjectives: Let $S_E$ be an $(n-1)$-dimensional end orbifold corresponding to a p-end $\tilde{E}$, and let $\mu$ be a holonomy homomorphism

$$\pi_1(\tilde{E}) \to \mathrm{PGL}(n+1, \mathbb{R}) \text{ (resp. } \mathrm{SL}_{\pm}(n+1, \mathbb{R}))$$

restricted from that of $\mathcal{O}$.

- Suppose that $\mu(\pi_1(\tilde{E}))$ acts on a (generalized) lens-shaped domain $K$ in $\mathbb{RP}^n$ (resp. in $\mathbb{S}^n$) with boundary a union of two open $(n-1)$-cells $A_1$ and $A_2$ and $\pi_1(\tilde{E})$ acts properly on $A_1$ and $A_2$. Then $\mu$ is said to be a *(generalized) lens-shaped* representation for $\tilde{E}$.
- $\mu$ is a *totally-geodesic* representation if $\mu(\pi_1(\tilde{E}))$ acts on a totally-geodesic subdomain.
- If $\mu(\pi_1(\tilde{E}))$ acts on a horoball $K$, then $\mu$ is said to be a *horospherical* representation. In this case, it follows $\mathrm{bd}K - \partial K = \{v_{\tilde{E}}\}$ for the p-end vertex $v_{\tilde{E}}$ of $\tilde{E}$.
- If $\mu(\pi_1(\tilde{E}))$ acts on a strictly joined domain, then $\mu$ is said to be a *strictly joined* representation.

Let $C'$ be a generalized lens and $L := \{v_{\tilde{E}}\} * C' - \{v_{\tilde{E}}\}$ be a generalized lens-cone over $C'$. A *concave p-end-neighborhood* is an imbedded p-end neighborhood of form $L - C'$.

DEFINITION 4.2. (Admissible ends) Let $\mathcal{O}$ be a real projective orbifold with the universal cover $\tilde{\mathcal{O}}$. Let $E$ be an R-end of $\mathcal{O}$ and $\tilde{E}$ be the corresponding p-end with the p-end fundamental group $\pi_1(\tilde{E})$.

- We say that the radial end $E$ of $\mathcal{O}$ is of *lens-type* if $\tilde{E}$ has a p-end neighborhood that is a lens-cone of form $L * \{v_{\tilde{E}}\} - \{v_{\tilde{E}}\}$ and $\pi_1(\tilde{E})$ acts on for its lens $L$.
- $E$ is of *generalized lens-type* if $\tilde{E}$ has a concave p-end neighborhood. Equivalently, a p-end neighborhood of $\tilde{E}$ is the interior of a generalized lens-cone of form $L * \{v_{\tilde{E}}\} - \{v_{\tilde{E}}\}$ and $\pi_1(\tilde{E})$ acts on the generalized lens $L$.

A p-R-end $\tilde{E}$ is *admissible* if the p-end fundamental group acts on $\tilde{S}_{\tilde{E}}$ as an admissible group and if $\tilde{E}$ is a horospherical or lens-type p-R-end.

A T-end $E$ is of *lens-type* if $E$ satisfies the lens-condition that the ideal boundary end orbifold $S_E$ has a lens-neighborhood $L$ in an ambient real projective orbifold containing $\mathcal{O}$. For a component $C_1$ of $L - S_E$ inside $\mathcal{O}$, $C_1 \cup S_E$ is said to be the *one-sided end neighborhood* of $S_E$. Given a p-end $\tilde{E}$, the orbifold $S_E$ is covered by a domain $\tilde{S}_{\tilde{E}}$ in the boundary of p-end neighborhood corresponding $\tilde{E}$ in a hyperspace.

A T-end is *admissible* if it is of lens-type and the p-end fundamental group for a p-end $\tilde{E}$ acts on the ideal boundary $\tilde{S}_{\tilde{E}}$ as an admissible group.

A p-end is *admissible in a generalized sense* if it is admissible or is a generalized lens-type p-R-end.

EXAMPLE 4.3. A model of a lens-type R-end can be made by a positive diagonal group acting on the standard simplex $T$ in $\mathbb{RP}^3$. We take a vertex $v$ to be $[1, 0, 0, 0]$

and we choose an abelian group $G$ of rank 3 acting on $T$ and properly and freely on the interior $F$ of the side of $T$ opposite $v$. We choose $G$ so that the eigenvalue at $v$ is not the largest or the smallest one for $g \in \Delta$ for the Zariski closure $\Delta$ of $G$. $(\{v\} * F - \{v\})/G$ is an end neighborhood of an ambient orbifold. The existence of lens follows by considering orbits of points under $\Delta$. (This follows by Theorem 5.1 of [**16**] since we can show that the uniform middle eigenvalue condition holds. See also Ballas [**1**], [**2**], and Ballas-Danciger-Lee [**4**] which include many graphics for ends.)

Example 2.6 give these examples by Proposition 4.6 of [**16**] or more generally by Theorem 6.5; that is, we show that these have to be admissible lens-type R-ends or horospherical R-ends. (Note also that these properties of the examples will hold during deformations as we will show later in Section 6.2.)

## 5. The relative hyperbolicity of $\pi_1(\mathcal{O})$ and the strict convexity

### 5.1. SPC-structures and its properties.

DEFINITION 5.1. For a strongly tame orbifold $\mathcal{O}$,

(IE) $\mathcal{O}$ or $\pi_1(\mathcal{O})$ satisfies the *infinite-index end fundamental group condition* (IE) if $[\pi_1(\mathcal{O}) : \pi_1(E)] = \infty$ for the end fundamental group $\pi_1(E)$ of each end $E$.

(NA) $\mathcal{O}$ or $\pi_1(\mathcal{O})$ satisfies the *nonannular property* (NA) if

$$\pi_1(\tilde{E}_1) \cap \pi_1(\tilde{E}_2)$$

is finite for two distinct p-ends $\tilde{E}_1, \tilde{E}_2$ of $\tilde{\mathcal{O}}$, and $\pi_1(\tilde{\mathcal{O}})$ does not contain a free abelian group of rank 2 not in the end fundamental groups.

(NA) implies that $\mathcal{O}$ contains no essential torus and also that $\pi_1(E)$ contains every element $g \in \pi_1(\mathcal{O})$ normalizing $\langle h \rangle$ for an infinite order $h \in \pi_1(E)$ for an end fundamental group $\pi_1(E)$ of an end $E$. These conditions are satisfied by complete hyperbolic manifolds with cusps and are group theoretical properties with respect to the end groups.

DEFINITION 5.2. An *SPC-structure* or *stable properly-convex real projective structure* on an $n$-orbifold is a real projective structure so that the orbifold is projectively diffeomorphic to a quotient orbifold of a properly convex domain in $\mathbb{RP}^n$ by a discrete group of projective automorphisms that is stable and irreducible.

DEFINITION 5.3. Suppose that $\mathcal{O}$ has an SPC-structure. Let $\tilde{U}$ be the inverse image in $\tilde{\mathcal{O}}$ of the union $U$ of some choice of a collection of disjoint end neighborhoods of $\mathcal{O}$ with compact $\mathrm{Cl}(U)$. If every straight arc and every non-$C^1$-point in $\mathrm{bd}\tilde{\mathcal{O}}$ are contained in the closure of a component of $\tilde{U}$, then $\mathcal{O}$ is said to be *strictly convex* with respect to the collection of the ends. And $\mathcal{O}$ is also said to have a *strict SPC-structure* with respect to the collection of ends.

Notice that the definition depends on the choice of $U$. However, we will show that if each component $U$ is required to be of lens-type or horospherical, then we show that the definition is independent of $U$ in [**19**].

THEOREM 5.4. *Let $\mathcal{O}$ be a noncompact strongly tame properly convex real projective manifold with generalized admissible ends and satisfies* (IE) *and* (NA). *Then any finite-index subgroup of the holonomy group is strongly irreducible and is not contained in a proper parabolic subgroup of* $\mathrm{PGL}(n+1, \mathbb{R})$ (*resp.* $\mathrm{SL}_\pm(n+1, \mathbb{R})$).

For proof, see Section 4 of [**19**].

**5.2. Bowditch's method.** There are results proved by Cooper, Long, and Tillman [**33**] and Crampon and Marquis [**34**] similar to below. However, the ends have to be horospherical in their work. We will use Bowditch's result [**12**] to show

THEOREM 5.5. *Let $\mathcal{O}$ be a noncompact strongly tame strict SPC-orbifold with generalized admissible ends $E_1, \dots, E_k$ and satisfies* (IE) *and* (NA). *Assume $\partial\mathcal{O} = \emptyset$. Let $\tilde{U}_i$ be the inverse image $U_i$ in $\tilde{\mathcal{O}}$ for a mutually disjoint collection of neighborhoods $U_i$ of the ends $E_i$ for $i = 1, \dots, k$. Then*

- *$\pi_1(\mathcal{O})$ is relatively hyperbolic with respect to the end fundamental groups*
$$\pi_1(E_1), \dots, \pi_1(E_k).$$
*Hence $\mathcal{O}$ is relatively hyperbolic with respect to $U_1 \cup \cdots \cup U_k$.*
- *If $\pi_1(E_{l+1}), \dots, \pi_1(E_k)$ are hyperbolic for some $1 \leq l \leq k$ (possibly some of the hyperbolic ones), then $\pi_1(\mathcal{O})$ is relatively hyperbolic with respect to the end fundamental group $\pi_1(E_1), \dots, \pi_1(E_l)$.*

For definitions and results on relative hyperbolicity of metric spaces, see Bowditch [**12**] or Farb [**37**].

The idea for proof is as follows: Let $U$ be a union of end neighborhoods of $\mathcal{O}$ diffeomorphic to an orbifold times an interval. $\mathcal{O} - U$ is a compact orbifold with boundary. We contract $\mathrm{Cl}(C) \cap \mathrm{bd}\tilde{\mathcal{O}}$ for each component $C$ of $p^{-1}(U)$ to a singleton to obtain a quotient space $X$. Then $X$ is homeomorphic to a compact metric space, i.e., a compactum. We demonstrate that the axioms of Bowditch are satisfied by analyzing the triples of points in $X$ in [**19**].

**5.3. Converse.** The converse to Theorem 5.5 is as follows:

THEOREM 5.6. *Let $\mathcal{O}$ be a noncompact strongly tame properly convex real projective orbifold with generalized admissible ends and satisfies* (IE) *and* (NA). *Assume $\partial\mathcal{O} = \emptyset$. Suppose that $\pi_1(\mathcal{O})$ is a relatively hyperbolic group with respect to the admissible end groups $\pi_1(E_1), ..., \pi_1(E_k)$. Then $\mathcal{O}$ is strictly SPC with respect to the admissible ends $E_1, \dots, E_k$.*

Let $U$ be as in the above section, and let $\tilde{U} = p^{-1}(U) \subset \tilde{\mathcal{O}}$. To give some idea of the proof, we take any segment in $\mathrm{bd}\tilde{\mathcal{O}}$ not contained in any component of $\mathrm{Cl}(\tilde{U}) \cap \mathrm{bd}\tilde{\mathcal{O}}$. Then we find a triangle $T$ with $\partial T \subset \mathrm{bd}\tilde{\mathcal{O}}$ using a sequence of points converging to an interior point of the segment. Also, we construct so that $\partial T$ is not in the closure of any p-end neighborhood.

Let us recall standard definitions in Section 3.1 of Drutu-Sapir [**36**]. An *ultrafilter* $\omega$ is a finite additive measure on $P(\mathbb{N})$ of $\mathbb{N}$ so that each subset has either measure 0 or 1 and all finite sets have measure 0. If a property $P(n)$ holds for all $n$ from a set with measure 1, we say that $P(n)$ holds *$\omega$-almost surely*.

Let $(X, d_X)$ be a metric space. Let $\omega$ be a nonprincipal ultrafilter over the set $\mathbb{N}$ of natural numbers. For a sequence $(x_i)_{i\in\mathbb{N}}$ of points of $X$, its *$\omega$-limit* is $x \in X$ if for every neighborhood $U$ of $x$ the property that $x_i \in U$ holds $\omega$-almost surely.

An *ultraproduct* $\prod X_n/\omega$ of a sequence of sets $(X_n)_{n\in\mathbb{N}}$ is the set of the equivalence classes of sequences $(x_n)$ where $(x_n) \sim (y_n)$ if $x_n = y_n$ holds for $\omega$-almost surely.

Given a sequence of metric spaces $(X_n, d_n)$, consider the ultraproduct $\prod X_n$ and an observation point $e = (e_n)$. Let $D(x, y) = \lim_\omega d_n(x_n, y_n)$. Let $\prod_e X_n/\omega$ denote the set of equivalence classes of sequences of bounded distances from $e$. The *$\omega$-limit* $\lim^\omega(X_n)_e$ is the metric space obtained from $\prod_e X_n/\omega$ by identifying all pair of points $x, y$ with $D(x, y) = 0$.

Given an ultrafilter $\omega$ over the set $\mathbb{N}$ of natural numbers, an observation point $e = (e_i)^\omega$, and sequence of numbers $\delta = (\delta_i)_{i\in\mathbb{N}}$ satisfying $\lim_\omega \delta_i = \infty$, the $\omega$-limit $\lim^\omega(X, d_X/\delta_i)_e$ is called the *asymptotic cone* of $X$. (See [**42**], [**43**] and Definitions 3.3 to 3.8 in [**36**].) We denote it by $Con^\omega(X, e, \delta)$.

For a sequence $(A_n)$ of subsets $A_n$ of $X$, we denote by $\lim^\omega(A_n)$ the subset of $Con^\omega(X, e, \delta)$ that consists of all elements $(x_n)$ where $x_n \in A_n$ $\omega$-almost surely. The asymptotic cone is always complete and $\lim^\omega(A_n)$ is closed.

Next, we choose a nonprincipal ultrafilter $\omega$ and a sequence $l_i \to \infty$. We use the $\omega$-limit $\tilde{\mathcal{O}}_\infty$ of $\frac{1}{l_i} d_{\tilde{\mathcal{O}}}$ on $\tilde{\mathcal{O}}$ with a constant base point $e_i = e \in \tilde{\mathcal{O}}$. This turns out to be a tree-graded space in the sense of Drutu and Sapir [**36**]. Let $T_i^o$ be $T^o$ with the metric $\frac{1}{l_i} d_{\tilde{\mathcal{O}}}|T^o$, which is a hex metric of de la Harpe [**35**].

The inverse image in $U$ of $\tilde{\mathcal{O}}$ of the union of disjoint end neighborhoods of $\mathcal{O}$. A piece in the limit tree-graded space is a limit of a sequence of components of $U$ by Proposition 7.26 of [**36**]. The sequence $\{T_i^o \subset \tilde{\mathcal{O}}\}$ converges to a triangle $T_\infty$ with the hex metric and we show that $T_\infty$ is not contained in a piece by a geometric argument. However, a triangle with a hex metric cannot be divided into more than one piece.

**5.4. Strict SPC-structures deform to strict SPC-structures.** By above Theorems 5.5 and 5.6, the property of strictness of the SPC-structures is topological. Hence, the strictness is a stable property among the set of the SPC-structures.

THEOREM 5.7. *Let $\mathcal{O}$ denote a noncompact strongly tame strict SPC-orbifold with admissible ends and satisfies* (IE) *and* (NA). *Assume $\partial\mathcal{O} = \emptyset$. Let*

$$E_1, \ldots, E_n, E_{n+1}, \ldots, E_k$$

*be the ends of $\mathcal{O}$ where $E_{n+1}, \ldots, E_k$ are some or all of the hyperbolic ends.*

- *Given a deformation through SPC-structures with generalized admissible ends of a strict SPC-orbifold with respect to admissible ends $E_1, \ldots, E_k$ to an SPC-structure with generalized admissible end, the SPC-structures remain strictly SPC with respect to $E_1, \ldots, E_k$.*
- *Given a deformation through SPC-structures with generalized admissible ends of a strict SPC-orbifold with respect to $E_1, \ldots, E_n$ to an SPC-structure with generalized admissible end, the SPC-structures remain strictly SPC with respect to admissible ends $E_1, \ldots, E_n$.*

## 6. The openness and closedness in character varieties

We will now begin to discuss the main aim of this papers. This is to identify the deformation spaces of convex real projective structures on a strongly tame orbifold $\mathcal{O}$ with end conditions with parts of character varieties of $\pi_1(\mathcal{O})$ with corresponding conditions on holonomy groups of ends. We mention that the uniqueness condition below simplifies the theory greatly. Otherwise, we need to use the sections picking the vertices and the totally geodesic planes fixed by the holonomy group of each end.

**6.1. The semi-algebraic properties of** $\mathrm{rep}^s(\pi_1(\mathcal{O}), \mathrm{PGL}(n+1,\mathbb{R}))$ **and related spaces.** We will now recall Section 2.1.8 and make it more precise.

A *parabolic subalgebra* $\mathfrak{p}$ is an algebra in a semisimple Lie algebra $\mathfrak{g}$ whose complexification contains a maximal solvable subalgebra of $\mathfrak{g}$ (p. 279–288 of [**59**]). A *parabolic subgroup* $P$ of a semisimple Lie group $G$ is the full normalizer of a parabolic subalgebra.

We recall from Section 2.1.8. Since $\mathcal{O}$ is the interior of a compact orbifold, there exists a finite set of generators $g_1, \dots, g_m$ with finitely many relators. First, $\mathrm{Hom}(\pi_1(\mathcal{O}), \mathrm{PGL}(n+1,\mathbb{R}))$ can be identified with a semi-algebraic subset of $\mathrm{PGL}(n+1,\mathbb{R})^m$ corresponding to the relators. Each end of $\mathcal{O}$ is assigned to be an $\mathcal{R}$-type end or a $\mathcal{T}$-type end.

Let $\mathrm{Hom}_{\mathcal{E}}(\pi_1(\mathcal{O}), \mathrm{PGL}(n+1,\mathbb{R}))$ denote the subspace of

$$\mathrm{Hom}(\pi_1(\mathcal{O}), \mathrm{PGL}(n+1,\mathbb{R}))$$

where the holonomy of each p-end fundamental group fixes a point of $\mathbb{RP}^n$ for an end of type $\mathcal{R}$ or acts on a subspace $P$ of codimension-one and on a lens meeting $P$ satisfying the lens-condition or a horoball tangent to $P$ for an end of type $\mathcal{T}$. Since there are only finitely many p-end fundamental groups up to conjugation by elements of $\pi_1(\mathcal{O})$, we obtain that

$$\mathrm{Hom}_{\mathcal{E}}(\pi_1(\mathcal{O}), \mathrm{PGL}(n+1,\mathbb{R}))$$

is a closed semi-algebraic subset provided that there is no $\mathcal{T}$-end. If there are $\mathcal{T}$-ends, then we obtain a union of open subsets of closed semi-algebraic subsets.

Since each end fundamental group is finitely generated, the conditions of having a common 1-dimensional eigenspace for each of a finite collection of finitely generated subgroups is a semi-algebraic condition.

Let $\rho \in \mathrm{Hom}_{\mathcal{E}}(\pi_1(E), \mathrm{PGL}(n+1,\mathbb{R}))$ where $E$ is a horospherical end. Then $\rho(\pi_1(E))$ is virtually abelian by Theorem 1.1 of [**15**]. Define

$$\mathrm{Hom}_{\mathcal{E},\mathbf{par}}(\pi_1(E), \mathrm{PGL}(n+1,\mathbb{R}))$$

to be the space of representations where an abelian group of finite index goes into a parabolic subgroup in a copy of $\mathrm{PO}(n,1)$. By Lemma 6.1,

$$\mathrm{Hom}_{\mathcal{E},\mathbf{par}}(\pi_1(E), \mathrm{PGL}(n+1,\mathbb{R}))$$

is a closed semi-algebraic set.

LEMMA 6.1. *Let $G$ be a finite extension of a finitely generated free abelian group $\mathbb{Z}^m$. Then $\mathrm{Hom}_{\mathcal{E},\mathbf{par}}(G, \mathrm{PGL}(n+1,\mathbb{R}))$ is a closed algebraic set.*

PROOF. Let $P$ be a maximal parabolic subgroup of a copy of $\mathrm{PO}(n+1,\mathbb{R})$ that fixes a point $x$. Then $\mathrm{Hom}(\mathbb{Z}^m, P)$ is a closed semi-algebraic set.

$$\mathrm{Hom}_{\mathcal{E},\mathbf{par}}(\mathbb{Z}^m, \mathrm{PGL}(n+1,\mathbb{R}))$$

equals a union

$$\bigcup_{g\in \mathrm{PGL}(n+1,\mathbb{R})} \mathrm{Hom}(\mathbb{Z}^m, gPg^{-1}),$$

another closed semi-algebraic set. Now $\mathrm{Hom}_{\mathcal{E},\mathbf{par}}(G, \mathrm{PGL}(n+1,\mathbb{R}))$ is a closed semi-algebraic subset of

$$\mathrm{Hom}_{\mathcal{E},\mathbf{par}}(\mathbb{Z}^m, \mathrm{PGL}(n+1,\mathbb{R})).$$

□

Let $E$ be an end orbifold of $\mathcal{O}$. Given

$$\rho \in \mathrm{Hom}_{\mathcal{E}}(\pi_1(E), \mathrm{PGL}(n+1, \mathbb{R})),$$

we define the following sets:

- Let $E$ be an end of type $\mathcal{R}$. Let

  $$\mathrm{Hom}_{\mathcal{E},\mathbb{RL}}(\pi_1(E), \mathrm{PGL}(n+1, \mathbb{R}))$$

  denote the space of representations $h$ of $\pi_1(E)$ where $h(\pi_1(E))$ acts on a lens-cone $\{p\} * L$ for a lens $L$ and $p \not\in \mathrm{Cl}(L)$ of a p-end $\tilde{E}$ corresponding to $E$ and the lens $L$ itself. Thus, it is an open subspace of the above semi-algebraic set $\mathrm{Hom}_{\mathcal{E}}(\pi_1(E), \mathrm{PGL}(n+1, \mathbb{R}))$.
- Let $E$ denote an end of type $\mathcal{T}$. Let

  $$\mathrm{Hom}_{\mathcal{E},\mathbb{TL}}(\pi_1(E), \mathrm{PGL}(n+1, \mathbb{R}))$$

  denote the space of totally geodesic representations $h$ of $\pi_1(E)$ satisfying the following condition:
  - $h(\pi_1(E))$ acts on an lens $L$ and a hyperspace $P$ where
  - $L^o \cap P \neq \emptyset$ and
  - $L/h(\pi_1(E))$ is a compact orbifold with two strictly convex boundary components.

  $$\mathrm{Hom}_{\mathcal{E},\mathbb{TL}}(\pi_1(E), \mathrm{PGL}(n+1, \mathbb{R}))$$

  again an open subset of the semi-algebraic set

  $$\mathrm{Hom}_{\mathcal{E}}(\pi_1(E), \mathrm{PGL}(n+1, \mathbb{R})).$$

  (This follows by the proof of Theorem 8.1 of [**15**].)

Let

$$R_E : \mathrm{Hom}(\pi_1(\mathcal{O}), \mathrm{PGL}(n+1, \mathbb{R})) \ni h \to h|\pi_1(E) \in \mathrm{Hom}(\pi_1(E), \mathrm{PGL}(n+1, \mathbb{R}))$$

be the restriction map to the p-end fundamental group $\pi_1(E)$ corresponding to the end $E$ of $\mathcal{O}$.

A *representative set* of p-ends of $\tilde{\mathcal{O}}$ is the subset of p-ends where each end of $\mathcal{O}$ has a corresponding p-end and a unique corresponding p-end. Let $\mathcal{R}_{\mathcal{O}}$ denote the representative set of p-ends of $\tilde{\mathcal{O}}$ of type $\mathcal{R}$, and let $\mathcal{T}_{\mathcal{O}}$ denote the representative set of p-ends of $\tilde{\mathcal{O}}$ of type $\mathcal{T}$. We define a more symmetric space:

$$\mathrm{Hom}^s_{\mathcal{E},ce}(\pi_1(\mathcal{O}), \mathrm{PGL}(n+1, \mathbb{R}))$$

to be

$$\begin{aligned}&\mathrm{Hom}^s(\pi_1(\mathcal{O}), \mathrm{PGL}(n+1, \mathbb{R})) \cap \\ &\left(\bigcap_{E \in \mathcal{R}_{\mathcal{O}}} R_E^{-1}\Big(\mathrm{Hom}_{\mathcal{E},\mathbf{par}}(\pi_1(E), \mathrm{PGL}(n+1, \mathbb{R})) \cup \mathrm{Hom}_{\mathcal{E},\mathbb{RL}}(\pi_1(E), \mathrm{PGL}(n+1, \mathbb{R}))\Big)\right) \cap \\ &\left(\bigcap_{E \in \mathcal{T}_{\mathcal{O}}} R_E^{-1}\Big(\mathrm{Hom}_{\mathcal{E},\mathbf{par}}(\pi_1(E), \mathrm{PGL}(n+1, \mathbb{R})) \cup \mathrm{Hom}_{\mathcal{E},\mathbb{TL}}(\pi_1(E), \mathrm{PGL}(n+1, \mathbb{R}))\Big)\right).\end{aligned}$$

Hence, this is a union of open subsets of semi-algebraic sets in

$$X := \mathrm{Hom}^s_{\mathcal{E}}(\pi_1(\mathcal{O}), \mathrm{PGL}(n+1, \mathbb{R})).$$

(We don't claim that the union is open in $X$. These definitions allow for changes between horospherical ends to lens-type radial ones and totally geodesic ones.)

Let $\mathrm{Hom}^s_{\mathcal{E},\mathrm{u}}(\pi_1(\mathcal{O}), \mathrm{PGL}(n+1,\mathbb{R}))$ denote the subspace of

$$\mathrm{Hom}^s_{\mathcal{E}}(\pi_1(\mathcal{O}), \mathrm{PGL}(n+1,\mathbb{R}))$$

where each element $h$ satisfies the following properties:

- $h|\pi_1(\tilde{E})$ fixes a unique point of $\mathbb{RP}^n$ corresponding to the common eigenspace of positive eigenvalues for lifts of elements of the p-end fundamental group $\pi_1(\tilde{E})$ of $\mathcal{R}$-type (recall Remark 2.2) and
- $h|\pi_1(\tilde{E})$ has a common null-space $P$ of an eigen-1-forms which is unique under the condition that
  - $\pi_1(\tilde{E})$ acts properly on a lens $L$ with $L \cap P$ with nonempty interior in $P$ or
  - $H - \{p\}$ for a horosphere $H$ tangent to $P$ at $p$

  for each p-end fundamental group $\pi_1(\tilde{E})$ of the end of $\mathcal{T}$-type.

We obtain the union of open subsets of semi-algebraic subsets since we need to consider finitely many generators of the fundamental groups of the ends again by Lemma 2.8.

REMARK 6.2. The lens condition is equivalent to the condition here. We repeat it here to put these into set theoretical terms.

Since

$$\mathrm{rep}^s_{\mathcal{E},\mathrm{u}}(\pi_1(\mathcal{O}), \mathrm{PGL}(n+1,\mathbb{R}))$$

is the Hausdorff quotient of the above set with the conjugation $\mathrm{PGL}(n+1,\mathbb{R})$-action, this is the union of open subsets of semi-algebraic subset by Proposition 1.1 of [**45**].

We define

$$\mathrm{Hom}^s_{\mathcal{E},\mathrm{u,ce}}(\pi_1(\mathcal{O}), \mathrm{PGL}(n+1,\mathbb{R}))$$

to be the subset

$$\mathrm{Hom}^s_{\mathcal{E},\mathrm{u}}(\pi_1(\mathcal{O}), \mathrm{PGL}(n+1,\mathbb{R})) \cap$$
$$\left(\bigcap_{E\in\mathcal{R}_{\mathcal{O}}} R_E^{-1}\Big(\mathrm{Hom}_{\mathcal{E},\mathbf{par}}(\pi_1(E), \mathrm{PGL}(n+1,\mathbb{R})) \cup \mathrm{Hom}_{\mathcal{E},\mathbb{RL}}\big(\pi_1(E), \mathrm{PGL}(n+1,\mathbb{R})\big)\Big)\right) \cap$$
$$\left(\bigcap_{E\in\mathcal{T}_{\mathcal{O}}} R_E^{-1}\Big(\mathrm{Hom}_{\mathcal{E},\mathbf{par}}(\pi_1(E), \mathrm{PGL}(n+1,\mathbb{R})) \cup \mathrm{Hom}_{\mathcal{E},\mathbb{TL}}\big(\pi_1(E), \mathrm{PGL}(n+1,\mathbb{R})\big)\Big)\right).$$

Similarly to the above,

$$\mathrm{rep}^s_{\mathcal{E},\mathrm{u,ce}}(\pi_1(\mathcal{O}), \mathrm{PGL}(n+1,\mathbb{R}))$$

is a union of open subsets of strata in

$$\mathrm{rep}^s_{\mathcal{E},\mathrm{u}}(\pi_1(\mathcal{O}), \mathrm{PGL}(n+1,\mathbb{R})).$$

EXAMPLE 6.3. The uniqueness as above holds automatically for convex real projective orbifolds where a set of orbifold singularities contains a leaf of a radial foliation of an R-end neighborhood as in Example 2.6. By Theorem 6.5, we obtain that these have lens-shaped radial ends only.

LEMMA 6.4. *Suppose that $\mathcal{O}$ is a strongly tame real projective orbifold with radial ends. Suppose the end fundamental group of an end $E$ is*

- *virtually generated by finite order elements and*
- *is virtually abelian or is hyperbolic.*

*Suppose that the end orbifold $\Sigma_E$ is convex. Then the end $E$ is either properly convex lens-type radial end or is horospherical.*

THEOREM 6.5. *Suppose that $\mathcal{O}$ is a strongly tame properly convex real projective orbifold with radial ends. Suppose that each end fundamental group is*

- *virtually generated by finite order elements and*
- *is virtually abelian or is hyperbolic.*

*Then the holonomy is in*

$$\mathrm{Hom}^s_{\mathcal{E},\mathrm{u,ce}}(\pi_1(\mathcal{O}), \mathrm{PGL}(n+1,\mathbb{R})).$$

We need the end classification results from [**16**], [**17**], and [**18**].

We immediately obtain:

COROLLARY 6.6. *Let $M$ be a real projective orbifold with radial ends. Suppose that $M$ admits finite-volume hyperbolic 3-orbifold with ends that are horospherical. Suppose that the end orbifold is either a small orbifold with cone points of orders 3 or a disk orbifold with corner reflectors orders 6. Then the ends must be of lens-type $R$-ends or horospherical $R$-ends.*

PROOF. Each end fundamental group is virtually abelian of rank 2. Hence, the end orbifold is finitely covered by a 2-torus. By the classification of real projective 2-torus [**6**], and the existence of order 3 or 6 singularities, the torus must be properly convex or complete affine. By the proof of Proposition 4.6 of [**16**], the end is either horospherical or of lens-type. (Here we just need that the end orbifold be convex.) □

This result may be generalized to higher dimensions but we lack the formulation.

6.1.1. *Main theorems.* We now state our main results:

- We define $\mathrm{Def}^s_{\mathcal{E},\mathrm{ce}}(\mathcal{O})$ to be the subspace of $\mathrm{Def}_{\mathcal{E}}(\mathcal{O})$ with real projective structures with generalized admissible ends and stable irreducible holonomy homomorphisms.
- We define $\mathrm{CDef}_{\mathcal{E},\mathrm{u,ce}}(\mathcal{O})$ to be the subspace of $\mathrm{Def}_{\mathcal{E},\mathrm{u}}(\mathcal{O})$ consisting of SPC-structures with generalized admissible ends.
- We define $\mathrm{SDef}_{\mathcal{E},\mathrm{u,ce}}(\mathcal{O})$ to be the subspace of $\mathrm{Def}_{\mathcal{E},\mathrm{u,ce}}(\mathcal{O})$ consisting of strict SPC-structures with admissible ends.

We remark that these spaces are dual to the same type of the spaces but switching the $\mathcal{R}$-end with $\mathcal{T}$-ends and vice versa by Proposition 3.7.

THEOREM 6.7. *Let $\mathcal{O}$ be a noncompact strongly tame $n$-orbifold with generalized admissible ends. Assume $\partial\mathcal{O} = \emptyset$. Suppose that $\mathcal{O}$ satisfies* (IE) *and* (NA). *Then the subspace*

$$\mathrm{CDef}_{\mathcal{E},\mathrm{u,ce}}(\mathcal{O}) \subset \mathrm{Def}^s_{\mathcal{E},\mathrm{u,ce}}(\mathcal{O})$$

*is open.*

*Suppose further that every finite-index subgroup of $\pi_1(\mathcal{O})$ contains no nontrivial infinite nilpotent normal subgroup. Then* hol *maps $\mathrm{CDef}_{\mathcal{E},\mathrm{u,ce}}(\mathcal{O})$ homeomorphically to a union of components of*

$$\mathrm{rep}^s_{\mathcal{E},\mathrm{u,ce}}(\pi_1(\mathcal{O}), \mathrm{PGL}(n+1,\mathbb{R})).$$

THEOREM 6.8. *Let $\mathcal{O}$ be a strict SPC noncompact strongly tame $n$-dimensional orbifold with admissible ends and satisfies* (IE) *and* (NA). *Assume $\partial\mathcal{O} = \emptyset$. Then*

- *$\pi_1(\mathcal{O})$ is relatively hyperbolic with respect to its end fundamental groups.*
- *The subspace $\mathrm{SDef}_{\mathcal{E},\mathrm{u,ce}}(\mathcal{O}) \subset \mathrm{Def}^s_{\mathcal{E},\mathrm{u,ce}}(\mathcal{O})$ of strict SPC-structures with admissible ends is open.*

*Suppose further that every finite-index subgroup of $\pi_1(\mathcal{O})$ contains no nontrivial infinite nilpotent normal subgroup. Then* hol *maps the deformation space $\mathrm{SDef}_{\mathcal{E},\mathrm{u,ce}}(\mathcal{O})$ of strict SPC-structures on $\mathcal{O}$ with admissible ends homeomorphically to a union of components of*

$$\mathrm{rep}^s_{\mathcal{E},\mathrm{u,ce}}(\pi_1(\mathcal{O}), \mathrm{PGL}(n+1, \mathbb{R})).$$

We prove Theorems 6.7 and 6.8 by dividing into the openness result in Section 6.2 and the closedness result in Section 6.3.

**6.2. Openness.** We will show the following by proving Theorem 6.10.

THEOREM 6.9. *Let $\mathcal{O}$ be a noncompact strongly tame real projective $n$-orbifold and satisfies* (IE) *and* (NA)*. Assume $\partial\mathcal{O} = \emptyset$. In $\mathrm{Def}^s_{\mathcal{E},\mathrm{u,ce}}(\mathcal{O})$, the subspace $\mathrm{CDef}_{\mathcal{E},\mathrm{u,ce}}(\mathcal{O})$ of SPC-structures with generalized admissible ends is open, and so is $\mathrm{SDef}_{\mathcal{E},\mathrm{u,ce}}(\mathcal{O})$.*

We are given a properly real projective orbifold $\mathcal{O}$ with ends $E_1, \dots, E_{e_1}$ of $\mathcal{R}$-type and $E_{e_1+1}, \dots, E_{e_1+e_2}$ of $\mathcal{T}$-type. Let us choose representative p-ends $\tilde{E}_1, \dots, \tilde{E}_{e_1}$ and $\tilde{E}_{e_1+1}, \dots, \tilde{E}_{e_1+e_2}$. Again, $e_1$ is the number of $\mathcal{R}$-type ends, and $e_2$ the number of $\mathcal{T}$-type ends of $\mathcal{O}$.

We define a subspace of $\mathrm{Hom}_{\mathcal{E}}(\pi_1(\mathcal{O}), \mathrm{PGL}(n+1, \mathbb{R}))$ to be as in Section 6.1.

Let $\mathcal{V}$ be an open subset of

$$\mathrm{Hom}^s_{\mathcal{E}}(\pi_1(\mathcal{O}), \mathrm{PGL}(n+1, \mathbb{R}))$$

invariant under the conjugation action of $\mathrm{PGL}(n+1, \mathbb{R})$ so that the following hold:

- one can choose a continuous section $s^{(1)}_{\mathcal{V}} : \mathcal{V} \to (\mathbb{RP}^n)^{e_1}$ sending a holonomy homomorphism to a common fixed point of $\Gamma_{\tilde{E}_i}$ for $i = 1, \dots, e_1$ and
- $s^{(1)}_{\mathcal{V}}$ satisfies

$$s^{(1)}_{\mathcal{V}}(gh(\cdot)g^{-1}) = g \cdot s^{(1)}_{\mathcal{V}}(h(\cdot)) \text{ for } g \in \mathrm{PGL}(n+1, \mathbb{R}).$$

$s^{(1)}_{\mathcal{V}}$ is said to be a *fixed-point section.* In these cases, we say that R-ends are *determined by $s^{(1)}_{\mathcal{V}}$.*

Again we assume that for the open subset $\mathcal{V}$ of

$$\mathrm{Hom}^s_{\mathcal{E}}(\pi_1(\mathcal{O}), \mathrm{PGL}(n+1, \mathbb{R}))$$

the following hold:

- one can choose a continuous section $s^{(2)}_{\mathcal{V}} : \mathcal{V} \to (\mathbb{RP}^{n*})^{e_2}$ sending a holonomy homomorphism to a common dual fixed point of $\pi_1(\tilde{E}_i)$ for $i = e_1 + 1, \dots, e_1 + e_2$,
- $s^{(2)}_{\mathcal{V}}$ satisfies $s^{(2)}_{\mathcal{V}}(gh(\cdot)g^{-1}) = (g^*)^{-1} \circ s^{(2)}_{\mathcal{V}}(h(\cdot))$ for $g \in \mathrm{PGL}(n+1, \mathbb{R})$, and

$s^{(2)}_{\mathcal{V}}$ is said to be a *dual fixed-point* section. In this case, we say that T-ends are *determined by $s^{(2)}_{\mathcal{V}}$.*

We define $s_{\mathcal{V}} : \mathcal{V} \to (\mathbb{RP}^n)^{e_1} \times (\mathbb{RP}^{n*})^{e_2}$ as $s^{(1)}_{\mathcal{V}} \times s^{(2)}_{\mathcal{V}}$ and call it a *fixing section.*

Let $\mathcal{V}$ and $s_{\mathcal{V}} : \mathcal{V} \to (\mathbb{RP}^n)^{e_1} \times (\mathbb{RP}^{n*})^{e_2}$ be as above.

- We define $\mathrm{Def}^s_{\mathcal{E},s_{\mathcal{V}},ce}(\mathcal{O})$ to be the subspace of $\mathrm{Def}_{\mathcal{E},s_{\mathcal{V}}}(\mathcal{O})$ of real projective structures with generalized admissible ends with end structures determined by $s_{\mathcal{V}}$, and stable irreducible holonomy homomorphisms in $\mathcal{V}$.
- We define $\mathrm{CDef}_{\mathcal{E},s_{\mathcal{V}},ce}(\mathcal{O})$ to be the subspace consisting of SPC-structures with generalized admissible ends and holonomy homomorphisms in $\mathcal{V}$ in $\mathrm{Def}^s_{\mathcal{E},s_{\mathcal{V}},ce}(\mathcal{O})$.
- We define $\mathrm{SDef}_{\mathcal{E},s_{\mathcal{V}},ce}(\mathcal{O})$ to be the subspace of consisting of strict SPC-structures with admissible ends and holonomy homomorphisms in $\mathcal{V}$ in $\mathrm{Def}^s_{\mathcal{E},s_{\mathcal{V}},ce}(\mathcal{O})$.

THEOREM 6.10. *Let $\mathcal{O}$ be a noncompact strongly tame real projective $n$-orbifold with generalized admissible ends and satisfies* (IE) *and* (NA). *Assume $\partial\mathcal{O} = \emptyset$. Choose an open* $\mathrm{PGL}(n+1,\mathbb{R})$*-conjugation invariant set*

$$\mathcal{V} \subset \mathrm{Hom}^s_{\mathcal{E}}(\pi_1(\mathcal{O}), \mathrm{PGL}(n+1,\mathbb{R})),$$

*and a fixing section $s_{\mathcal{V}} : \mathcal{V} \to (\mathbb{RP}^n)^{e_1} \times (\mathbb{RP}^{n*})^{e_2}$.*

*Then* $\mathrm{CDef}_{\mathcal{E},s_{\mathcal{V}},ce}(\mathcal{O})$ *is open in* $\mathrm{Def}^s_{\mathcal{E},s_{\mathcal{V}},ce}(\mathcal{O})$, *and so is* $\mathrm{SDef}_{\mathcal{E},s_{\mathcal{V}},ce}(\mathcal{O})$.

By Theorems 6.9 and 2.10, we obtain:

COROLLARY 6.11. *Let $\mathcal{O}$ be a noncompact strongly tame real projective $n$-orbifold with generalized admissible ends and satisfies* (IE) *and* (NA). *Assume $\partial\mathcal{O} = \emptyset$. Then*

$$\mathrm{hol} : \mathrm{CDef}_{\mathcal{E},\mathrm{u},\mathrm{ce}}(\mathcal{O}) \to \mathrm{rep}^s_{\mathcal{E},\mathrm{u},\mathrm{ce}}(\pi_1(\mathcal{O}), \mathrm{PGL}(n+1,\mathbb{R}))$$

*is a local homeomorphism.*

*Furthermore, if $\mathcal{O}$ has a strict SPC-structure with admissible ends, then so is*

$$\mathrm{hol} : \mathrm{SDef}_{\mathcal{E},\mathrm{u},\mathrm{ce}}(\mathcal{O}) \to \mathrm{rep}^s_{\mathcal{E},\mathrm{u},\mathrm{ce}}(\pi_1(\mathcal{O}), \mathrm{PGL}(n+1,\mathbb{R})).$$

We just give a heuristic idea for proof of Corollary 6.11 for $\mathrm{SDef}_{\mathcal{E},\mathrm{u},\mathrm{ce}}(\mathcal{O})$ here as in [**19**]. We begin by taking a properly convex open cone $C$ in $\mathbb{R}^{n+1}$ corresponding to $\tilde{\mathcal{O}}$. Let $\mathcal{O}$ be a properly convex real projective orbifold with generalized admissible ends. Let $\Gamma = h(\pi_1(\mathcal{O}))$ in $\mathrm{PGL}(n+1,\mathbb{R})$ be the holonomy group. Then $\Gamma$ acts on $C$. There is a Koszul-Vinberg function on $C$. The Hessian will be a $\Gamma$-invariant metric. Thus, $\mathcal{O}$ has an induced metric $\mu$ by using sections. For each p-end $\tilde{E}$, we obtain a p-end neighborhood $U$. We approximate the p-end neighborhood $U$ very close to $\tilde{\mathcal{O}}$ in the Hausdorff metric $\mathbf{d}^H$. Let $C_U \subset \mathbb{R}^{n+1}$ denote the open cone associated with $U$. We choose a Koszul-Vinberg Hessian metric on $C_U$ approximating that of $C$.

Let $h_t$ be a parameter of representations in the appropriate character variety with $h_0 = h$. Let $\Gamma_t = h_t(\pi_1(\mathcal{O}))$ where $\Gamma_0 = \Gamma$.

By Theorem 2.10, a convex real projective structure on $\mathcal{O}$ with radial or totally geodesic ends has the holonomy homomorphism $h_t$ and developing maps $\mathbf{dev}_t : \tilde{\mathcal{O}} \to \mathbb{RP}^n$. $\mathbf{dev}_0$ can be considered as the inclusion $\tilde{\mathcal{O}} \to \mathbb{RP}^n$. $\mathbf{dev}_t$ may not be an inclusion in general. However, by the generalized admissibility of the ends, $\mathbf{dev}_t|U$ is an imbedding for $0 < t < \epsilon$ for sufficiently small $\epsilon > 0$. Here, a key point is that $U_t$ can be chosen to be properly convex when deforming. If $\Gamma_t|\pi_1(E)$ is not virtually abelian, it cannot have any horospherical representation. Thus, by the stability of the lens condition, we have openness for the associated end neighborhood. If $\pi_1(E)$

is virtually abelian, $\Gamma_0(\pi_1(E))$ can be horospherical. If $\Gamma_t|\pi_1(E)$ is horospherical for $t > 0$, then this case is straightforward. We consider the case when $\Gamma_t|\pi_1(E)$, $t > 0$, becomes diagonalizable by our lens condition for holonomy homomorphisms. By the classification of Benoist [**5**] of projective structures with diagonalizable holonomy groups, we have brick decompositions for the transversal orbifold structure $\Sigma_{E,t}$ for $E$ and $\mathbf{dev}_t$. The transversal projective structure on $\Sigma_{E,t}$ has to have only one brick. Otherwise, we can find a fundamental domain of a finite cover of $\Sigma_{E,t}$ which under $\mathbf{dev}_t$ is noninjective. Since during the deformation the brick number doesn't change, we obtain a contradiction that $\mathbf{dev}_t|F_t$ cannot converge to a map with compact image as $t \to 0$. (In other words, a sequence of real projective structures with more than one bricks cannot converge to a complete affine structure.) Therefore, $\Sigma_{E,t}$ is properly convex for $\mathbf{dev}_t, t > 0$. Since the holonomy is for the lens type ones, we will have lens-type ends. (See Proposition 6.5 of [**19**].) Let $C_{U_t} \subset \mathbb{R}^{n+1}$ denote the convex open cone corresponding to $U_t = \mathbf{dev}_t(U) \subset \mathbb{RP}^n$.

The affine space $\mathbb{R}^{n+1}$ is compactified as $\mathbb{RP}^{n+1}$. We show that $\mathrm{Cl}(C_{U_t})$ changes in a continuous manner under the Hausdorff metric in $\mathbb{RP}^{n+1}$ containing $\mathbb{R}^{n+1}$. Also, the Hessian metric $\mu$ in the complement of the union of end neighborhoods varies continuously to Hessian metrics $\mu_t$. Then we patch these metrics together to obtain a Hessian metric for $\mathcal{O}$. The existence of Hessian metrics and by generalized admissibility of ends, we can show the proper convexity. (The author learned that Cooper, Long, and Tillman [**32**] came up with the similar arguments in slightly different settings. Also, they use different topology using developing maps of ends neighborhood. This makes thing simpler and maybe more clear.)

**6.3. The closedness of convex real projective structures.** We recall

$$\mathrm{rep}^s_{\mathcal{E}}(\pi_1(\mathcal{O}), \mathrm{PGL}(n+1, \mathbb{R}))$$

the subspace of stable irreducible characters of

$$\mathrm{rep}_{\mathcal{E}}(\pi_1(\mathcal{O}), \mathrm{PGL}(n+1, \mathbb{R}))$$

which is shown to be the union of open subsets of semi-algebraic subsets in Section 6.1, and denote by $\mathrm{rep}^s_{\mathcal{E},\mathrm{u,ce}}(\pi_1(\mathcal{O}), \mathrm{PGL}(n+1, \mathbb{R}))$ the subspace of stable irreducible characters of $\mathrm{rep}_{\mathcal{E},\mathrm{u,ce}}(\pi_1(\mathcal{O}), \mathrm{PGL}(n+1, \mathbb{R}))$, an open subset of a semialgebraic set.

In this section, we will need to discuss $\mathbb{S}^n$ but only inside a proof.

THEOREM 6.12. *Let $\mathcal{O}$ be a noncompact strongly tame SPC $n$-orbifold with generalized admissible ends and satisfies* (IE) *and* (NA). *Assume $\partial\mathcal{O} = \emptyset$, and that the nilpotent normal subgroups of every finite-index subgroup of $\pi_1(\mathcal{O})$ are trivial. Then the following hold*:

- *The deformation space* $\mathrm{CDef}_{\mathcal{E},\mathrm{u,ce}}(\mathcal{O})$ *of SPC-structures on $\mathcal{O}$ with generalized admissible ends maps under* hol *homeomorphically to a union of components of* $\mathrm{rep}^s_{\mathcal{E},\mathrm{u,ce}}(\pi_1(\mathcal{O}), \mathrm{PGL}(n+1, \mathbb{R}))$.
- *The deformation space* $\mathrm{SDef}_{\mathcal{E},\mathrm{u,ce}}(\mathcal{O})$ *of strict SPC-structures on $\mathcal{O}$ with admissible ends maps under* hol *homeomorphically to the union of components of* $\mathrm{rep}^s_{\mathcal{E},\mathrm{u,ce}}(\pi_1(\mathcal{O}), \mathrm{PGL}(n+1, \mathbb{R}))$.

We will give some general idea behind the proofs here and show only the closedness. We will use the developing maps to $\mathbb{S}^n$. Given a sequence $\mu_i$ of properly convex real projective structures on $\mathcal{O}$ satisfying some boundary conditions, let $h_i$ denote the corresponding holonomy homomorphism with developing maps $\mathbf{dev}_i$.

Let $K_i \subset \mathbb{S}^n$ denote the closure of the images $\mathbf{dev}_i(\tilde{\mathcal{O}})$. We may choose a subsequence so that $h_i \to h$ for a representation $h : \pi_1(\mathcal{O}) \to \mathrm{PGL}(n+1, \mathbb{R})$ and $K_i \to K$ for a compact convex domain $K$.

If $K_i$ geometric converges to a convex domain $K$ with empty interior, then $h$ is reducible since $K$ is contain in a proper subspace. Hence, $h$ is not in the target character space.

Suppose that $K_i$ geometrically converges to a convex domain $K$ with nonempty interior $K^o$ by choosing a subsequence. Suppose that $K$ is not properly convex. Then $K$ contain a pair of antipodal points. We take the maximal great sphere $\mathbb{S}^i$ in $K$ for $i \geq 0$. The limiting holonomy group acts on $\mathbb{S}^i$ and hence is reducible. Thus, $h$ is not in the target character subspace.

Hence, $K$ is properly convex, and $K^o/\Gamma$ is a properly convex real projective orbifold $\mathcal{O}'$. We can show that $\mathcal{O}'$ is diffeomorphic to $\mathcal{O}$. By [**40**], we can show that $h_i$ converges to a faithful representation $h : \pi_1(\mathcal{O}) \to \mathrm{SL}_{\pm}(n+1, \mathbb{R})$.

**6.4. Nicest cases.** Theorems 6.5 and 6.12 imply the following:

COROLLARY 6.13. *Let $\mathcal{O}$ be a strongly tame SPC $n$-dimensional real projective orbifold with only radial ends and satisfies* (IE) *and* (NA). *Suppose that each end fundamental group is generated by finite order elements and is virtually abelian or hyperbolic. Assume $\partial\mathcal{O} = \emptyset$, and that the nilpotent normal subgroups of every finite-index subgroup of $\pi_1(\mathcal{O})$ are trivial. Then* hol *maps the deformation space* $\mathrm{CDef}_{\mathcal{E}}(\mathcal{O})$ *of SPC-structures on $\mathcal{O}$ homeomorphically to a union of components of*

$$\mathrm{rep}^s_{\mathcal{E},\mathrm{u},\mathrm{ce}}(\pi_1(\mathcal{O}), \mathrm{PGL}(n+1, \mathbb{R})).$$

*The same can be said for* $\mathrm{SDef}_{\mathcal{E}}(\mathcal{O})$.

These types of deformations from structures with cusps to ones with lens-type ends are realized in our main examples as stated in Section 1.2. We need the restrictions on the target space since the convexity of $\mathcal{O}$ is not preserved under the hyperbolic Dehn surgery deformations of Thurston, as pointed out by Cooper at ICERM in September 2013.

Strongly tame properly convex Coxeter orbifolds admitting complete hyperbolic structures will satisfy the premise. Also, $2h\_1\_1$ and the double of the simplex orbifold do also.

For Coxeter orbifolds, this simplifies further.

COROLLARY 6.14. *Let $\mathcal{O}$ be a strongly tame Coxeter $n$-dimensional real projective orbifold, $n \geq 3$, with only radial ends admitting a complete hyperbolic structure. Then* $\mathrm{SDef}_{\mathcal{E},\mathrm{u},\mathrm{ce}}(\mathcal{O})$ *is homeomorphic to the union of components of*

$$\mathrm{rep}^s_{\mathcal{E}}(\pi_1(\mathcal{O}), \mathrm{PGL}(n+1, \mathbb{R})).$$

*Finally,*

$$\mathrm{SDef}_{\mathcal{E},\mathrm{u},\mathrm{ce}}(\mathcal{O}) = \mathrm{SDef}_{\mathcal{E}}(\mathcal{O}).$$

We give a sketch of the proof. Consider a component $C$ of

$$\mathrm{rep}^s_{\mathcal{E},\mathrm{u},\mathrm{ce}}(\pi_1(\mathcal{O}), \mathrm{PGL}(n+1, \mathbb{R}))$$

corresponding to a component of $\mathrm{SDef}_{\mathcal{E},\mathrm{u},\mathrm{ce}}(\mathcal{O})$. Let $C'$ be the inverse image of $C$ in

$$\mathrm{Hom}^s_{\mathcal{E},\mathrm{u},\mathrm{ce}}(\pi_1(\mathcal{O}), \mathrm{PGL}(n+1, \mathbb{R})).$$

Then we claim that $C'$ is open in

$$\mathrm{Hom}^s_{\mathcal{E}}(\pi_1(\mathcal{O}), \mathrm{PGL}(n+1, \mathbb{R})) :$$

Let $h : \pi_1(\mathcal{O}) \to \mathrm{PGL}(n+1, \mathbb{R})$ be a representation in $C'$. By Theorem 2.10, there is a neighborhood $J$ of $h$ realized by orbifold $\mathcal{O}_b$ diffeomorphic to $\mathcal{O}$ for each $b \in J$. Each end is convex since a compact projective Coxeter $(n-1)$-orbifold, $n-1 \geq 2$, admitting a Euclidean structure, is always convex by Vinberg [**62**]. By Lemma 6.4, the end is properly convex of lens-type or is horospherical. Hence, $h$ is in

$$\mathrm{Hom}^s_{\mathcal{E},\mathrm{u,ce}}(\pi_1(\mathcal{O}), \mathrm{PGL}(n+1, \mathbb{R})).$$

The closedness follows as in Section 6.3.

## References


[1] S. A. Ballas, *Deformations of noncompact projective manifolds*, Algebr. Geom. Topol. **14** (2014), no. 5, 2595–2625, DOI 10.2140/agt.2014.14.2595. MR3276842

[2] S. A. Ballas, *Finite volume properly convex deformations of the figure-eight knot*, Geom. Dedicata **178** (2015), 49–73, DOI 10.1007/s10711-015-0043-2. MR3397481

[3] Samuel A. Ballas, Daryl Cooper, and Arielle Leitner, *A classification of generalized cusps on properly convex projective n-manifolds*, in preparation.

[4] Samuel A. Ballas, Jeffrey Danciger, and Gye-Seon Lee, *Convex projective structures on non-hyperbolic three-manifolds*, arXiv:1508.04794.

[5] Y. Benoist, *Nilvariétés projectives* (French), Comment. Math. Helv. **69** (1994), no. 3, 447–473, DOI 10.1007/BF02564497. MR1289337

[6] Y. Benoist, *Tores affines* (French, with English summary), Crystallographic groups and their generalizations (Kortrijk, 1999), Contemp. Math., vol. 262, Amer. Math. Soc., Providence, RI, 2000, pp. 1–37, DOI 10.1090/conm/262/04166. MR1796124

[7] Y. Benoist, *Convexes divisibles* (French, with English and French summaries), C. R. Acad. Sci. Paris Sér. I Math. **332** (2001), no. 5, 387–390, DOI 10.1016/S0764-4442(01)01878-X. MR1826621

[8] Yves Benoist, *Convexes divisibles. I*, Algebraic groups and arithmetic, Tata Inst. Fund. Res., Mumbai, 2004, pp. 339–374. MR2094116

[9] Y. Benoist, *Convexes divisibles. III* (French, with English and French summaries), Ann. Sci. École Norm. Sup. (4) **38** (2005), no. 5, 793–832, DOI 10.1016/j.ansens.2005.07.004. MR2195260

[10] Y. Benoist, *Convexes divisibles. IV. Structure du bord en dimension 3* (French, with English summary), Invent. Math. **164** (2006), no. 2, 249–278, DOI 10.1007/s00222-005-0478-4. MR2218481

[11] N. Bergeron and T. Gelander, *A note on local rigidity*, Geom. Dedicata **107** (2004), 111–131, DOI 10.1023/B:GEOM.0000049122.75284.06. MR2110758

[12] B. H. Bowditch, *A topological characterisation of hyperbolic groups*, J. Amer. Math. Soc. **11** (1998), no. 3, 643–667, DOI 10.1090/S0894-0347-98-00264-1. MR1602069

[13] M. R. Bridson and A. Haefliger, *Metric spaces of non-positive curvature*, Grundlehren der Mathematischen Wissenschaften [Fundamental Principles of Mathematical Sciences], vol. 319, Springer-Verlag, Berlin, 1999. MR1744486

[14] R. D. Canary, D. B. A. Epstein, and P. L. Green, *Notes on notes of Thurston [MR0903850]*, Fundamentals of hyperbolic geometry: selected expositions, London Math. Soc. Lecture Note Ser., vol. 328, Cambridge Univ. Press, Cambridge, 2006, pp. 1–115. With a new foreword by Canary. MR2235710

[15] S. Choi, *A classification of radial and totally geodesic ends of properly convex real projective orbifolds*, arXiv:1304.1605.

[16] ———, *A classification of radial or totally geodesic ends of real projective orbifolds. I: a survey of results*, arXiv:1501.00348, to appear in Advanced Studies in Pure Mathematics, Math. Soc. Japan.

[17] ———, *A classification of radial or totally geodesic ends of real projective orbifolds. II: properly convex ends*, arXiv:1501.00352.

[18] ______, *A classification of radial or totally geodesic ends of real projective orbifolds. III: nonproperly convex convex ends*, arXiv:1507.00809.

[19] ______, *The convex real projective orbifolds with radial or totally geodesic ends: The closedness and openness of deformations*, arXiv:1011.1060.v3.

[20] ______, *$\mathbb{RP}^n$-orbifolds with ends and their deformation spaces*, in preparation as a monograph.

[21] ______, *The mathematica file to compute tetrahedral real projective orbifolds*, `http://mathsci.kaist.ac.kr/~schoi/coefficientsol8.nb`.

[22] S. Choi, *Convex decompositions of real projective surfaces. II. Admissible decompositions*, J. Differential Geom. **40** (1994), no. 2, 239–283. MR1293655

[23] S. Choi, *Geometric structures on orbifolds and holonomy representations*, Geom. Dedicata **104** (2004), 161–199, DOI 10.1023/B:GEOM.0000022859.74165.44. MR2043960

[24] S. Choi, *The deformation spaces of projective structures on 3-dimensional Coxeter orbifolds*, Geom. Dedicata **119** (2006), 69–90, DOI 10.1007/s10711-006-9050-7. MR2247648

[25] S. Choi, *Geometric structures on 2-orbifolds: exploration of discrete symmetry*, MSJ Memoirs, vol. 27, Mathematical Society of Japan, Tokyo, 2012. MR2962023

[26] S. Choi and W. M. Goldman, *The deformation spaces of convex $\mathbb{RP}^2$-structures on 2-orbifolds*, Amer. J. Math. **127** (2005), no. 5, 1019–1102. MR2170138

[27] S. Choi, R. Greene, G. Lee, and L. Marquis, *Projective deformations of hyperbolic coxeter 3-orbifolds of finite volume*, in preparation.

[28] S. Choi, C. D. Hodgson, and G.-S. Lee, *Projective deformations of hyperbolic Coxeter 3-orbifolds*, Geom. Dedicata **159** (2012), 125–167, DOI 10.1007/s10711-011-9650-8. MR2944525

[29] S. Choi, G. Lee, and L. Marquis, *Convex real projective structures on manifolds and orbifolds*, arXiv:1605.02548, to appear in the Handbook of Group Actions, (L. Ji, A. Papadopoulos, S.-T. Yau, eds.), Higher Education Press and International Press.

[30] D. Cooper, D. Long, and M. Thistlethwaite, *Computing varieties of representations of hyperbolic 3-manifolds into* $\mathrm{SL}(4, \mathbb{R})$, Experiment. Math. **15** (2006), no. 3, 291–305. MR2264468

[31] D. Cooper, D. D. Long, and M. B. Thistlethwaite, *Flexing closed hyperbolic manifolds*, Geom. Topol. **11** (2007), 2413–2440, DOI 10.2140/gt.2007.11.2413. MR2372851

[32] D. Cooper, D. Long, and S. Tillmann, *Deforming convex projective manifolds*, arXiv:1511.06206.

[33] D. Cooper, D. D. Long, and S. Tillmann, *On convex projective manifolds and cusps*, Adv. Math. **277** (2015), 181–251, DOI 10.1016/j.aim.2015.02.009. MR3336086

[34] M. Crampon and L. Marquis, *Finitude géométrique en géométrie de Hilbert* (French, with English and French summaries), Ann. Inst. Fourier (Grenoble) **64** (2014), no. 6, 2299–2377, DOI 10.5802/aif.2914. MR3331168

[35] P. de la Harpe, *On Hilbert's metric for simplices*, Geometric group theory, Vol. 1 (Sussex, 1991), London Math. Soc. Lecture Note Ser., vol. 181, Cambridge Univ. Press, Cambridge, 1993, pp. 97–119, DOI 10.1017/CBO9780511661860.009. MR1238518

[36] C. Druţu and M. Sapir, *Tree-graded spaces and asymptotic cones of groups*, Topology **44** (2005), no. 5, 959–1058, DOI 10.1016/j.top.2005.03.003. With an appendix by Denis Osin and Mark Sapir. MR2153979

[37] B. Farb, *Relatively hyperbolic groups*, Geom. Funct. Anal. **8** (1998), no. 5, 810–840, DOI 10.1007/s000390050075. MR1650094

[38] W. M. Goldman, *Geometric structures on manifolds and varieties of representations*, Geometry of group representations (Boulder, CO, 1987), Contemp. Math., vol. 74, Amer. Math. Soc., Providence, RI, 1988, pp. 169–198, DOI 10.1090/conm/074/957518. MR957518

[39] W. M. Goldman, *Convex real projective structures on compact surfaces*, J. Differential Geom. **31** (1990), no. 3, 791–845. MR1053346

[40] W. M. Goldman and J. J. Millson, *Local rigidity of discrete groups acting on complex hyperbolic space*, Invent. Math. **88** (1987), no. 3, 495–520, DOI 10.1007/BF01391829. MR884798

[41] R. Greene, *The deformation theory of discrete reflection groups and projective structures*, ProQuest LLC, Ann Arbor, MI, 2013. Thesis (Ph.D.)–The Ohio State University. MR3322033

[42] M. Gromov, *Groups of polynomial growth and expanding maps*, Inst. Hautes Études Sci. Publ. Math. **53** (1981), 53–73. MR623534

[43] M. Gromov, *Hyperbolic groups*, Essays in group theory, Math. Sci. Res. Inst. Publ., vol. 8, Springer, New York, 1987, pp. 75–263, DOI 10.1007/978-1-4613-9586-7_3. MR919829

[44] D. Heard, C. Hodgson, B. Martelli, and C. Petronio, *Hyperbolic graphs of small complexity*, Experiment. Math. **19** (2010), no. 2, 211–236. MR2676749

[45] D. Johnson and J. J. Millson, *Deformation spaces associated to compact hyperbolic manifolds*, Discrete groups in geometry and analysis (New Haven, Conn., 1984), Progr. Math., vol. 67, Birkhäuser Boston, Boston, MA, 1987, pp. 48–106, DOI 10.1007/978-1-4899-6664-3_3. MR900823

[46] J.-L. Koszul, *Déformations de connexions localement plates* (French), Ann. Inst. Fourier (Grenoble) **18** (1968), no. fasc. 1, 103–114. MR0239529

[47] N. H. Kuiper, *On convex locally-projective spaces*, Convegno Internazionale di Geometria Differenziale, Italia, 1953, Edizioni Cremonese, Roma, 1954, pp. 200–213. MR0063115

[48] A. Leitner, *A classification of subgroups of $SL(4, \mathbb{R})$ isomorphic to $\mathbb{R}^3$ and generalized cusps in projective 3 manifolds*, Topology Appl. **206** (2016), 241–254, DOI 10.1016/j.topol.2016.04.005. MR3494445

[49] A. Leitner, *Conjugacy limits of the diagonal Cartan subgroup in $SL_3(\mathbb{R})$*, Geom. Dedicata **180** (2016), 135–149, DOI 10.1007/s10711-015-0095-3. MR3451461

[50] A. Leitner, *Limits under conjugacy of the diagonal subgroup in $SL_n(\mathbb{R})$*, Proc. Amer. Math. Soc. **144** (2016), no. 8, 3243–3254, DOI 10.1090/proc/12959. MR3503693

[51] W. L. Lok, *Deformations of Locally Homogeneous Spaces and Kleinian Groups (Teichmueller, Hyperbolic)*, ProQuest LLC, Ann Arbor, MI, 1984. Thesis (Ph.D.)–Columbia University. MR2633813

[52] L. Marquis, *Espace des modules de certains polyèdres projectifs miroirs* (French, with English summary), Geom. Dedicata **147** (2010), 47–86, DOI 10.1007/s10711-009-9438-2. MR2660566

[53] I. Moerdijk and D. A. Pronk, *Orbifolds, sheaves and groupoids*, $K$-Theory **12** (1997), no. 1, 3–21, DOI 10.1023/A:1007767628271. MR1466622

[54] D. W. Morris, *Introduction to arithmetic groups*, Deductive Press, 2015. MR3307755

[55] J. R. Munkres, *Topology: a first course*, Prentice-Hall, Inc., Englewood Cliffs, N.J., 1975. MR0464128

[56] J. Porti and S. Tillmann, *Projective structures on a hyperbolic 3-orbifold*, working title, in preparations, 2016.

[57] W. Thurston, *Geometry and topology of 3-manifolds*, mimeographed notes, available from `http://library.msri.org/books/gt3m/`, 1984.

[58] W. P. Thurston, *Three-dimensional geometry and topology. Vol. 1*, Princeton Mathematical Series, vol. 35, Princeton University Press, Princeton, NJ, 1997. Edited by Silvio Levy. MR1435975

[59] V. S. Varadarajan, *Lie groups, Lie algebras, and their representations*, Graduate Texts in Mathematics, vol. 102, Springer-Verlag, New York, 1984. Reprint of the 1974 edition. MR746308

[60] J. Vey, *Une notion d'hyperbolicité sur les variétés localement plates* (French), C. R. Acad. Sci. Paris Sér. A-B **266** (1968), A622–A624. MR0236837

[61] È. B. Vinberg, *The theory of homogeneous convex cones* (Russian), Trudy Moskov. Mat. Obšč. **12** (1963), 303–358. MR0158414

[62] È. B. Vinberg, *Discrete linear groups that are generated by reflections* (Russian), Izv. Akad. Nauk SSSR Ser. Mat. **35** (1971), 1072–1112. MR0302779

[63] A. Weil, *On discrete subgroups of Lie groups*, Ann. of Math. (2) **72** (1960), 369–384, DOI 10.2307/1970140. MR0137792

Department of Mathematical Sciences, KAIST, Daejeon 305-701, South Korea
*E-mail address*: `schoi@math.kaist.ac.kr`

Contemporary Mathematics
Volume **696**, 2017
http://dx.doi.org/10.1090/conm/696/14017

# A gluing theorem and applications in subspaces of the universal Teichmüller space

Jinhua Fan and Jun Hu

ABSTRACT. Let $f$ and $g$ be two quasiconformal homeomorphisms of the open unit disk $\mathbb{D} = \{z : |z| < 1\}$ with $f = g$ on the boundary $\mathbb{S}^1$ of $\mathbb{D}$. We show that for any compact subset $\Lambda$ of $\mathbb{D}$ and any $\epsilon > 0$, there exists a quasiconformal homeomorphism $h$ of $\mathbb{D}$ such that (i) $h = f$ on a disk $D = \{z : |z| < r\}$ containing $\Lambda$ (where $0 < r < 1$), (ii) $h = g$ on an annulus $\{z : r' < |z| < 1\}$ (where $r < r' < 1$), and (iii) the maximal dilatation $K(h|A)$ of $h$ on the gap $A = \{z : r \leq |z| \leq r'\}$ is less than $\max\{K(f|_A), K(g|_A)\} + \epsilon$. We find this result summarizes the essence of the main step or provides a direct way to obtain several known results concerning the inner metric and the Kobayashi metric on three subspaces of the universal Teichmüller space.

## 1. Introduction

Let $\mathbb{D}$ be the open unit disk on the complex plane centered at the origin and $\mathbb{S}^1$ the boundary of $\mathbb{D}$. Each quasiconformal homeomorphism $f$ of $\mathbb{D}$ extends to a homeomorphism of the closed unit disk $\overline{\mathbb{D}}$. We denote by $f|_{\mathbb{S}^1}$ the restriction of this extension to $\mathbb{S}^1$. Given positive real numbers $r$, $r_1$ and $r_2$ with $r_1 < r_2$, we denote by $D_r = \{z : |z| < r\}$ and $D_{r_1,r_2} = \{z : r_1 < |z| < r_2\}$. In this paper, we first show the following gluing theorem.

THEOREM 1. *Let $f$ and $g$ be two quasiconformal homeomorphisms of $\mathbb{D}$ with $f|_{\mathbb{S}^1} = g|_{\mathbb{S}^1}$. Then for any $0 < r_0 < 1$ and any $\epsilon > 0$, there exist $r_0 < r < 1$ and a quasiconformal homeomorphism $F$ of $\mathbb{D}$ such that*
*(1) $F = f$ on $D_{r_0}$,*
*(2) $F = g$ on $D_{r,1}$, and*
*(3) $K(F|_{D(r_0,r)}) \leq \max\{K(f|_{D(r_0,r)}), K(g|_{D(r_0,r)})\} + \epsilon$.*

Before giving applications of this theorem, we briefly introduce the universal Teichmüller space and three subspaces.

Two quasiconformal homeomorphisms $f$ and $g$ of $\mathbb{D}$ are said to be equivalent if $f|_{\mathbb{S}^1}$ and $g|_{\mathbb{S}^1}$ differ by postcomposition by a conformal isometry of $\mathbb{D}$. Denote by $[f]$ the equivalence class of $f$. Then the universal Teichmüller space can be defined

2010 *Mathematics Subject Classification.* Primary 30C75, 30F60.
*Key words and phrases.* The universal Teichmüller space, VMOA Teichmüller space, Weil-Petersson Teichmüller space, Teichmüller metric, path metric, symmetric homeomorphism.
The work of the first author was supported by NNSF of China (No. 11201228 and 11571172). The work of the second author was partially supported by PSC-CUNY grants.

as

$$T(\mathbb{D}) = \{[f] : f \text{ is a quasiconformal homeomorphism of } \mathbb{D}\}.$$

The Teichmüller metric on $T(\mathbb{D})$ is defined as

$$d_T([f],[g]) = \frac{1}{2} \inf \log K(g \circ f^{-1}),$$

where $K(g \circ f^{-1})$ is the maximal dilatation of $g \circ f^{-1}$ and the infimum is taken over all $f$ and $g$ in their equivalence classes $[f]$ and $[g]$.

A quasiconformal homeomorphism $f$ of $\mathbb{D}$ is said to be asymptotically conformal if its Beltrami coefficient $\mu_f$ is vanishing near the boundary of $\mathbb{D}$; that is, for any $\epsilon > 0$, there exists $0 < r < 1$ such that $||\mu_f|_{\mathbb{D}\setminus\mathbb{D}_r}||_\infty < \epsilon$. One subspace of $T(\mathbb{D})$ is defined as

$$T_0(\mathbb{D}) = \{[f] \in T(\mathbb{D}) : f \text{ is an asymptotically conformal homeomorphism of } \mathbb{D}\}.$$

A quasiconformal homeomorphism $f$ of $\mathbb{D}$ is said to have an integrable Beltrami coefficient if

$$\int_{\mathbb{D}} |\mu_f(z)|^2 \rho_{\mathbb{D}}^2(z) dxdy < \infty,$$

where $\rho_{\mathbb{D}}(z) = 1/(1-|z|^2)$ denotes the Poincaré density on $\mathbb{D}$. Another subspace of $T(\mathbb{D})$ is defined as

$$T_2(\mathbb{D}) = \{[f] \in T(\mathbb{D}) : \mu_f \text{ is integrable }\},$$

which is called the integrable Teichmüller space in [**2**] or the Weil-Petersson Teichmüller space in [**20**].

A positive measure $\lambda$ defined on a simply connected domain $\Omega$ of $\widehat{\mathbb{C}}$ is called a Carleson measure if

$$\|\lambda\|_c^2 = \sup\{\frac{\lambda(\Omega \cap D(z,r))}{r} : z \in \partial\Omega, 0 < r < \text{diameter}(\Omega)\} < \infty,$$

where $D(z,r) = \{w : |w-z| < r\}$. A Carleson measure $\lambda$ is said to be vanishing if

$$\lim_{r \to 0} \frac{\lambda(\Omega \cap D(z,r))}{r} = 0$$

uniformly on $z \in \partial\Omega$. The third subspace of $T(\mathbb{D})$ is defined as

$$T_v(\mathbb{D}) = \{[f] \in T(\mathbb{D}) : |\mu_f|^2 \rho_{\mathbb{D}}(z) \text{ is a vanishing Carleson measure on } \mathbb{D}\},$$

which is called the VMOA Teichmüller space.

It is shown in [**2**] that $T_2(\mathbb{D}) \subset T_0(\mathbb{D})$ and in [**21**] that $T_v(\mathbb{D}) \subset T_0(\mathbb{D})$ (see Section 3 for the outlines of the proofs).

The Teichmüller metric $d_T$ on the universal Teichmüller space $T(\mathbb{D})$ induces metrics on three spaces $T_0(\mathbb{D})$, $T_2(\mathbb{D})$ and $T_v(\mathbb{D})$. Questions can be raised for these metrics. A challenging open question is to characterize the geodesics under the induced metric in each space. A closely related question is to ask: given two points in $T_0(\mathbb{D})$ (resp. $T_2(\mathbb{D})$ or $T_v(\mathbb{D})$), does the geodesic in $T(\mathbb{D})$ connecting them stay in $T_0(\mathbb{D})$ (resp. $T_2(\mathbb{D})$ or $T_v(\mathbb{D})$)? As far as we know, these questions remain open. On the other hand, if there are positive answers to these questions, then one can conclude that the induced metric on each subspace is an inner distance and the Kobayashi metric coincides with the Teichmüller metric on each subspace. The goal of this paper is to use the gluing theorem to obtain these metric properties on the three subspaces without knowing whether or not there are positive answers to those questions.

Let $(X, d)$ be a metric space. A path $\gamma : [a, b] \to X$ is said to be rectifiable with respect to $d$ if the $d-$length of $\gamma$

$$\text{Length}_d(\gamma) = \sup\{\sum_{j=1}^{n} d(\gamma(t_{j-1}), \gamma(t_j)) : a = t_0 < t_1 < t_2 < \cdots < t_n = b\}$$

is finite. We say that $d$ is an inner metric if

$$d(x_1, x_2) = \inf_{\gamma} \text{Length}_d(\gamma)$$

for all $x_1, x_2 \in X$, where the infimum is taken over all rectifiable paths connecting $x_1$ and $x_2$ with respect to $d$.

It is well known that the metric space $(T(\mathbb{D}), d_T)$ is path-connected. Using the existence of geodesic between any two points in $(T(\mathbb{D}), d_T)$, we can easily see that $d_T$ is an inner distance on $T(\mathbb{D})$. Using solutions to Beltrami differential equations depending on parameters, we can see that the three spaces $(T_0(\mathbb{D}), d_T)$, $(T_2(\mathbb{D}), d_T)$ and $(T_v(\mathbb{D}), d_T)$ are also path-connected. As the first application of the gluing theorem, we obtain a positive result on the inner metric. See Remark 2 for why we call it a folklore theorem.

THEOREM 2 (Folklore Theorem). *The restriction of the Teichmüller metric $d_T$ on each of the three spaces $T_0(\mathbb{D})$, $T_2(\mathbb{D})$ and $T_v(\mathbb{D})$ is an inner distance.*

Teichmüller spaces are complex analytic manifolds on which Kobayashi's pseudo metrics can be defined. A natural question is to study whether or not the Kobayashi pseudo metric and the Teichmüller metric coincide with each other on a Teichmüller space. This is proved to be true by Royden on finite-dimensional Teichmüller spaces [**19**] and by Gardiner on infinite-dimensional Teichmüller spaces [**11**]. An easy proof of the coincidence of the two metrics on all Teichmüller spaces was given in [**6**]. The same problem was studied on several subspaces of the universal Teichmüller space $T(\mathbb{D})$. The same result is obtained in [**7**] for $T_0(\mathbb{D})$ (see [**14**] for a direct proof), in [**16**] for $T_2(\mathbb{D})$, and in [**9**] for $T_v(\mathbb{D})$.

It is shown in [**2**] that $T_2(\mathbb{D}) \subset T_0(\mathbb{D})$ and in [**21**] that $T_v(\mathbb{D}) \subset T_0(\mathbb{D})$ (see Section 3 for the outlines of the proofs). It is used in [**14**] that the inclusion map from $T_0(\mathbb{D})$ to $T(\mathbb{D})$ is holomorphic. The inclusion maps from $T_2(\mathbb{D})$ into $T(\mathbb{D})$ and from $T_v(\mathbb{D})$ into $T(\mathbb{D})$ are also holomorphic (see [**9**]). Using the facts that any holomorphic map between complex analytic manifolds does not increase distance under Kobayashi's metrics and the Kobayashi metric is as the same as the Teichmüller metric on $T(\mathbb{D})$, one can conclude that the Kobayashi metric is greater than or equal to the Teichmüller metric on each of $T_0(\mathbb{D})$, $T_2(\mathbb{D})$ and $T_v(\mathbb{D})$. The remaining step to reach the coincidence of the two metrics on each of them is to obtain the reversed inequality between them. The gluing theorem leads to an easy proof of that property.

THEOREM 3 ([**7**], [**16**] and [**9**]). *The Kobayashi metric is less than or equal to the Teichmüller metric on $T_0(\mathbb{D})$, $T_2(\mathbb{D})$ or $T_v(\mathbb{D})$.*

REMARK 1. Note that if there is an affirmative answer to the open question on whether or not the geodesic in T(D) between any two points in $T_0(\mathbb{D})$ (resp. $T_2(\mathbb{D})$ or $T_v(\mathbb{D})$) is contained in $T_0(\mathbb{D})$ (resp. $T_2(\mathbb{D})$ or $T_v(\mathbb{D})$), then one can conclude immediately Theorems 2 and 3 for $T_0(\mathbb{D})$ (resp. $T_2(\mathbb{D})$ or $T_v(\mathbb{D})$).

REMARK 2. It is known that the Kobayashi metric on any complex analytic manifold is an inner metric (see [**10**]). Since the coincidence of the Kobayashi metric and the Teichmüller metric on $T_0(\mathbb{D})$, $T_2(\mathbb{D})$ and $T_v(\mathbb{D})$ has been obtained in [**7**], [**16**] and [**9**] respectively, Theorem 2 is a known consequence. This is why we state it as a folklore theorem. Our goal is to emphasize that it can be proved by using definition and the gluing theorem without using the coincidence of the two metrics on those spaces and the known relationship between the Kobayashi metric and the inner metric on complex analytic manifolds.

The paper is arranged as follows. In section 2, we show the gluing theorem; in section 3, we introduce more background for the three subspaces $T_0(\mathbb{D})$, $T_2(\mathbb{D})$ and $T_v(\mathbb{D})$ and then show Theorems 2 and 3.

*Acknowledgement.* Both authors wish to thank the referee for comments, suggestions and corrections of typos.

## 2. Proof of the gluing theorem

Before we start to show our gluing theorem, we recall some background such as the definition of boundary dilatation, Strebel's frame mapping theorem [**22**] and the main inequality of Reich and Strebel [**18**].

Let $\Omega$ be a plane domain with at least two boundary points. For a quasiconformal homeomorphism $f$ from $\Omega$ onto a domain on the complex plane, denote by $h$ the restriction of $f$ on the boundary of $\Omega$. The boundary dilatation of $f$ is defined as

$$H^*(f) = \inf\{K(f|_{\Omega\backslash E}) : E \text{ is a compact subset of } \Omega\},$$

where $K(f|_{\Omega\backslash E})$ is the maximum dilatation of $f$ on $\Omega\backslash E$.

Let $T(\Omega)$ be the Teichmüller space of $\Omega$ (see [**12**]). Given each element $[f] \in T(\Omega)$, the boundary dilatation of the equivalence class $[f]$ is defined as

$$H(h) = H([f]) = \inf\{H^*(g) : g \in [f]\}.$$

The extremal maximum dilatation $K_0([f])$ of $[f]$ is defined as

$$K_0([f]) = \inf\{K(g) : g \in [f]\}.$$

It is obvious that

$$H([f]) \leq K_0([f]).$$

Now we state the version of Strebel's frame mapping theorem [**22**] in our setting, which is used in the proof of our Theorem 1.

**Theorem A** [**22**]. *Given $[f] \in T(\Omega)$, if $H([f]) < K_0([f])$, then $[f]$ is represented by a unique extremal map $f$ with the Beltrami coefficient $\mu_f = k_0 \frac{\overline{\varphi}}{|\varphi|}$, where $0 < k_0 < 1$ and $\varphi(z)$ is an associated holomorphic quadratic differential with*

$$\|\varphi\| = \int_\Omega |\varphi| = 1.$$

The following Main Inequality of Reich and Strebel [**18**] is a main tool to show our Theorem 1.

**Theorem B** [**18**]. *Let $f$ and $g$ be quasiconformal homeomorphisms between Riemann surfaces $R$ and $R'$ which are homotopic modulo the boundary. Then for any integrable holomorphic quadratic differential $\varphi = \varphi(z)dz^2$, we have*

$$\|\varphi\| \leq \int_R |\varphi(z)| \frac{|1-\mu_f \frac{\varphi}{|\varphi|}|^2}{1-|\mu_f|^2} D_{g^{-1}}(f(z))dxdy,$$

*where $D_{g^{-1}}(w) = \frac{1+|\mu_{g^{-1}}(w)|}{1-|\mu_{g^{-1}}(w)|}$ is the maximum dilatation of $g^{-1}$ at $w$ and $\mu_f, \mu_g$ and $\mu_{g^{-1}}$ are the Beltrami coefficients of the quasiconformal homeomorphisms $f, g$ and $g^{-1}$ respectively.*

Proof of Theorem 1. Note first that there exits $r_0 < \tilde{r} < 1$ such that

$$g(|z| = \tilde{r}) \subset \mathbb{D} \backslash \overline{f(D_{r_0})}.$$

For any $r$ with $\tilde{r} < r < 1$, let

$$h_r = \begin{cases} f(z), & |z| = r_0, \\ g(z), & |z| = r. \end{cases}$$

It is clear that $h_r$ is a homeomorphism between the boundaries of two annuli. We choose $r_1$ and $r_2$ with $r_0 < r_1 < r_2 < r$ such that $f(\overline{D}_{r_0,r_1}) \cap g(\overline{D}_{r_2,r}) = \emptyset$. Let

$$h_{r_1,r_2} = \begin{cases} f(z), & |z| = r_1, \\ g(z), & |z| = r_2. \end{cases}$$

Now we show that $h_{r_1,r_2}$ extends to a quasiconformal extension $f_{r_1,r_2}$ defined on $D_{r_1,r_2}$. Let $\mathbb{S}^1_t$ be the circle centered at the origin and of radius $t$, where $t > 0$. Let $A$ be the annulus bounded by two Jordan curves $f(\mathbb{S}^1_{r_1})$ and $g(\mathbb{S}^1_{r_2})$. Then there is a conformal map $\Phi$ that maps $A$ to a round annulus bounded by $\mathbb{S}^1_{r_1}$ and $\mathbb{S}^1_{r_3}$ for some $r_3 > 0$. Since $f$ and $g$ are quasiconformal maps on $\mathbb{D}$, we can show that $\Phi \circ f$ is a quasisymmetric map from $\mathbb{S}^1_{r_1}$ to $\mathbb{S}^1_{r_1}$ and $\Phi \circ g$ is a quasisymmetric map from $\mathbb{S}^1_{r_2}$ to $\mathbb{S}^1_{r_3}$. By Lemma 1 in [**15**], there exists a quasiconformal map $\Psi$ from $D_{r_1,r_2}$ to $D_{r_1,r_3}$ such that $\Psi$ is equal to $\Phi \circ f$ on $\mathbb{S}^1_{r_1}$ and $\Phi \circ g$ on $\mathbb{S}^1_{r_2}$. Then $\Phi^{-1} \circ \Psi$ is a quasiconformal extension of $h_{r_1,r_2}$ to $D_{r_1,r_2}$.

Let

$$F_{r_0,r} = \begin{cases} f(z), & z \in D_{r_0,r_1}, \\ f_{r_1,r_2}(z), & z \in D_{r_1,r_2}, \\ g(z), & z \in D_{r_2,r}. \end{cases}$$

By definition,

$$H(h_r) \leq H^*(F_{r_0,r}) \leq \max\{K(f|_{D_{r_0,r_1}}), K(g|_{D_{r_2,r}})\}.$$

Clearly,

$$\max\{K(f|_{D_{r_0,r_1}}), K(g|_{D_{r_2,r}})\} \leq \max\{K(f|_{D_{r_0,r}}), K(g|_{D_{r_0,r}})\}.$$

Thus

$$H(h_r) \leq \max\{K(f|_{D_{r_0,r}}), K(g|_{D_{r_0,r}})\}. \tag{2.1}$$

Now we let $f_r$ be an extremal quasiconformal extension of $h_r$ to the annulus $D_{r_0,r}$. If

$$K(f_r) \leq \max\{K(f|_{D_{r_0,r}}), K(g|_{D_{r_0,r}})\} + \varepsilon$$

for some $r_0 < r < 1$, then

$$F_r = \begin{cases} f(z), & z \in D_{r_0}, \\ f_r(z), & z \in D_{r_0,r}, \\ g(z), & z \in D_{r,1}. \end{cases}$$

satisfies the conditions of Theorem 1.

In the following, we show that such a constant $r$ exists. Suppose not, that means for all $\tilde{r} < r < 1$,

$$K(f_r) > \max\{K(f|_{D_{r_0,r}}), K(g|_{D_{r_0,r}}))\} + \varepsilon. \tag{2.2}$$

Then $K(f_r) > H(h_r)$. By Theorem A, $f_r$ is a Teichmüller map with $\mu_{f_r} = k_r \frac{\overline{\varphi_r}}{|\varphi_r|}$, where $0 < k_r < 1$ and $\varphi_r dz^2$ is a holomorphic quadratic differential with

$$\|\varphi_r\| = \int_{D_{r_0,r}} |\varphi_r| = 1.$$

We obtain a contradiction in the following two steps.

*Step 1. We show that $\varphi_r \to 0$ uniformly on any compact of $D_{r_0,1}$ as $r \to 1$.*

Suppose this is not true. Then there exist a non-zero holomorphic map $\phi_0$ defined on $D_{r_0,1}$ and an increasing sequence $\{r_n\}$ of real numbers between $\tilde{r}$ and 1 such that as $n \to \infty$, $r_n \to 1$ and $\varphi_{r_n} \to \varphi_0$, where $\varphi_{r_n}$ is the associated quadratic differential of the Teichmüller map $f_{r_n}$. Since $\{K(F_{r_n})\}_{n=1}^{\infty}$ is bounded, we can choose a subsequence of $\{f_{r_n}\}$, denoted again by $\{f_{r_n}\}$ for simplicity, such that as $n \to \infty$, $k_{r_n} \to k_0$ and the Beltrami coefficient $\mu_{r_n}$ of $f_{r_n}$ converges to $\mu_0 = k_0\overline{\varphi_0}/|\varphi_0|$ uniformly on every compact subset of $D_{r_0,1}$.

Since all $F_{r_n}$ are normalized to fix three points and $\{K(F_{r_n})\}_{n=1}^{\infty}$ is bounded, by passing to a subsequence we may assume further that as $n \to \infty$, $F_{r_n}$ converges to a quasiconformal map $F_0$ uniformly on $\overline{\mathbb{D}}$ and the Beltrami coefficient of $F_{r_n}$ converges to the Beltrami coefficient of $F_0$. Then the restriction of the Beltrami coefficient $\mu_{F_0}$ to $D_{r_0,1}$ is equal to $\mu_0 = k_0\overline{\varphi_0}/|\varphi_0|$. Now Thus $F_0|_{D_{r_0,1}}$ is an extremal Teichmüller map. By the assumption (2.2), we obtain that

$$K(F_0|_{D_{r_0,1}}) \geq K(f|_{D_{r_0,1}}) + \varepsilon. \tag{2.3}$$

From the construction of $F_{r_n}$ and $[f] = [g]$, we obtain $F_0 \in [f]$. Thus (2.3) contradicts that $F_0|_{D_{r_0,1}}$ is an extremal Teichmüller map. Therefore, $\varphi_r$ converges to 0 uniformly on any compact of $D_{r_0,1}$ as $r \to 1$ .

*Step 2. We show that there exists $\tilde{r} < r < 1$ such that*

$$K(f_r) \leq \max\{K(f|_{D_{r_0,1}}), K(g|_{D_{r_0,1}}))\} + \varepsilon.$$

Let $r'$ be a fixed number in $(\tilde{r}, 1)$. It follows from (2.1) that there exist a compact subset $E$ of $D_{r_0,r'}$ and a quasiconformal extension $\tilde{f}_{r'}$ of $h_{r'}$ to $D_{r_0,r'}$ such that

$$K(\tilde{f}_{r'}|_{D_{r_0,r'}\setminus E}) < \max\{K(f|_{D_{r_0,r'}}), K(g|_{D_{r_0,r'}})\} + \frac{\varepsilon}{2}. \tag{2.4}$$

Let

$$\tilde{F}_{r'} = \begin{cases} f(z), & z \in D_{r_0}, \\ \tilde{f}_{r'}(z), & z \in D_{r_0,r'}, \\ g(z), & z \in D_{r',1}. \end{cases}$$

For any $r' < r < 1$, $f_r|_{\partial D_{r_0,r}} = \tilde{F}_{r'}|_{\partial D_{r_0,r}}$. Applying Theorem B to $f = f_r, g = \tilde{F}_{r'}$ and $R = D_{r_0,r}$, we obtain

$$\begin{aligned} 1 = \|\varphi_r\| &\leq \int_{D_{r_0,r}} |\varphi_r(z)| \frac{|1-\mu_{f_r}\frac{\varphi_r}{|\varphi_r|}|^2}{1-|\mu_{f_r}|^2} D_{\tilde{F}_{r'}^{-1}}(f_r(z))dxdy \\ &= \int_{D_{r_0,r}} \frac{|\varphi_r(z)|}{K(f_r)} D_{\tilde{F}_{r'}^{-1}}(f_r(z))dxdy. \end{aligned}$$

Thus

$$\begin{aligned} K(f_r) &\leq \int_{D_{r_0,r}} |\varphi_r(z)| D_{\tilde{F}_{r'}^{-1}}(f_r(z))dxdy \\ &= \int_{f_r^{-1}\circ\tilde{F}_{r'}(E)} |\varphi_r(z)| D_{\tilde{F}_{r'}^{-1}}(f_r(z))dxdy \\ &+ \int_{D_{r_0,r}\backslash f_r^{-1}\circ\tilde{F}_{r'}(E)} |\varphi_r(z)| D_{\tilde{F}_{r'}^{-1}}(f_r(z))dxdy. \end{aligned} \tag{2.5}$$

From the definitions of $f_r$, $F_r$ and $\widetilde{F}_{r'}$, we know that $K(F_r^{-1}\circ\widetilde{F}_{r'})$ is uniformly bounded on the closure of $D_{r_0,1}$ for each $r$ with $r' < r < 1$. Clearly, $F_r^{-1}\circ\widetilde{F}_{r'}$ fixes every point on $\mathbb{S}^1$ for each $r$ with $r' < r < 1$. Thus, $f_r^{-1}\circ\widetilde{F}_{r'}(E) = F_r^{-1}\circ\widetilde{F}_{r'}(E)$ is contained in a compact subset $E_c$ of $D_{r_0,1}$ for all $r$ with $r' < r < 1$. Since $\{\varphi_r\}$ is a degenerating sequence,

$$\int_{f_r^{-1}\circ\tilde{F}_{r'}(E)} |\varphi_r(z)| D_{\tilde{F}_{r'}^{-1}}(f_r(z))dxdy \leq \int_{E_c} |\varphi_r(z)| D_{\tilde{F}_{r'}^{-1}}(f_r(z))dxdy < \frac{\varepsilon}{2} \tag{2.6}$$

if $r$ is sufficiently close to 1.

Now we show for each $r$ with $r' < r < 1$,

$$\int_{D_{r_0,r}\backslash f_r^{-1}\circ\tilde{F}_{r'}(E)} |\varphi_r(z)| D_{\tilde{F}_{r'}^{-1}}(f_r(z))dxdy \leq \max\{K(f|_{D_{r_0,r}}), K(g|_{D_{r_0,r}})\} + \frac{\varepsilon}{2}. \tag{2.7}$$

For any $z \in D_{r_0,r}\backslash f_r^{-1}\circ\tilde{F}_{r'}(E)$, we let $w = f_r(z)$ and $\zeta = \tilde{F}_{r'}^{-1}(w)$. Then

$$w \in f_r(D_{r_0,r})\backslash\tilde{F}_{r'}(E)$$

and hence

$$\zeta \in \tilde{F}_{r'}^{-1}\circ f_r(D_{r_0,r})\backslash E = D_{r_0,r}\backslash E = D_{r',r} \cup (\overline{D}_{r_0,r'}\backslash E),$$

which means $\zeta$ is either contained in $D_{r',r}$ or $\overline{D}_{r_0,r'}\backslash E$. Clearly,

$$D_{\tilde{F}_{r'}^{-1}}(f_r(z)) = D_{\tilde{F}_{r'}^{-1}}(w) = D_{\tilde{F}_{r'}}(\zeta).$$

Using the definition of $\tilde{F}_{r'}$ and (2.4), we can see that for any $z \in D_{r_0,r}\backslash f_r^{-1}\circ\tilde{F}_{r'}(E)$,

$$D_{\tilde{F}_{r'}^{-1}}(f_r(z)) \leq \max\{K(f|_{D_{r_0,r}}), K(g|_{D_{r_0,r}})\} + \frac{\varepsilon}{2},$$

which implies (2.7) since $\int_{D_{r_0,r}} |\varphi_r(z)|dxdy = 1$.

Using (2.5), (2.6) and (2.7), we conclude that when $r$ is sufficiently close to 1,

$$K(f_r) \leq \max\{K(f|_{D_{r_0,r}}), K(g|_{D_{r_0,r}})\} + \varepsilon.$$

This is a contradiction to the assumption (2.2). Thus the assumption (2.2) fails, and hence the conclusion of Theorem 1 follows. □

REMARK 3. For a different purpose, a theorem of gluing two quasiconformal maps in a different pattern is developed in [**3**].

Our Theorem 1 implies the following corollary.

COROLLARY 1. *For every element* $[f] \in T_0(\mathbb{D})$ *and any* $\epsilon > 0$, *there exists an asymptotically conformal homeomorphism* $g$ *of* $\mathbb{D}$ *such that* $[g] = [f]$ *and*

$$K(g) < K_0([f]) + \epsilon,$$

*where* $K_0([f])$ *stands for the maximal dilatation of the extremal representative of* $[f]$.

Let $R$ be a Riemann surface and $T_0(R)$ be the collection of the points in the Teichmüller space $T(R)$ having asymptotically conformal representatives. Corollary 2 on page 298 [**12**] states that for every $[f] \in T_0(R)$ and $\epsilon > 0$, there exists an asymptotically conformal representative $[g]$ of $[f]$ with $K_0([g]) < K_0([f]) + \epsilon$. Therefore, our Corollary 1 is a special case of Corollary 2 in [**12**], in which the Riemann surface is the open unit disk.

## 3. Proofs of Theorems 2 and 3

We first introduce more background for the three subspaces $T_0(\mathbb{D})$, $T_2(\mathbb{D})$ or $T_v(\mathbb{D})$.

Let $L^\infty(\mathbb{D})$ denote the Banach space of essentially bounded measurable complex-valued functions defined on the open unit disk $\mathbb{D}$, and let $M(\mathbb{D})$ be the open unit ball in $L^\infty(\mathbb{D})$ centered at the 0-constant function. For each element $\mu \in M(\mathbb{D})$, there is a unique quasiconformal homeomorphism $f^\mu$ of $\mathbb{D}$ with Beltrami coefficient $\mu$ and normalized to fix three points $1, -1$ and $i$. Two elements $\mu$ and $\nu$ of $M(\mathbb{D})$ are equivalent if $f^\mu|_{\mathbb{S}^1} = f^\nu|_{\mathbb{S}^1}$, where $\mathbb{S}^1 = \partial\mathbb{D}$. Denote by $[\mu]$ or $[f^\mu]$, $[f^\mu|_{\mathbb{S}^1}]$ the equivalence class of $\mu$. Then

$$T(\mathbb{D}) = \{[\mu] : \mu \in M(D)\}.$$

Let $\widehat{\mathbb{C}} = \mathbb{C} \cup \{\infty\}$ denote the Riemann sphere and $\mathbb{D}^* = \widehat{\mathbb{C}} \backslash \overline{\mathbb{D}}$. Given each element $\mu \in M(\mathbb{D})$, let $f_\mu$ be a quasiconformal homeomorphism of $\widehat{\mathbb{C}}$ with Beltrami coefficient equal to $\mu$ on $\mathbb{D}$ and 0 on $\mathbb{D}^*$, and furthermore let $f_\mu$ be normalized on $\mathbb{D}^*$ in the following form

$$f_\mu(z) = z + \frac{b_1}{z} + \cdots \quad \text{as } z \to \infty.$$

Then $f_\mu|_{\mathbb{D}^*} = f_\nu|_{\mathbb{D}^*}$ if and only if $f^\mu|_{\mathbb{S}^1} = f^\nu|_{\mathbb{S}^1}$.

Let $S_{f_\mu|_{\mathbb{D}^*}}$ be the Schwarzian derivative of $f_\mu|_{\mathbb{D}^*}$ and let $B(\mathbb{D}^*)$ be the Banach space of holomorphic quadratic differentials on $\mathbb{D}^*$ with the following norm

$$\|\phi\|_{B(\mathbb{D}^*)} = \sup_{z \in \mathbb{D}^*} |\phi(z)| \rho_{\mathbb{D}^*}^{-2}(z) < \infty,$$

where $\rho_{\mathbb{D}^*}(z) = 1/(|z|^2 - 1)$ denotes the Poincaré density on $\mathbb{D}^*$. The map

$$\beta : T(\mathbb{D}) \to B(\mathbb{D}^*) \text{ given by } \beta([\mu]) \mapsto S_{f_\mu|_{\mathbb{D}^*}}$$

is the Bers embedding.

Now we briefly recall how to characterize the three subspaces through the Bers embedding and recall some properties of the three subspaces.

Let $M_0(\mathbb{D})$ be the collection of the elements of $M(\mathbb{D})$ vanishing near the boundary of $\mathbb{D}$. Using the definition of $T_0(\mathbb{D})$ in the introduction, we see

$$T_0(\mathbb{D}) = \{[\mu] : \mu \in M_0(\mathbb{D})\}.$$

It is proved in [**13**] that $T_0(\mathbb{D})$ is a normal subgroup of $T(\mathbb{D})$ under the composition. The quotient space $T(\mathbb{D})/T_0(\mathbb{D})$ is called the asymptotic Teichmüller space of $\mathbb{D}$ and denoted by $AT(\mathbb{D})$ in [**12**]. Furthermore, it is shown in [**8**] that $[\mu] \in T_0(\mathbb{D})$ if and only if $\beta([\mu]) \in B_0(\mathbb{D}^*)$, where $B_0(\mathbb{D}^*)$ is the subset of $B(\mathbb{D}^*)$ consisting of all $\phi \in B(\mathbb{D}^*)$ such that

$$\lim_{r\to 1^+} \sup_{|z|=r} |\phi(z)|\rho_{\mathbb{D}^*}^{-2}(z) = 0. \tag{3.1}$$

Let $M_2(\mathbb{D})$ be the subset of $M(\mathbb{D})$ consisting of all $\mu \in M(\mathbb{D})$ such that

$$\int_{\mathbb{D}} |\mu(z)|^2 \rho_{\mathbb{D}}^2(z)dxdy < \infty,$$

where $\rho_{\mathbb{D}}(z) = 1/(1-|z|^2)$ denotes the Poincaré density on $\mathbb{D}$. Then

$$T_2(\mathbb{D}) = \{[\mu] : \mu \in M_2(\mathbb{D})\}.$$

It is proved in [**2**] that $[\mu] \in T_2(\mathbb{D})$ if and only if $\beta([\mu]) \in B_2(\mathbb{D}^*)$, where $B_2(\mathbb{D}^*)$ is the subset of $B(\mathbb{D}^*)$ consisting of all $\phi \in B(\mathbb{D}^*)$ such that

$$\int_{\mathbb{D}^*} |\phi(z)|^2 \rho_{\mathbb{D}^*}^{-2}(z)dxdy < \infty. \tag{3.2}$$

It is also proved in Lemma 2 of [**2**] that (3.2) implies (3.1). Therefore,

$$T_2(\mathbb{D}) \subset T_0(\mathbb{D}). \tag{3.3}$$

Furthermore, it is proved in [**2**] that $T_2(\mathbb{D})$ is a group under the composition.

Let $M_v(\mathbb{D})$ be the subset of $M(\mathbb{D})$ consisting of all $\mu \in M(\mathbb{D})$ such that $|\mu(z)|^2\rho_{\mathbb{D}}(z)$ is a vanishing Carleson measure. Then

$$T_v(\mathbb{D}) = \{[\mu] : \mu \in M_v(\mathbb{D})\}.$$

It is proved in [**21**] that $[\mu] \in T_v(\mathbb{D})$ if and only if $\beta([\mu]) \in B_v(\mathbb{D}^*)$, where $B_v(\mathbb{D}^*)$ is the subset of $B(\mathbb{D}^*)$ consisting of all $\phi \in B(\mathbb{D}^*)$ such that

$$|\phi(z)|^2 \rho_{\mathbb{D}^*}^{-3}(z) \text{ is a vanishing Carleson measure on } \mathbb{D}^*. \tag{3.4}$$

The proof of Lemma 2 in [**2**] can be modified to show that (3.4) implies (3.1) [**21**]. Therefore,

$$T_v(\mathbb{D}) \subset T_0(\mathbb{D}). \tag{3.5}$$

Furthermore, it is shown in [**21**] that $T_v(\mathbb{D})$ is also a group under the composition. For more studies of $T_v(\mathbb{D})$, we refer to [**1**] and [**17**].

Before we start to show Theorem 2, we develop several lemmas. Note first that $T_2(\mathbb{D}) \subset T_0(\mathbb{D})$ doesn't mean $M_2(\mathbb{D}) \subset M_0(\mathbb{D})$ and $T_v(\mathbb{D}) \subset T_0(\mathbb{D})$ doesn't mean $M_v(\mathbb{D}) \subset M_0(\mathbb{D})$. In order to prove Theorem 2, we select special Beltrami coefficients representing the points in $T_2(\mathbb{D})$ or $T_v(\mathbb{D})$. Given an element $[f] \in T_2(\mathbb{D})$, let $\Phi(f|_{\mathbb{S}^1})$ be the Douady-Earle extension of $f|_{\mathbb{S}^1}$ ([**5**]). It is proved in [**2**] and reproved in [**16**] that the Beltrami coefficient $\mu(\Phi(f|_{\mathbb{S}^1}))$ of $\Phi(f|_{\mathbb{S}^1})$ belongs to $M_2(\mathbb{D})$. Since $T_2(\mathbb{D}) \subset T_0(\mathbb{D})$, $[f] \in T_0(\mathbb{D})$. Using a result in [**8**] concerning the Douady-Earle extensions of the elements in $T_0(\mathbb{D})$, we also know $\mu(\Phi(f|_{\mathbb{S}^1})) \in M_0(\mathbb{D})$. We summarize the properties of $\mu(\Phi(f|_{\mathbb{S}^1}))$ in the following.

LEMMA 1. *For any* $[\mu] \in T_2(\mathbb{D})$, *there exists* $\nu \in [\mu]$ *such that* $\nu \in M_2(\mathbb{D}) \cap M_0(\mathbb{D})$.

A similar lemma for the elements in $T_v(\mathbb{D})$ is obtained in [**9**].

LEMMA 2. *For any* $[\mu] \in T_v(\mathbb{D})$, *there exists* $\nu \in [\mu]$ *such that* $\nu \in M_v(\mathbb{D}) \cap M_0(\mathbb{D})$.

PROOF OF THEOREM 2. The proofs of the conclusion on the three spaces are very similar. We only present the proof for $T_2(\mathbb{D})$.

Since $T_2(\mathbb{D})$ is a group and the right translation map $\alpha([f]) = [f \circ g^{-1}]$ is an isometry of $T_2(\mathbb{D})$ with respect to $d_T$, it suffices to show

$$d_T([0],[\mu]) = \inf_{\gamma} \mathrm{Length}_{d_T}(\gamma)$$

for every $[\mu] \in T_2$, where the infimum is taken over all rectifiable paths $\gamma \subset T_2(\mathbb{D})$ connecting $[0]$ and $[\mu]$ under $d_T$. For any rectifiable path $\gamma$ connecting $[0]$ and $[\mu]$, we have

$$\mathrm{Length}_{d_T}(\gamma) \geq d_T([0],[\mu]).$$

To obtain Theorem 2, we only need to show

$$\inf_{\gamma} \mathrm{Length}_{d_T}(\gamma) \leq d_T([0],[\mu]).$$

Note first that $T_2(\mathbb{D}) \subset T_0(\mathbb{D})$. By the frame mapping theorem, each point $[\mu]$ of $T_2(\mathbb{D}) \setminus \{[0]\}$ has a unique extremal Beltrami coefficient $\mu = k_0 \frac{\overline{\varphi}}{|\varphi|}$, where $k_0 \in (0,1)$ and $\varphi$ is a holomorphic quadratic differential on $\mathbb{D}$ with $\|\varphi\| = 1$. By Lemma 1, there exists $\nu \in [\mu]$ such that $\nu \in M_2(\mathbb{D}) \cap M_0(\mathbb{D})$. Let $0 < r_0 < 1$ be sufficiently close to 1 such that

$$\frac{1 + \|\nu|_{D_{r_0,1}}\|_\infty}{1 - \|\nu|_{D_{r_0,1}}\|_\infty} < \frac{1+k_0}{1-k_0}. \tag{3.6}$$

Now let $f$ and $g$ be quasiconformal homeomorphisms of $\mathbb{D}$ with Beltrami coefficients equal to $\mu$ and $\nu$ respectively such that $[f] = [g] = [\mu]$. Using Theorem 1, we know that for any $0 < \epsilon < 1 - k_0$, there exist $r_0 < r < 1$ and a quasiconformal homeomorphism $F$ of $\mathbb{D}$ such that $F = f$ on $D_{r_0}$, $F = g$ on $D_{r,1}$, and

$$K(F|_{D(r_0,r)}) \leq \max\{K(f|_{D(r_0,r)}), K(g|_{D(r_0,r)})\} + \epsilon.$$

Let $\eta_\varepsilon(z) = \mu_F$. Then

$$\eta_\varepsilon(z) = \begin{cases} \mu(z), & z \in D_{r_0} \\ \mu_F(z), & z \in D_{r_0,r} \\ \nu(z), & z \in D_{r,1}. \end{cases}$$

Thus, $\frac{1+\|\eta_\varepsilon\|_\infty}{1-\|\eta_\varepsilon\|_\infty} \leq \frac{1+k_0}{1-k_0} + \varepsilon$. Clearly,

$$g(t) = [\frac{t\eta_\varepsilon}{\|\eta_\varepsilon\|_\infty}], t \in \mathbb{D}$$

is a holomorphic map from $\mathbb{D}$ to $T(\mathbb{D})$ and it is also a curve staying in $T_2(\mathbb{D})$. Let $\gamma(t) = g(t), t \in [0,1]$. Then $\gamma$ is a path in $T_2(\mathbb{D})$ connecting $[0]$ and $[\mu]$. Now given any partition $0 = t_0 < t_1 < t_2 < \cdots < t_{n-1} < t_n = 1$ of $[0,1]$, let $s_k = t_k \|\eta_\varepsilon\|_\infty$ for each $0 \leq k \leq n$. Then

$$d_T(\gamma(t_{k-1}), \gamma(t_k)) = \frac{1}{2} \log \frac{1 + \frac{s_k - s_{k-1}}{1 - s_k s_{k-1}}}{1 - \frac{s_k - s_{k-1}}{1 - s_k s_{k-1}}} = \int_{s_{k-1}}^{s_k} \frac{1}{1-x^2} dx,$$

where $k = 1, 2, \cdots, n$. It follows that

$$\sum_{k=1}^{n} d_T(\gamma(t_{k-1}), \gamma(t_k)) = \int_0^{s_n} \frac{1}{1-x^2} dx = \frac{1}{2} \log \frac{1+s_n}{1-s_n} = \frac{1}{2} \log \frac{1+\|\eta_\varepsilon\|_\infty}{1-\|\eta_\varepsilon\|_\infty}.$$

Applying the definition of $\mathrm{Length}_{d_T}(\gamma)$, we obtain

$$\mathrm{Length}_{d_T}(\gamma) \leq \frac{1}{2}\log(\frac{1+k_0}{1-k_0} + \varepsilon).$$

Letting $\varepsilon \to 0$, we conclude

$$\inf_{\gamma} \mathrm{Length}_{d_T}(\gamma) \leq \frac{1}{2}\log\frac{1+k_0}{1-k_0} = d_T([0],[\mu]).$$

□

PROOF OF THEOREM 3. The approaches to obtain the inequality on the three spaces are similar, which are even shorter than the one to prove Theorem 3. For an example, we present the proof for $T_v(\mathbb{D})$.

Denote by $d_{K,T_v}$ the Kobayashi metric on $T_v(\mathbb{D})$. Since $T_v(\mathbb{D})$ is a group and the right translation map $\alpha([f]) = [f \circ g^{-1}]$ is isometric respect to $d_T$ and $d_{K,T_v}$, it is sufficient to show

$$d_{K,T_v}([0],[\mu]) \leq d_T([0],[\mu])$$

for any point $[\mu] \in T_v(\mathbb{D})$.

Note that $T_v(\mathbb{D}) \subset T_0(\mathbb{D})$. By the frame mapping theorem, each point $[\mu]$ of $T_v(\mathbb{D}) \setminus \{[0]\}$ has a unique extremal Beltrami coefficient $\mu = k_0\frac{\overline{\varphi}}{|\varphi|}$, where $k_0 \in (0,1)$ and $\varphi$ is a holomorphic quadratic differential on $\mathbb{D}$ with $\|\varphi\| = 1$. By Lemma 2, there exists $\nu \in [\mu]$ such that $\nu \in M_v(\mathbb{D}) \cap M_0(\mathbb{D})$. Let $0 < r_0 < 1$ be sufficiently close to 1 such that

$$\frac{1+\|\nu|_{D_{r_0,1}}\|_\infty}{1-\|\nu|_{D_{r_0,1}}\|_\infty} < \frac{1+k_0}{1-k_0}. \tag{3.7}$$

Now let $f$ and $g$ be quasiconformal homeomorphisms of $\mathbb{D}$ with Beltrami coefficients equal to $\mu$ and $\nu$ respectively such that $[f] = [g] = [\mu]$. Using Theorem 1, we know that for any $0 < \epsilon < 1 - k_0$, there exist $r_0 < r < 1$ and a quasiconformal homeomorphism $F$ of $\mathbb{D}$ such that $F = f$ on $D_{r_0}$, $F = g$ on $D_{r,1}$, and

$$K(F|_{D(r_0,r)}) \leq \max\{K(f|_{D(r_0,r)}), K(g|_{D(r_0,r)})\} + \epsilon.$$

Let $\eta_\varepsilon(z) = \mu_F$. Then

$$\eta_\varepsilon(z) = \begin{cases} \mu(z), & z \in D_{r_0} \\ \mu_F(z), & z \in D_{r_0,r} \\ \nu(z), & z \in D_{r,1}. \end{cases}$$

Thus, $\frac{1+\|\eta_\varepsilon\|_\infty}{1-\|\eta_\varepsilon\|_\infty} \leq \frac{1+k_0}{1-k_0} + \varepsilon$. Clearly,

$$g(t) = [\frac{t\eta_\varepsilon}{\|\eta_\varepsilon\|_\infty}], t \in \mathbb{D}$$

is a holomorphic map from $\mathbb{D}$ to $T_v(\mathbb{D})$. Using the definition of the Kobayashi metric, we obtain

$$d_{K,T_v}([0],[\mu]) \leq \rho_{\mathbb{D}}(0, \|\eta_\varepsilon\|_\infty) \leq \frac{1}{2}\log\frac{1+(k_0+\varepsilon)}{1-(k_0+\varepsilon)}.$$

Letting $\varepsilon \to 0$, we conclude $d_{K,T_v}([0],[\mu]) \leq d_T([0],[\mu])$. □

Using Lemmas 1 and 2 and the map $F$ constructed in the proof of Theorem 2 or 3, we obtain the following corollary as well.

COROLLARY 2. *For every element* $[f] \in T_2(\mathbb{D})$ *(resp.* $T_v(\mathbb{D})$*) and any* $\epsilon > 0$, *there exists a quasiconformal homeomorphism* $F$ *of* $\mathbb{D}$ *such that* $[F] = [f]$ *and*

$$K(F) < K_0([f]) + \epsilon,$$

*where* $K_0([f])$ *is the extremal maximum dilatation of* $[f]$.

## References


[1] K. Astala and M. Zinsmeister, *Teichmüller spaces and BMOA*, Math. Ann. **289** (1991), no. 4, 613–625, DOI 10.1007/BF01446592. MR1103039

[2] G. Cui, *Integrably asymptotic affine homeomorphisms of the circle and Teichmüller spaces*, Sci. China Ser. A **43** (2000), no. 3, 267–279, DOI 10.1007/BF02897849. MR1766456

[3] G. Cui and Y. Qi, *Local boundary dilatation of quasiconformal maps in the disk*, Proc. Amer. Math. Soc. **130** (2002), no. 5, 1383–1389, DOI 10.1090/S0002-9939-01-06353-5. MR1879961

[4] G. Cui and M. Zinsmeister, *BMO-Teichmüller spaces*, Illinois J. Math. **48** (2004), no. 4, 1223–1233. MR2114154

[5] A. Douady and C. J. Earle, *Conformally natural extension of homeomorphisms of the circle*, Acta Math. **157** (1986), no. 1-2, 23–48, DOI 10.1007/BF02392590. MR857678

[6] C. J. Earle, I. Kra, and S. L. Krushkal′, *Holomorphic motions and Teichmüller spaces*, Trans. Amer. Math. Soc. **343** (1994), no. 2, 927–948, DOI 10.2307/2154750. MR1214783

[7] C. J. Earle, F. P. Gardiner, and N. Lakic, *Asymptotic Teichmüller space. II. The metric structure*, In the tradition of Ahlfors and Bers, III, Contemp. Math., vol. 355, Amer. Math. Soc., Providence, RI, 2004, pp. 187–219, DOI 10.1090/conm/355/06452. MR2145063

[8] C. J. Earle, V. Markovic, and D. Saric, *Barycentric extension and the Bers embedding for asymptotic Teichmüller space*, Complex manifolds and hyperbolic geometry (Guanajuato, 2001), Contemp. Math., vol. 311, Amer. Math. Soc., Providence, RI, 2002, pp. 87–105, DOI 10.1090/conm/311/05448. MR1940165

[9] J. Fan and J. Hu, *Holomorphic contractibility and other properties of the Weil-Petersson and VMOA Teichmüller spaces*, Ann. Acad. Sci. Fenn. Math. **41** (2016), no. 2, 587–600, DOI 10.5186/aasfm.2016.4137. MR3525386

[10] T. Franzoni and E. Vesentini, *Holomorphic maps and invariant distances*, Notas de Matemática [Mathematical Notes], vol. 69, North-Holland Publishing Co., Amsterdam-New York, 1980. MR563329

[11] F. P. Gardiner, *Approximation of infinite-dimensional Teichmüller spaces*, Trans. Amer. Math. Soc. **282** (1984), no. 1, 367–383, DOI 10.2307/1999593. MR728718

[12] F. P. Gardiner and N. Lakic, *Quasiconformal Teichmüller theory*, Mathematical Surveys and Monographs, vol. 76, American Mathematical Society, Providence, RI, 2000. MR1730906

[13] F. P. Gardiner and D. P. Sullivan, *Symmetric structures on a closed curve*, Amer. J. Math. **114** (1992), no. 4, 683–736, DOI 10.2307/2374795. MR1175689

[14] J. Hu, Y. P. Jiang, and Z. Wang, *Kobayashi's and Teichmüller's metrics on the Teichmüller space of symmetric circle homeomorphisms*, Acta Math. Sin. (Engl. Ser.) **27** (2011), no. 3, 617–624, DOI 10.1007/s10114-011-9045-7. MR2770830

[15] Y. Jiang and Y. Qi, *A gluing theorem for quasiconformal mappings*, Kodai Math. J. **35** (2012), no. 3, 415–424, DOI 10.2996/kmj/1352985446. MR2997472

[16] M. Yanagishita, *Teichmüller distance and Kobayashi distance on subspaces of the universal Teichmüller space*, Kodai Math. J. **36** (2013), no. 2, 209–227. MR3081243

[17] Ch. Pommerenke, *On univalent functions, Bloch functions and VMOA*, Math. Ann. **236** (1978), no. 3, 199–208, DOI 10.1007/BF01351365. MR0492206

[18] E. Reich and K. Strebel, *Extremal quasiconformal mappings with given boundary values*, Contributions to analysis (a collection of papers dedicated to Lipman Bers), Academic Press, New York, 1974, pp. 375–391. MR0361065

[19] H. L. Royden, *Automorphisms and isometries of Teichmüller space*, Advances in the Theory of Riemann Surfaces (Proc. Conf., Stony Brook, N.Y., 1969), Ann. of Math. Studies, No. 66. Princeton Univ. Press, Princeton, N.J., 1971, pp. 369–383. MR0288254

[20] Y. Shen, *Weil-Petersson Teichmüller space*, arXiv:1304.3197.

[21] Y. Shen and H. Wei, *Universal Teichmüller space and BMO*, Adv. Math. **234** (2013), 129–148, DOI 10.1016/j.aim.2012.10.017. MR3003927

[22] K. Strebel, *On the existence of extremal Teichmueller mappings*, J. Analyse Math. **30** (1976), 464–480, DOI 10.1007/BF02786734. MR0440031

Department of Applied Mathematics, Nanjing University of Science and Technology, Nanjing 210094, People's Republic of China

*E-mail address*: jinhuafan@hotmail.com

Department of Mathematics, Brooklyn College of CUNY, Brooklyn, New York 11210 – and – Ph.D. Program in Mathematics, Graduate Center of CUNY, 365 Fifth Avenue, New York, New York 10016

*E-mail address*: junhu@brooklyn.cuny.edu, JHu1@gc.cuny.edu

Contemporary Mathematics
Volume **696**, 2017
http://dx.doi.org/10.1090/conm/696/14018

# Extremal length and uniformization

Frederick P. Gardiner

## Introduction

The classical theorem that identifies Riemann surfaces of negative Euler characteristic with surfaces carrying a hyperbolic structure is called uniformization. For surfaces that have complex structure it gives simultaneously a common parametrization by a parameter varying in a simply connected domain. For topological reasons the covering domain is connected and simply connected and, this being so, the uniformization theorem says that it must be conformal to either the Riemann sphere, the complex plane or the unit disc.

Since this theorem was first proved more than a century ago by Koebe, there have been many accounts, nearly all of them relying on some form of potential theory and the classification of open surfaces into those that have and those that do not have Green's functions. The exposition given here differs in that it avoids potential theory replacing it by the notion of extremal length. The proof involves finding moduli of curve families and the existence of conformal mappings from abstract doubly connected Riemann surfaces to planar annuli, that is, regions in the plane of the form $\{z : R_1 < |z| < R_2\}$.

Our proof starts out the same way as it does in most classical expositions. It is assumed there is given a simply connected Riemann surface $X$ with a local parameter $z$ vanishing at a point $p_0 \in X$ and mapping a neighborhood of $p_0$ onto the unit disc in the $z$-plane. Then one forms the disc $D_\epsilon(p_0) = \{p \in X : |z(p)| \leq \epsilon\}$ and tries to uniformize the annular surface $X_\epsilon = X - D_\epsilon(p_0)$. In our approach we consider moduli $M(A_n)$ of annuli $A_n$ contained in $X_{1/n}$ with one boundary component equal to the boundary of $D_{1/n}(p_0)$. Then we let $M(A_n)$ to be the extremal length of the family curves contained in $A_n$ and homotopic to $A_n$. We first establish that for such annuli with non-zero modulus there exists conformal maps $z_n$ that map $A_n$ onto a ring domains in the $z$-plane lying between $|z| = 1$ and $|z| = R_n$ where $M(A_n) = 2\pi/\log R_n$. There are two cases:

Case 1. There exists a sequence of annuli $A_n$ such that $M(A_n)$ decreases to 0 as $n \to \infty$. In this case we show $X$ is conformal either to the complex plane or to the Riemann sphere.

---

2010 *Mathematics Subject Classification.* Primary 32G15, 30C70, 30C75.

The author is grateful to Zhe Wang and Ara Basmajian for help in the preparation of this manuscript.

Case 2. For any such sequence of annuli $A_n$ the moduli $M(A_n)$ are bounded below by a positive number independently of n. In this case we show $X$ is conformal to the unit disc.

In both cases we find uniformizing annuli or uniformizing punctured discs for $X_\epsilon$. We repeat the same constructions for the surfaces $X_\epsilon$ with decreasing values of $\epsilon$ and obtain conformal maps from $X_\epsilon$ onto annuli in the complex plane. Then we take the limit as $\epsilon$ approaches 0 and the crux of the argument is to show that these limits exist. The underlying entity that controls the limiting process is a quadratic differential, holomorphic on $X - \{p_0\}$ and with a double pole at $p_0$ whose regular trajectories are closed and fill $X - \{p_0\}$ except for possibly one additional point.

Our proof could be given in one or two pages but we take the opportunity to prove results with other applications and to develop the following ideas.

(1) weak measured foliations and the Dirichlet integral,
(2) conjugate differentials and conformal metrics,
(3) pulling back measured foliations,
(4) intersection number and the Cauchy-Schwarz inequality,
(5) comparison of moduli of annuli to moduli of curve families,
(6) generalized locally $L^2$ derivatives and Weyl's lemma,
(7) integrable holomorphic quadratic differentials,
(8) Strebel's variational technique,
(9) extremal lengths as moduli for Teichmüller space,
(10) Rado's theorem on the separability of all Riemann surfaces.

We use a technique that involves an alternative expression for absolute intersection numbers of simple closed curves and prove several results that we believe have not been previously noticed. In particular, we introduce the idea of intersection number of weak measured foliations[1] and show how this kind of intersection applies to homotopy intersection of simple closed curves. For two closed curves $\alpha$ and $\beta$ by definition the intersection $i(\alpha, \beta)$ is the total minimal number of intersections, all counted positively, of the unoriented curves $\alpha'$ and $\beta'$ where $\alpha'$ and $\beta'$ are any closed curves in the surface $X$ freely homotopic to $\alpha$ and $\beta$, respectively. If we let $A$ and $B$ be arbitrary annuli in $X$ in the homotopy classes of $\alpha$ and $\beta$, we can define functions $u_A$ and $u_B$ that are harmonic on $A$ and $B$ and that are equal to 0 and 1 on the two sides of $A$ and $B$, respectively. Then it turns out that

$$i(\alpha, \beta) = \inf \left\{ \int\int_{A \cap B} |du_A \wedge du_B| \right\}$$

where the infimum is taken over all such annuli $A$ and $B$. This inequality combined with the Cauchy-Schwarz inequality

$$\left( \int\int_{A \cap B} |du_A \wedge du_B| \right)^2 \leq Dir(|du_A|) Dir(|du_B|)$$

gives a way to estimate extremal lengths of measured foliations in terms of Dirichlet integrals. It also brings in Riemann's idea for showing the existence of harmonic functions, which is to minimize the Dirichlet integral with side conditions. In our set-up we minimize the Dirichlet integral of a weak measured foliation subject to

[1]In earlier papers I have used the terminology partial measured foliation instead of weak measured foliation. The improved terminology was suggested to me by Dylan Thurston.

the side conditions on its heights. The significance of intersection numbers in Teichmüller theory is first developed in Thurston's theory of measured foliations, [**10**] [**27**]. See also [**21**].

The paper is organized into ten sections. In the first and second we define Riemann surfaces, weak measured foliations, differentials, the Dirichlet integral, conjugate differentials and conformal metrics. In section 3 we show that the Dirichlet integral of the pull back by a K-quasiconformal map of a weak measured foliation is $K$-invariant. In section 4 we give a method for computing intersection number and show how it can be estimated from above by Dirichlet integrals. In section 5 we define modulus of an annulus, observe that it is a conformal invariant and show that the modulus $M(A)$ of the closed curve family $\Gamma_A$ in $A = \{z : 1 < |z| < R\}$ that separates its two boundary contours is $2\pi/\log R$. Section 6 presents several equivalent ways to define the modulus of a free homotopy class of simple closed curves on a Riemann surface. This is a key result that not only permits an elementary proof of Koebe's uniformization theorem but also provides a family of invariants that permit the construction the extremal length embedding of any Teichmüller space [**13**]. In section 7 we present Strebel's variational method which together with Weyl's lemma presented in section 9 provides the holomorphic quadratic differential whose unique existence is the key to the equivalence of the four different formulas for extremal length given in Theorem 1. In section 8 we prove the Riemann mapping theorem and the uniformization theorem by the same method. In section 9 we prove Weyl's lemma and finally, in section 10, we discuss the one-dimensional $C^\infty$ manifold structure on the *long line* and the corresponding 2-dimensional manifold structure on the *long cylinder* and observe how our proof of uniformization shows why this 2-dimensional smooth surface cannot have subordinate complex structure. This example is featured to clarify a theorem of Rado that says any Riemann surface is metrizable by a metric that has a countable basis for its topology. Once the uniformization theorem is proved Rado's theorem is an obvious corollary. In studying classical proofs of uniformization it is often difficult to see why the existence of a complex structure forces the underlying topological space to be separable.

In the last analysis it turns out that the uniformizing coordinate for a simply connected Riemann surface $X$ can be found by solving two differently stated extremal problems that apply to a given point on any Riemann surface $X$. These two problems both lead to an underlying cylindrical quadratic differential that has a double pole at one point with a quadratic residue equal to $-1$. One fixes a local parameter $w$ that vanishes at a point $p_0 \in X$. The first extremal problem involves the family $\mathcal{F}$ of univalent functions that map the unit disc into $X$ and that map 0 to $p_0$ and it is to find a function $f \in \mathcal{F}$ that realizes the infimum

$$U = \inf\left\{\frac{1}{|(w \circ f)'(0)|} : f \in \mathcal{F}\right\}.$$

The second extremum problem involves something called the reduced extremal length $Q$ at the point $p_0$ in $X$. One first finds the extremal distance in $X_\epsilon$ from the perimeter of a small disc of radius $\epsilon$ centered at $p_0$ to the ideal boundary of $X_\epsilon$. The reduced length $Q$ is obtained by correcting this constant by subtracting the number $(1/2\pi)\log(1/\epsilon)$ and taking the limit as $\epsilon \to 0$. It turns out that by using Koebe's distortion theorem for univalent functions holomorphic in the unit disc [**3**] one can show that $U$ and $Q$ are related by the formula $U = \exp(-2\pi Q)$, [**20**]. Case 1 occurs when $U = 0$ and $Q = \infty$, and in this case $X$ is conformal to the Riemann sphere or the complex plane. Case 2 occurs when $U > 0$ and $Q < \infty$, and in this case $X$ is conformal to the unit disc. One sees that for a simply connected Riemann surface $X$ finding the function inverse to the uniformizing function $z$ amounts to finding a univalent function holomorphic in the unit disc which realizes the extremal value $U$. These ideas are developed in [**20**].

The uniformization theorem, a form of which is also known as Hilbert's 22nd problem [**6**], was developed over a period of time that starts with Riemann in the mid nineteenth century and goes to the present. A summary of contributions up to 1945 is given in Herman Weyl's book, *Der Idee die Riemann Fläche* and translated into English *The Concept of a Riemann Surface*. Most textbooks written for first semester courses in complex analysis contain the Riemann mapping theorem which is preliminary to uniformization, but few prove uniformization.

Since the publication of Weyl's book in 1954 there have been many other accounts. Several of these are nicely summarized in Henry Kwan's Ph.D. 2004 thesis [**17**] written under the supervision of Yun-Tong Siu at Harvard University. However, few address the possibility that the Riemann surface might be non-separable and none of them take advantage of the possibility of a proof based on extremal length.

The following authors are among those that include a statement of the theorem and sometimes a proof in notes or books: Abikoff [**1**], Bers [**5**], Ahlfors and Sario [**4**], Donaldson [**8**], Farkas and Kra [**9**], Imayoshi and Taniguchi [**15**], Kwan [**17**], Pommerenke [**22**], G. Springer [**25**], Schiffer and Spencer [**24**], Strebel [**26**], and Tsuji [**28**].

## 1. Riemann surfaces, weak measured foliations and the Dirichlet integral

DEFINITION 1. A *Riemann surface* is a connected topological Hausdorff space $X$ together with a system of local homeomorphisms (called local parameters) $z_j$ defined on neighborhoods $N_j$ contained in $X$ and mapping into $\mathbb{C}$ such that

(1) the union of the neighborhoods $N_j$ over $j$ covers $X$,
(2) in any non-empty overlap $N_j \cap N_k$ the composition $z_k \circ (z_j)^{-1}$ is holomorphic.

DEFINITION 2. A *chart* on a Riemann surface $X$ is a homeomorphism $w$ from an open set in $X$ onto an open subset of $\mathbb{C}$ for which $w \circ z_j^{-1}$ is holomorphic for every local parameter $z_j$ for which points in the domain of $z_j$ overlap with points in the domain of $w$.

Note that with these definitions local parameters are also charts.

DEFINITION 3. A continuous function $u$ defined on a set $\Omega \subset \mathbb{C}$ has generalized first partial derivatives $u_x$ and $u_y$ if for every $C^\infty$ function $\phi$ with compact support

$$\int\int u\phi_x dxdy = -\int\int u_x\phi dxdy$$

and

$$\int\int u\phi_y dxdy = -\int\int u_y\phi dxdy.$$

DEFINITION 4. Let $z_j$ be a set of charts defined on open subsets $N_j$ of a Riemann surface $X$. A *weak measured foliation* $|du|$ is a system of continuous, real valued functions $u_j$ defined on open sets $z_j(N_j)$ in $\mathbb{C}$ such that

(i) for any two of the sets $N_j$ and $N_k$ with $N_j \cap N_k$ non-empty,

$$u_j(z_j(p)) = \pm u_k(z_k(p)) + constant$$

for all $p \in N_j \cap N_k$,

(ii) in each open set $N_j$, $u_j$ has generalized first numerator derivatives with respect to $x$ and $y$, $(u_j)_x$ and $(u_j)_y$, where $z = x + iy$, and

(iii) these generalized first partial derivatives are locally square integrable, that is, for each $j$,

$$\int\int_{z(N_j)} \left((u_j)_x^2 + (u_j)_y^2\right) dxdy < \infty.$$

Note that in this definition we *do not* assume $N = \bigcup_j N_j$ is equal to $X$.

DEFINITION 5. The *Dirichlet integral* of a weak measured foliation $|du|$ on a Riemann surface $X$ is equal to

$$Dir(|du|) = \int\int \left(u_x^2 + u_y^2\right) dxdy.$$

Here, one consider the domain of integration to be either $N = \bigcup_k N_j$ or to be $X$, bearing in mind that by definition the integrand is identically zero in $X - N$.

Let $\gamma(t)$ be a continuous curve in $X$ defined for $t \in [0,1]$ and $p_k = \gamma(t_k)$ where $0 = t_0 < t_1 < \cdots t_n = 1$. Let $mesh(\{p_k\}) = \max_{1 \leq k \leq n} t_k - t_{k-1}$.

DEFINITION 6. The height of $\gamma$ with respect to $|du|$ is given by

$$ht(\gamma, |du|) = \limsup_{mesh(\{p_k\}) \to 0} \sum_{j=1}^{n} |u_k(p_j) - u_k(p_{j-1})| = \int_\gamma |du|$$

where we put $|u_k(p_j) - u_k(p_{j-1})| = 0$ if either $p_j$ or $p_{j-1}$ is not in $N$ and where we assume that if both successive points $p_{j-1}$ and $p_j$ lie $N$ then for sufficiently small value of $mesh(\{p_k\})$ there is a value of $k$ for which they both lie in the same set $N_k$.

## 2. Conformal metrics and conjugate differentials

DEFINITION 7. A conformal metric on a Riemann surface $X$ is a system of non-negative measurable functions $\rho_j$ defined on open sets $z_j(N_j)$ in $\mathbb{C}$ such that in each overlapping set $N_j \cap N_k$,

$$\rho_j(z_j) = \rho_k(z_k) \left|\frac{dz_k}{dz_j}\right| \quad \text{a.e.}$$

We note that in the definition of a Riemann surface we require that the domains of local parameters be defined on a family of open sets $N_j$ that cover all of the Riemann surface $X$, but we do not make this requirement in the definitions of a weak measured foliation $|du|$ or of a metric $\rho(z)|dz|$. At any points where a conformal metric $\rho(z)|dz|$ or the weak measured foliation $|du|$ are not defined by convention we put $\rho$ and $|du|$ equal to zero.

Given a conformal metric $\rho(z)|dz|$, where $\rho$ is either upper semicontinuous or lower semicontinuous, we define the length of a piecewise differentiable arc $\gamma(t), t_0 \leq t \leq t_1$ to be the line integral

$$L(\rho, \gamma) = \int_{t_0}^{t_1} \rho(\gamma(t))|\gamma'(t))|dt$$

and we define the area of $\rho$ to be the double integral

$$area(\rho) = \int\int_X \rho^2(z)dxdy.$$

DEFINITION 8. Assume $z = x+iy$ is a local coordinate on a Riemann surface $X$. For any differential form $Pdx + Qdy$, where $P$ and $Q$ are locally square integrable, the conjugate differential form is

$$^*(Pdx + Qdy) = -Qdx + Pdy.$$

DEFINITION 9. A *quadratic differential* $q(z)(dz)^2$ on $X$ is an assignment of a continuous function $q_j$ on sets $z_j(N_j)$ on each open set $N_j$ forming a covering of $X$ such that for any $p \in N_j \cap N_k$,

$$q_j(z_j(p)) = q_k(z_k(p)) \left(\frac{dz_k(p)}{dz_j(p)}\right)^2.$$

$q$ is *holomorphic* if each of the functions $q_j$ is holomorphic.

LEMMA 1. *A weak measured foliation $|du|$ with finite Dirichlet integral on a Riemann surface $X$ determines a conformal metric $\rho$ by putting*

$$\rho(z)|dz| = |du + i * du| \tag{1}$$

*and a quadratic differential $\psi$ by putting*

$$\psi = (du + i * du)^2,$$

*where we put both $\rho$ and $\psi$ identically equal to 0 in $X - N$. Moreover,*

$$area(\rho) = \int\int_X \rho^2 dxdy = Dir(|du|). \tag{2}$$

PROOF. Since

$$\rho^2 = (du + i * du)(du - i * du)$$
$$= u_x^2dx^2 + 2u_xu_ydxdy + u_y^2dy^2 + u_y^2dx^2 - 2u_xu_ydxdy + u_x^2dy^2$$
$$= (u_x^2 + u_y^2)(dx^2 + dy^2),$$

$\rho = (u_x^2 + u_y^2)^{1/2}|dz|$ is an infinitesimal conformal metric. For such a metric we define its area to be

$$area(\rho) = \int\int_X \rho(z)^2 dx \wedge dy.$$

Then for $\rho(z)|dz| = |\psi(z)|^{1/2}|dz|$ defined by (1),

$$area(|\psi|^{1/2}) = \int\int_X |\psi(z)|dx \wedge dy.$$

Note that $dx \wedge dy = (1/2)|dz \wedge d\overline{z}|$. Finally, we have

$$\int\int |\psi|dx \wedge dy = (1/2) \int\int_X |(du + i * du) \wedge (du - i * du)| = \int\int_X du \wedge *du$$

$$= \int\int_X (u_xdx + u_ydy) \wedge (-u_ydx + u_xdy) = \int\int_X (u_x^2 + u_y^2)dx \wedge dy = Dir(|du|).$$

□

The quadratic differential $\psi$ is not necessarily holomorphic and by definition in $N = \bigcup_j N_j$ both $\psi$ and $\rho$ are identically equal to zero.

LEMMA 2. *Any holomorphic quadradic differential $q$ on $X$ determines a one-parameter family of weak measured foliations by the formula*

$$v = \text{ the imaginary part of } \left(\int (e^{i\theta} q(z_j))^{1/2} dz_j\right) \tag{3}$$

*where $v$ is defined only on the complement of the set of zeros of $q$ in $X$.*

PROOF. Since we only consider points where $q$ is nonzero, the ambiguity in equation (3) involves the choice of a plus or minus sign in taking a square root and the choice of an additive constant in taking the antiderivative. Thus $|dv|$ is well defined on the complement in $X$ of the set of zeros of $q$. □

When $\theta = 0$, $|dv|$ is called the vertical foliation of $q$ and when $\theta = \pi$, $|dv|$ is called the horizontal foliation of $q$. We note that a parameterized differentiable arc $\gamma(t) \in X$ for which

$$q(\gamma(t))\gamma'(t)^2 > 0$$

is an arc contained in a horizontal leaf of the vertical foliation of $q$. Similarly if

$$q(\gamma(t))\gamma'(t)^2 < 0$$

then the image of $\gamma$ is contained in a vertical leaf of the horizontal foliation. Segments of these types are called regular horizontal or regular vertical trajectories of $q$. Points where $q$ is equal to zero are called critical points and at these points $|dv|$ is undefined. The trajectory structure in a neighborhood of a critical point $p$ where $q = 0$ has a zero of order $k$ is conformally equivalent to the trajectory structure of $z^k(dz)^2$. At these points $q$ has $k + 2$ trajectories that meet at $p$ and the angle between any two neighboring trajectories is $2\pi/(k + 2)$ radians. These are called $k + 2$-pronged singularities.

## 3. Pulling back a weak measured foliaton

In this section we show how the Dirichlet integral of a weak measured foliation is changed when changing coordinates by a $K$-quasiconformal map.

DEFINITION 10. Let $w = f(z) = u(z) + iv(z)$ be a homeomorphism from a Riemann surface $X$ onto a Riemann surface $Y$ and assume $u$ and $v$ have locally square integrable first partial distributional derivatives. Then $f$ is called *quasiconformal* if there is a constant $k < 1$ such that

$$|f_{\bar{z}}(z)| \le k|f_z(z)|, \ a.e.,$$

where

$$\frac{\partial}{\partial z} = (1/2)\left(\frac{\partial}{\partial x} - i\frac{\partial}{\partial y}\right) \text{ and}$$

$$\frac{\partial}{\partial \bar{z}} = (1/2)\left(\frac{\partial}{\partial x} + i\frac{\partial}{\partial y}\right).$$

LEMMA 3. *Let $f : X \to Y$ be quasiconformal and $|du|$ be a weak measured foliation on $Y$ with finite Dirichlet integral. Then $|d(u \circ f)|$ is a weak measured foliation on $X$ with finite Dirichlet integral and*

$$Dir_X(|d(u \circ f|)| \le K Dir_Y(|du|),$$

*where $K = \frac{1+k}{1-k}$.*

PROOF. We put $w = f(z)$ and note that

$$(u \circ f)_z = (u_w \circ f) f_z + (u_{\overline{w}} \circ f) \overline{f}_z.$$

Since $|\overline{f}_z| = |f_{\overline{z}}|$, $u$ is real-valued and defined up to plus or minus sign and an additive constant, $|u_w| = |u_{\overline{w}}|$. Thus we can imitate the calculation given at the end of Chapter 1 in [**3**]. Putting $w = f(z)$ we have

$$|(u \circ f)_z| = |(u_w \circ f) f_z + (u_{\overline{w}} \circ f) \overline{f}_z|$$

$$\leq (|u_w| \circ f)(|f_z| + |\overline{f}_z|) = (|u_w| \circ f)(|f_z| + |f_{\overline{z}}|), \text{ and}$$

$$Dir_X(|d(u \circ f)|) = 4 \int\int_X |(u \circ f)_z|^2 \left| \frac{dz \wedge d\overline{z}}{2} \right|$$

$$\leq 2 \int\int_X (|u_w| \circ f)^2 (|f_z| + |f_{\overline{z}}|)^2 \, |dz \wedge d\overline{z}|$$

$$= 2 \int\int_X (|u_w| \circ f)^2 (|f_z|^2 - |f_{\overline{z}}|^2) \cdot \frac{(|f_z| + |f_{\overline{z}}|)^2}{|f_z|^2 - |f_{\overline{z}}|^2} \, |dw \wedge d\overline{w}|$$

$$\leq 4K \int\int_Y |u_w|^2 \frac{|dw \wedge d\overline{w}|}{2} = K \; Dir_Y(|du|).$$

□

## 4. Intersection number and the Cauchy-Schwarz inequality

If two curves on a surface intersect transversally at several points, their intersection number is the total number of essential intersections. This differs from the homology notion of intersection where the curves are oriented and where you count an intersection positively or negatively according to its orientation. The intersection $i(\alpha, \beta)$ of the homotopy classes of two closed curves $\alpha$ and $\beta$ is the minimum number of intersections of $\alpha'$ and $\beta'$ where $\alpha'$ and $\beta'$ are freely homotopic to $\alpha$ and $\beta$.

There is a more devious way to view homotopical intersection of simple closed curves on oriented surfaces. For a closed curve $\alpha$, first slightly fatten it so it becomes a ribbon (or annulus) $A_\alpha$ and then form a harmonic function $u_\alpha$ on $A_\alpha$ that is equal to zero on one side of $A_\alpha$ and equal to one on the other side. Then it turns out that the intersection number is equal to the infimum of the double integrals

$$\int\int |du_\alpha \wedge du_\beta|, \tag{4}$$

where the domain of integration is the common intersection of the two ribbons for $\alpha$ and $\beta$ and the infimum is taken over all curves in the same free homotopy class and all possible ribbons. The absolute value is placed under the integral sign in (4) to ensure that all intersections are counted positively.

Here we show how this notion of intersection is related to Dirichlet integrals. First we note an application of the Cauchy-Schwarz inequality to the wedge product of two weak measured foliations.

LEMMA 4. *Let $|du|$ and $|dv|$ be weak measured foliations on a Riemann surface $X$. Then*

$$\left( \int\int_X |du \wedge dv| \right)^2 \leq Dir(|du|) Dir(|dv|). \tag{5}$$

PROOF. The following inequality for real numbers is elementary:

$$\left(det\begin{pmatrix} a & b \\ c & d \end{pmatrix}\right)^2 \le (a^2+b^2)(c^2+d^2).$$

Adding the terms indexed by pairs of integers $(j,k)$ leads to the generalization

$$\left(\sum_j\sum_k |\alpha_j\delta_k - \beta_j\gamma_k|\right)^2 \le \left(\sum_j \alpha_j^2+\beta_j^2\right)\left(\sum_k \gamma_k^2+\delta_k^2\right), \tag{6}$$

which is a discrete version of (5). Interpreting the double integrals in formula (5) as Lebesgue integrals of $L^1$-functions, the discrete version (6) implies (5). □

DEFINITION 11. A closed curve curve in a surface $X$ is a continuous map $\gamma$ from the unit circle into $X$. It is called simple if $\gamma$ is injective. A simple closed curve contained in an annulus is called a core curve if it separates the two boundary components.

DEFINITION 12. For any two simple closed curves $\alpha$ and $\beta$ on a Riemann surface $X$, $i(\alpha,\beta)$ is the number of essential intersections of $\alpha$ with $\beta$. That is, $i(\alpha,\beta)$ is the minimum number of intersections $\alpha'$ with $\beta'$ where $\alpha'$ and $\beta'$ are closed curves freely homotopic in $X$ to $\alpha$ and $\beta$, respectively.

LEMMA 5. **(The intersection inequality)** *Let $A_\alpha$ be an annulus with smooth boundary containing an essential simple close curve $\alpha$ and $u_\alpha$ be a function harmonic in $A_\alpha$ with one-sided limiting values equal to 0 and 1 on the two sides of $A_\alpha$. Let $u_\beta$ have the same corresponding properties for an annulus $A_\beta$ corresponding to a simple closed curve $\beta$. Finally, let $i(\alpha,\beta)$ be the number of essential intersections $\alpha$ and $\beta$. Then*

$$i(\alpha,\beta) \le \int\int |du_\alpha \wedge du_\beta|. \tag{7}$$

PROOF. Note that we can assume any simple closed curve $\alpha$ is an essential closed curve contained in some annulus $A_\alpha \subset X$. The fattening can be trimmed in such a way as to have a differentiable boundary; because $X$ is orientable the boundary will have two components that are homotopic simple closed curves and there will be a function $u_\alpha$ that is equal to 0 on one of these boundary curves and equal to 1 on the other.

The integrand $|du_\alpha \wedge du_\beta|$ is nonnegative where the annulus $A_\alpha$ intersects the annulus $A_\beta$ and zero elswhere. To estimate the integral in (7) note that since the level curves of $u_\alpha$ are homotopic to $\alpha$ and similarly those of $u_\beta$ are homotopic to $\beta$, they necessarily intersect at least $i(\alpha,\beta)$ times. Thus the double integral can be estimated by integrating by parts and we obtain

$$\int\int_X |du_\alpha \wedge du_\beta| \ge \left|i(\alpha,\beta)\int_0^1 [u_\alpha \wedge du_\beta]_{u_\alpha=0}^{u_\alpha=1}\right| = i(\alpha,\beta)\left|\int_0^1 du_\beta\right| = i(\alpha,\beta).$$

□

## 5. Extremal length of annuli

In this section we assume $A$ is an annulus in a given homotopy class on a Riemann surface $X$. Also, except for Lemma 8, we assume $X$ contains only one free homotopy class of simple closed curve. We let $\gamma$ be one of the closed curves in this class and $\Gamma$ be the set of all such curves in this class.

DEFINITION 13. The extremal length of a curve family $\Gamma$ is given by

$$M(\Gamma) = \sup_{\rho} \frac{L(\rho, \Gamma)^2}{area(\rho)}, \tag{8}$$

where $\rho(z)|dz|$ is a continuous conformal metric on $X$ and where $L(\rho, \Gamma)$ is the infimum of lengths of closed curves $\gamma$ in $\Gamma$ and $area(\rho) = \int\int_X \rho^2$.

The Riemann surface $X$ is called annular if it has only one such class $\Gamma$. In this case we usually write $A$ instead of $X$ and since there is only one such class we write $M(A)$ instead of $M(\Gamma)$. If $A$ is an annular Riemann surface we also define the conjugate curve family $\Gamma^*$ and the conjugate modulus $M^*(A)$. $\Gamma^*$ is the family of arcs in $A$ that join its two boundary components and $M^*(A) = M(\Gamma^*)$. This number is often called the modulus of $A$ and is given by the following definition.

DEFINITION 14.

$$M^*(A) = \sup_{\rho} \frac{L(\rho, \Gamma^*)^2}{area(\rho)}. \tag{9}$$

In both (8) and (9) the supremum is over all conformal metrics $\rho$ defined on $X$.

LEMMA 6. *If* $A = \{z : 1 < |z| < R\}$ *then* $M(A) = 2\pi/\log R$ *and* $M^*(A) = (\log R)/2\pi$.

PROOF. Any closed curve $\gamma$ in $A$ that winds once around the origin satisfies

$$\pm 1 = \frac{1}{2\pi i} \int_{\gamma} \frac{dz}{z}$$

and so any closed curve $\gamma \in \Gamma(A)$ satisfies

$$1 \leq \int_{\gamma} \frac{1}{2\pi r} |dz| = \int_{\gamma} \rho_0(z)|dz|,$$

where $\rho_0(z)|dz| = 1/(2\pi r)|dz|$. This inequality is an equality for any of the concentric closed curves $re^{it}, 0 \leq t \leq 2\pi$ so $L(\rho_0) = 1$. On the other hand,

$$area(\rho_0) = (1/2\pi)^2 \int\int_A (1/r^2) r dr d\theta = (\log R)/2\pi$$

and so $M^*(A) \geq 2\pi/\log R$.

For the reverse inequality, observe that for any allowable metric $\rho|dz|$,

$$L(\rho) \leq \int_0^{2\pi} \rho(re^{it}) r dt,$$

$$L(\rho)/r \leq \int_0^{2\pi} \rho(re^{it}) dt.$$

Integrating both sides from $r = 1$ to $r = R$ gives

$$L(\rho) \log R \leq \int\int_A \rho dr dt = \int\int_A \rho r^{1/2} \cdot \frac{1}{r^{1/2}} dr dt$$

and applying Schwarz's inequality we get

$$L(\rho)^2 (\log R)^2 \leq \int\int_A \rho^2 r dr dt \int\int_A \frac{1}{r} dr dt,$$

$$L(\rho)^2/area(\rho) \leq 2\pi/(\log R) \text{ for every allowable } \rho.$$

We conclude $\rho_0$ realizes the value $M(A)$ which is equal to $2\pi/(\log R)$.

For the second equality, the same metric $\rho_0$ is extremal, in the sense that it realizes the value $M^*(A)$. To see this note that for the radial arcs $\gamma_0^*(t) = te^{i\theta}, 1 \leq t \leq R$,

$$\int_{\gamma_0^*} \rho_0|dz| = (1/2\pi)\int_1^R (1/t)dt = \frac{1}{2\pi}\log R$$

and for any arc $\gamma^*$ in $A$ joining its inner to outer boundary,

$$(1/2\pi)\int_{\gamma^*}(1/t)dt \geq \frac{1}{2\pi}\log R$$

so $L(\rho_0, \Gamma^*) = \frac{1}{2\pi}\log R$. Also $area(\rho_0) = \left(\frac{1}{2\pi}\right)\log R$ so

$$L(\rho_0, \Gamma^*)^2/area(\rho_0) = (\log R)/(2\pi) \text{ and}$$

$$M^*(A) \geq (\log R)/(2\pi). \tag{10}$$

To see that $\rho_0$ realizes the maximal value for this curve family note that any arc $\gamma^*(t)$ joining $|z| = 1$ to $|z| = R$, the definition of $L(\rho)$ implies

$$L(\rho) \leq \int_{\gamma^*} \rho(z)|dz|.$$

Integrating along the arc $te^{i\theta}, 1 \leq t \leq R$ for each $\theta$ and then integrating with respect to $\theta$ from 0 to $2\pi$ gives

$$2\pi L(\rho) \leq \int_{\theta=0}^{\theta=2\pi}\int_{|z|=1}^{|z|=R} \rho(z)drd\theta.$$

Thus,

$$2\pi L(\rho) \leq \int\int_A \rho(z)r^{1/2}\cdot\frac{1}{r^{1/2}}drd\theta, \text{ and by Schwarz's inequality,}$$

$$(2\pi L(\rho))^2 \leq \int\int_A \rho(z)^2 rdrd\theta \cdot \int\int_A \frac{1}{r}drd\theta,$$

$$(2\pi L(\rho))^2 \leq area(\rho)2\pi\log R \text{ and so}$$

$$\frac{L(\rho)^2}{area(\rho)} \leq \frac{\log R}{2\pi}.$$

we conclude that $M^*(A) \leq \log R/2\pi$ and so from (10) $M^*(A) = \log R/2\pi$. □

There are several observations we wish to make about the result of this lemma:

(1) the same metric $\rho_0$ is extremal for the supremum problems involving both $\Gamma$ and $\Gamma^*$, and $\rho_0 = |q|^{1/2}$ where

$$q(z)(dz)^2 = -\left(\frac{1}{2\pi}\cdot\frac{dz}{z}\right)^2$$

is a holomorphic quadratic differential on $A$,

(2) the curve families $\Gamma$ and $\Gamma^*$ contain subfamilies $\Gamma_0$ and $\Gamma_0^*$, which are the horizontal and vertical trajectories of $q$,

(3) $\Gamma_0$ and $\Gamma_0^*$ are minimally extremal for $\rho_0$ in the sense that $\rho_0$ is extremal for the extremal problems $M(\Gamma_0)$ and $M(\Gamma_0^*)$. This observation is called Beurling's lemma, (see [**2**], chapter 4, section 7).
(4) from Lemma 3 and a change of variable it follows that $M(A)$ is a conformal invariant,
(5) $q^{1/2}dz = \pm(i/2\pi)(d\log z) = \pm(i/2\pi)(d\log|z| + id\theta) = i(du + idv)$, where $u = \log|z|$ is a harmonic function in $A$ equal to 0 on $|z| = 1$ and equal to $\log R$ on $|z| = R$,
(6) the level lines defined by $u =$ *a constant* and $v =$ *a constant* comprise the curve families $\Gamma_0$ and $\Gamma_0^*$,
(7) similar observations hold for the vertical and horizontal foliations of any integrable holomorphic quadratic differential on any Riemann surface
(8) for the extremal problem $M(\Gamma)$ on any Riemann surface $X$ if $M(\Gamma) > 0$ then the quadratic differential $q$ is integrable and holomorphic on $X$ and $q$ expresses the first variation of $\log M(\Gamma)$ when the complex structure on $X$ is allowed to vary.

Since it is not necessary for the proof of the uniformization theorem we do not include justification of statement (8) in this article. It can be found in the last pages of [**12**].

LEMMA 7. *For any annular Riemann surface $A$ conformal to $\{z : 1 < |z| < R\}$, $M(A) = 2\pi/\log R$ and there is a unique harmonic function $u$ defined on $A$ with the property that $u = 0$ on one boundary component of $A$ and is equal to $(1/2\pi)\log R$ on the other boundary component.*

PROOF. $u$ is unique by the maximum principle. If $z : A \to \{z : 1 < |z| < \log R\}$ is conformal, then $u = (1/2\pi)\log|z|$ is harmonic and has the prescribed boundary values. □

LEMMA 8. *Suppose there is a homeomorphism $h$ from $\{z : 1 \leq |z| \leq 2\}$ onto a domain $A$ in a Riemann surface $X$ and*

$$\gamma_0 = h(\{z : |z| = 1\}) \text{ and } \gamma_1 = h(\{z : |z| = 2\})$$

*are its two boundary contours. Then there exists a unique harmonic function $u$ defined on $A$ which is equal to 0 on $\gamma_0$ and equal to 1 on $\gamma_1$. Moreover, $u$ determines a conformal map $z = \exp\left(2\pi\int(du + i^*du)\right)$ from $A$ to a region of the form $\{z : 1 < |z| < R\}$ and a quadratic differential*

$$q_A = -\left(\frac{1}{2\pi}\cdot\frac{dz}{z}\right)^2 \tag{11}$$

*defined on $A$ whose horizontal and vertical trajectories are, respectively, the concentric circles* $|z| =$ a constant *and the rays* $\arg z =$ a constant.

PROOF. One considers the family of all $C^1$ functions defined on $A$ with the given boundary values on the two simple closed curves that form the boundary of $A$ and minimizes the Dirichlet integral. One uses $L^2$ estimates on first partial derivatives and Lemma 11, Weyl's lemma, to show that the minimum $u$ exists and is harmonic. Moreover, if $p_n$ is a convergent sequence of points in $A$ and if $u(p_n) \to 1$, then $p_n$ must converge to a point $\gamma_1$. Also, $u(p_n)$ converging to 0 implies $p_n$ converges to a point in $\gamma_0$. This is because otherwise the harmonic function $u$

would take its minimum or maximum value at an interior point of $A$, which would contradict the maximum principle for harmonic functions.

After constructing $u$ we form $U = (\log R)u$ where $R$ is chosen so that $M(A) = 2\pi/\log R$. Then $U$ is harmonic in $A$ with boundary value 0 on $\gamma_0$ and $\log R$ on $\gamma_1$ and $\zeta = U + i^*U$ maps $A$ to a rectangle in the $\zeta$-plane with width $\log R$ and height $2\pi$ and $z = \exp(i\zeta)$ maps $A$ to the region between two concentric circles in the $z$-plane with inner radius 1 and outer radius $R$. □

The essence of Lemma 8 is the statement that every topological annulus embedded in a Riemann surface is conformal to a round annulus and the horizontal and vertical trajectories of the quadratic differential $q$ in (11) are realized by concentric circles and radial segments in the $z$-plane.

## 6. Moduli of curve families on a Riemann surface

This section concerns moduli of homotopy classes of simple closed curves on an arbitrary Riemann surface $X$. Although it has several equivalent definitions, we choose to define the modulus of the homotopy class of $\Gamma$ in $X$ in the following way.

DEFINITION 15. The modulus of the curve family $\Gamma$ in $X$ is equal to

$$M(\Gamma) = \sup_{\rho} \left\{ \frac{L(\Gamma, \rho)^2}{\int\int_A \rho^2 dxdy} \right\}, \tag{12}$$

where the supremum is taken over all conformal metrics $\rho(z)|dz|$ on $X$ and

$$L(\rho, \Gamma) = \inf_{\gamma \in \Gamma} \int_{\gamma} \rho(z)|dz|.$$

In this definition of $M(\Gamma)$ there is no particular annulus on which to apply the Grötzsch length-area argument given in Lemma 6. However, because of the next theorem, one can single out a unique annulus on which the argument can be applied. It is the annulus embedded in $X$ in the same homotopy class for which $M(\Gamma)$ is as small as possible. Moreover, when $0 < M(\Gamma)$ this annulus is uniquely associated to a particular integrable quadratic differential $q$ defined and holomorphic on $X$. Even when $M(\Gamma) = 0$, there is also a corresponding domain $A$ and a corresponding holomorphic quadratic differential but which fails to be integrable. In the non-integrable case the domain $A$ is conformal either to a punctured disc or a doubly punctured Riemann sphere. All of the regular horizontal trajectories of $q$ are closed and in the homotopy class of $\Gamma$. Moreover, except for the critical trajectories of $q$ which comprise a union of analytic arcs of zero area, these regular closed horizontal trajectories fill $X$.

We make the following definitions:

(i) $M_1 = \inf_A \{M(A)\}$, where the infimum is over all annuli $A$ homotopic to $\gamma$.

(ii) $M_2 = \inf_{|du|} \{Dir(|du|)\}$, where the infimum is over all $C^1$ functions $u$ defined on an arbitrary annulus $A$ homotopic to $\gamma$ in $X$ and equal to 1 on one of its boundary contours and equal to 0 on the other.

(iii) $M_3 = \sup_{|dv|} \left\{ \frac{1}{Dir(|dv|)} \right\}$, where the supremum is over all weak measured foliations $|dv|$ for which $\int_{\gamma} |dv| \geq 1$ for all $\gamma \in \Gamma$.

THEOREM 1. *Suppose $\gamma \in \Gamma$ and there is a closed curve $\beta$ contained in $X$ such that $i(\gamma, \beta) > 0$. Then*

$$M(\Gamma) = M_1 = M_2 = M_3.$$

*Moreover, there is a unique annulus $A$ in $X$ that realizes the infimum for $M_1$. It is the characteristic annulus of a quadratic differential*

$$q_A(z)(dz)^2 = -\left(\frac{dz}{z}\right)^2$$

*on $A$. Moreover $q_A$ is defined and holomorphic on $X$ and $A$ is equal to the union of the closed horizontal trajectores of $q_A(z)(dz)^2$. The square root $q_A(z)^{1/2}dz$ is equal to $\pm(du + iM(\Gamma)dv)$, where $|du|$ and $|dv|$ are the two unique foliations that realize the extremal values $M_2$ and $M_3$. Up to scalar multiple the metric $\rho$ that uniquely realizes the supremum for $M(\Gamma)$ is given by $\rho = |q_A|^{1/2}$. Finally, if $A = X$ then $z$ is a univalent map the maps $A$ to a round annulus $\{z : 1/R < |z| < 1\}$ where $M(\Gamma) = 2\pi/\log R$.*

REMARK. For a family $\Gamma$ of closed curves on a surface the quantities $M_2$ and $M_3$ give effective ways to estimate from below and from above the extremal length $M(\Gamma)$ defined in (15). This observation was pointed out to me by Dylan Thurston.

PROOF. For a given annulus $A$ in $X$ with core curves in the homotopy class of $\Gamma$ we choose the harmonic function $u_A$ given by Lemma 8. For this $u_A$ the Dirichlet integral $D(|du_A|)$ is equal to $2\pi/\log R$ which is equal to $M(A)$ so $M_1 \leq M_2$. On the other hand, every such harmonic function is by definition supported on an annulus $A$ and so $M_2 \leq M_1$.

From the intersection inequality (5), the Cauchy-Schwarz inequality (4) and the definitions of $|du|$ and $|dv|$ in $M_2$ and $M_3$ we get

$$\frac{1}{Dir(|dv|)} \leq Dir(|du|), \tag{13}$$

which shows that $M_3 \leq M_2$. More generally, let $A$ be an annulus with core curve in the homotopy class of $\Gamma$ and $u$ a harmonic function of the type described in (ii). If there is a curve $\beta$ that intersects $\gamma$, then $i(\beta, \gamma) > 0$ and

$$0 < \frac{i(\gamma, \beta)^2}{Dir(|dv|)} \leq Dir(|du|), \tag{14}$$

where $v$ is any $C^1$ function on an annulus $B$ in the homotopy class of $\beta$ which is equal to 0 and 1 on the two sides of $B$. Thus by (14) the infimum taken in the definition of $M_1$ is the infimum of a set which is bounded below. Given an annulus in the set we know by Lemma 8 there is a univalent function $1/z_A$ from $A$ onto $\{z : 1/R < |z| < 1\}$. If we let $c_A$ be the univalent function which is inverse to $z_A$ then as $M(A)$ decreases to the infimum $M_1$ the functions $c_A$ have increasing domains of definition. The functions $z_A$ determine quadratic differentials

$$q_A = -\left(\frac{1}{2\pi}\right)^2 \cdot \left(\frac{dz_A}{z_A}\right)^2, \tag{15}$$

which have $L_1$-norm bounded by $1/M_1$. Since the $L_1$-norms of $q_A$ are bounded we can select a sequence $A_n$ such that $q_{A_n}$ converge uniformly on compact subsets. The limit of such a sequence determines an annulus $A$ that realizes the extremal value $M_1$. By Strebel's variational technique, Lemma 9, and Weyl's Lemma, Lemma 11, for this $A$ the quadratic differential $q_A$ given by formula (15) is holomorphic on $X$ and its $L_1$-norm is equal to $1/M_1$.

The closed trajectories for this $q_A$ sweep out a unique extremal annulus in $X$. Any closed curve $\gamma \in \Gamma$, even if while staying in $X$ it veers outside of $A$, measured in the metric $|q_A|^{1/2}$ it is at least as long as any of the closed trajectories of $q_A$. This is because $q$ is holomorphic and closed horizontal trajectories of a holomorphic quadratic differential minimize length in the metric $|q|^{1/2}$ among all closed curves in the same homotopy class. For this reason, one can apply Grötzsch's argument on the annulus $A$ to show that $M(\Gamma) = M_1$.

But then from Lemma 1 the supremum in $M_3$ is realized by a measured foliation equal to a constant times the conjugate measured foliation $|^*du|$. That is, the $|dv|$ that realizes this supremum is given by $(1/M(\Gamma))|^*du|$. The holomorphic quadratic differential

$$q_A = (du + i^*du)^2 = (du + iM(\Gamma)dv)^2$$

where $|du|$ realizes $M_2$ and $|dv|$ realizes $M_3$ gives the metric $|q_A|^{1/2}$ extremal metric for $M(\Gamma)$. □

## 7. Strebel's variational technique

LEMMA 9. *Suppose $M(\Gamma) > 0$. Then there is a unique annulus $A$ in $X$ homotopic to $\Gamma$ for which $M(A)$ is minimum. This annulus is is filled by closed horizontal trajectories of a quadratic differential $q_\gamma$ holomorphic on $X$ and the complement of $A$ in $X$ consists of the critical trajectories of $q$ and*

$$\int\int_X |q_\gamma| = (2\pi)^2/M(\Gamma).$$

*Finally, for this particular annulus $A$, $M(A) = M(\Gamma)$.*

PROOF. Let $z$ be a local parameter that vanishes at $p$ and defined in a neighborhood $N$ of $p$ and put

$$w = \begin{cases} z + \epsilon h(z) & \text{in } N \\ z & \text{outside of } N, \end{cases}$$

where $h(z)$ has compact support in $N$ and continuous first partial derivatives. Then the Beltrami coefficient $\nu(z)$ of $w$ is given by

$$\nu(z) = \frac{w_{\bar{z}}(z)}{w_z(z)} = \frac{\epsilon h_{\bar{z}}(z)}{1 + \epsilon h_z(z)}. \tag{16}$$

Provided $\epsilon$ is small enough, $||\nu||_\infty < 1$, and this will guarantee that $z \mapsto w$ is a homeomorphism. Note that $h$ can be any continuous function with continuous first partial derivatives and with compact support in the simply connected open set $N$.

From Lemma 8 we can assume $\zeta = \xi + i\eta$ maps $A$ to a rectangle

$$R = \{0 \leq \xi \leq a, 0 \leq \eta \leq b\}.$$

This map carries any horizontal segment $\alpha$ in $R$ defined by $y$ equal to a constant to closed curve in the class $\Gamma$. Also, let $A' = w(A)$ and assume $\zeta'$ is a conformal parameter that takes $A'$ to a rectangle

$$R' = \{0 \leq \xi' \leq a', 0 \leq \eta' \leq b, \}.$$

Then

$$a' \leq \int_{w(\alpha)} |d\zeta'| = \int_\alpha |w_\zeta d\zeta + w_{\bar\zeta} d\bar\zeta| = \int_\alpha |w_\zeta||1+\nu(\zeta)|d\xi.$$

Thus,

$$a'b \leq \int\int_R |w_\zeta||1+\nu| d\xi d\eta, \text{ and}$$

by introducing factor of $(1-|\nu|^2)^{1/2}$ in the numerator and denominator and apply Schwarz's inequality we obtain

$$(a'b)^2 \leq \int\int_R |w_\zeta|^2(1-|\nu|^2) d\xi d\eta \int\int_R \frac{|1+\nu|^2}{1-|\nu|^2} d\xi d\eta.$$

Noting that by the Jacobian change of variable formula the first integral is equal to the area of $R'$, that is, $a'b$, we get

$$a'b \leq \int\int_R \frac{|1+\nu|^2}{1-|\nu|^2} d\xi d\eta. \tag{17}$$

We put $q = (d\zeta)^2$ in $R$ and identically equal to zero on $X - R$ and similarly we put $q' = (d\zeta')^2$ in $R'$ and identically equal to zero on $X - R'$. Also, we view $\nu, q$ and $q'$ as invariant differential forms. Then equation (17) becomes

$$\int\int_X |q'| dxdy \leq \int\int_X \frac{\left|1+\nu q/|q|\right|^2}{1-|\nu|^2} |q| dxdy. \tag{18}$$

Since $M(A)$ realizes the smallest possible value of moduli of annuli in the homotopy class $\Gamma$, $q$ has the smallest norm in this class and we have $\int\int_X |q| dxdy \leq \int\int_X |q'| dxdy$, from (18) we obtain

$$\int\int_X |q| dxdy \leq \int\int_X \frac{\left|1+\nu q/|q|\right|^2}{1-|\nu|^2} |q| dxdy. \tag{19}$$

By squaring out the numerator on the right hand side of (19) and dividing by 2 yields

$$0 \leq \text{Re} \int\int_X \frac{\nu q}{1-|\nu|^2} dxdy + \int\int_X \frac{|\nu|^2|q|}{1-|\nu|^2}. \tag{20}$$

The first order consequence of (16) and (20) is that for every $C^1$ function $h$ with compact support in $N \subset X$,

$$\pm \int\int_N h_{\bar z}(z) q(z)\ dxdy \geq 0$$

which implies

$$\int\int_N h_{\bar z}(z) q(z)\ dxdy = 0.$$

By Weyl's lemma, Lemma 11 in section 10, this implies $q$ is holomorphic at every point $p \in X$ up to the addition of a function that is almost everywhere equal to zero.

In summary, there is a global holomorphic quadratic differential $q$ defined on $X$ and an extremal annulus $A$ contained in $X$ and a univalent holomorhic function $z$ mapping $A$ to the region between two concentric circles $|z| = R_1$ and $|z| = R_2$ in the complex plane such that the restriction of $q$ to $A$ satisfies

$$q(z)(dz)^2 = -(1/2\pi)^2(dz/z)^2.$$

□

From the minimum norm principle in [**11**], [**12**] and [**18**] we know that up to a multiplicative constant with absolute value equal to one $q$ is unique among holomorphic quadratic differentials $q'$ that minimize $\int\int_X |q'|dxdy$ and satisfy $\int_\gamma |q'|^{1/2} \geq 1$ for all $\gamma \in \Gamma$.

The application of Lemma 9 that we use to prove Theorems 1 and 2 to a very particular curve family $\Gamma$. First we choose a local parameter$w$ that vanishes at a point $p_0$ on a Riemann surface $X$ and let

$$D_\epsilon(p) = \{p \in X : |w(p)| \leq \epsilon\}.$$

Then we consider the curve family $\Gamma_\epsilon$ in $X_\epsilon = X - D_\epsilon(p_0)$. The closure of the extremal annulus $A$ in $Y_\epsilon$ must contain the boundary of $D_\epsilon$. If it did not, by enlarging $A$ to contain the part of $X_\epsilon$ exterior to $A$ and contiguous to the boundary of $D_\epsilon$ one decreases $M(A)$.

## 8. Uniformization

Before proceeding to the main theorem we define something that we call a tongue. The standard tongue $T_0$ is the domain $\{z = x+iy : -1 \leq x \leq 1, 0 \leq y \leq 2\}$ together with its marked spine $J_0 = \{z : x = 0, 0 \leq y \leq 1\}$.

DEFINITION 16. A tongue $T$ is any image of the standard tongue $T_0$ by a quasiconformal map $h$. That is, $T = h(T_0)$.

We give names to the parts of a tongue according to the following terminology. We call, respectively, the parts $J = h(J_0)$, $L = h(L_0)$, $R = h(R_0)$, and $B = h(B_0)$ the spine, the left hand base, the right hand base and the base of $T$ where

$J_0 = \{z : x = 0, 0 \leq y \leq 1\}$,
$L_0 = \{z : y = 0, -1 \leq x \leq 0\}$,
$R_0 = \{z : y = 0, 0 \leq x \leq\}$,
$B_0 = L_0 \cup R_0$.

Note that for any tongue $T$ contained in a Riemann surface $X$ there is an associated weak measured foliation $|dv_T|$ on $X$ with support equal to $T$ such that $v_T$ is equal to 0 on its spine and equal to 1 its left and right hand sides and on its top. It is easy to construct $v_{T_0}$ on the standard tongue and one can put $v_T = v_{T_0} \circ h^{-1}$ where $h$ is a quasiconformal map from $T_0$ onto $T$.

In the next lemma we assume $X$ is any Riemann surface, $z$ is a local parameter vanishing at $p_0$ defined in a neighborhood of $p_0$, $D_\epsilon(p_0) = \{p : |z(p)| \leq \epsilon\}$ and

$$X_\epsilon = X - D_\epsilon(p_0).$$

Finally, we let $\Gamma_\epsilon$ be the family of simple closed curves in $X_\epsilon$ that are homotopic to $\gamma_\epsilon = \{p : |z(p)| = \epsilon\}$.

LEMMA 10. **(The tongue lemma)** *Assume the extremal length* $M(\Gamma_\epsilon) = 0$ *and* $C$ *is a sequentially compact subset of* $X_\epsilon$. *Then there exists a sequence of annuli* $A_n \subset X_\epsilon$ *in the homotopy class of* $\{p : |z(p)| = \epsilon\}$ *with the following properties:*

*1) the sequence* $M(A_n)$ *decreases to* 0,
*2)* $A_n$ *is ascending in the sense that* $A_k \subset A_{k+1}$ *for each integer* $k$ *and*
*3)* $C \subset \bigcup_n A_n$.

PROOF. We know from Theorem 1 there is a sequence of annuli $A_n$ with properties 1) and 2). For two successive annuli $A_k$ and $A_{k+1}$ in the sequence $A_n$ we can enlarge $A_{k+1}$ to an annular domain $\tilde{A}_{k+1}$ in such a way that $A_k \subset \tilde{A}_{k+1}$.

To see this consider the family $\mathcal{F}$ of simple closed curves $\gamma$ contained in $A_k \cup A_{k+1}$. Note that every simple closed curve $\gamma \in \mathcal{F}$ separates $X_\epsilon$ into two components. We shall refer to the component adjacent to $\{p : |z(p) - z(p_0)| = \epsilon\}$ as the inside and to the other component as the outside. Form the function $v_\gamma$ equal 0 on $\{p : |z(p) - z(p_0)| = \epsilon\}$ and equal to 1 on $\gamma$ and outside of $\gamma$ and which is harmonic inside $\gamma$. Then let $A_\gamma = \{p : v(p) < 1\}$ and

$$\tilde{A}_{k+1} = \{p : \text{there exists } \gamma \in \mathcal{F} \text{ such that } p \in A_\gamma\}.$$

Inductively we construct the new sequence $\tilde{A}_n$ of annuli with property 2). Since $A_n \subset \tilde{A}_n$, the sequence also has property 1).

For any point $p \in X_\epsilon$ let $\beta$ be an arc in $X_\epsilon$ that joins $\gamma_\epsilon = \{p : |z(p)| = \epsilon\}$ to $p$ and let $T$ be a tongue in $X_\epsilon$ with spine $\beta$ and base contained in $\gamma_\epsilon$. Also let $|dv_T|$ be the tongue described above. If a point $p$ is not in $A_n$ then any core curve in $A_n$ must cross the spine of $T$ and so from Lemmas 4 and 5

$$4/Dir(|dv_T|) \leq Dir(|du_{A_n}),$$

where $u_{A_n}$ is a function equal to zero on one side of $A_n$ and equal to 1 on the other side. Then by Theorem 1 it is impossible for $M(A_n)$ to approach 0.

This provides a countable open covering of $C$ of points for which $M(A_n)$ cannot approach 0 and implies property 3). □

The next theorem is the uniformization theorem and its corollary is the Riemann mapping theorem.

THEOREM 2. **(Uniformization)** *Any simply connected Riemann surface $X$ is conformal either to the Riemann sphere $\overline{\mathbb{C}}$ or to the complex plane $\mathbb{C}$ or to the unit disc $\mathbb{D} = \{z : |z| < 1\}$.*

PROOF. Take a point $p_0 \in X$ and a local parameter z defined and a value of $\epsilon$ small enough so that $D_{\epsilon_0}(p_0) = \{p : |z(p) - z(p_0)| \leq \epsilon_0\} \subset X$. Then consider the doubly connected Riemann surface $X_\epsilon = X - D_{\epsilon(p_0)}$ where $0 < \epsilon < \epsilon_0$. Put $\Gamma_\epsilon$ equal to the family of closed curves in $X_\epsilon$ homotopic to the boundary of $D_\epsilon(p_0)$ and divide the proof into two cases,

Case 1: $M(\Gamma_\epsilon) = 0$, and

Case 2: $M(\Gamma_\epsilon) > 0$.

In Case 1 we construct the squence of annuli $A_n$ with the properties of Lemma 10 and consider the subset $\bigcup_n A_n$ of $X_\epsilon$. It can omit at most one point of $X_\epsilon$ because otherwise the extremal length of the family of the family $\Gamma_\epsilon$ would be positive. If it has exactly one element then $X_\epsilon$ is a conformal to the exterior of $D_\epsilon$ in $\mathbb{C}$ by a mapping $z_\epsilon$. The limit as $\epsilon \to 0$ of these univalent mappings converges uniformly on compact subsets to a conformal map from $X$ onto $\mathbb{C}$. If it has no elements then $X$ is conformal to $\overline{\mathbb{C}}$.

In Case 2 by Theorem 1 there is an annulus $A_\epsilon$ with $M(A_\epsilon) > 0$ with the property that $D_{\epsilon_0}(p_0) \cup A_\epsilon$ fills $X$. Letting $\epsilon \to 0$, one can obtain a subsequence of these uniformizing parameters for these $A_\epsilon$ that converges to a conformal mapping from $X$ onto a disc of finite radius. □

COROLLARY 1. **(The Riemann mapping theorem)** *Any simply connected domain $X$ contained in $\overline{\mathbb{C}}$ is conformal either to the Riemann sphere $\overline{\mathbb{C}}$ or to the complex plane $\mathbb{C}$ or to the unit disc $\mathbb{D} = \{z : |z| < 1\}$.*

PROOF. The conclusion is exactly the same as the conclusion of Theorem 2 but the hypothesis is weaker. □

In many presentations the proof of this corollary starts with an arbitrary simply connected domain in $X \subset \overline{\mathbb{C}}$ omitting two or more points. This assumption implies it is conformal to a domain in $\mathbb{C}$ omitting one or more points. The further assumption that it is simply connected implies that one can apply a branch of a square root to obtain a conformal image of $X$ that lies in a bounded domain in $\mathbb{C}$. After this step the extremal length method of Theorem 1 yields the Riemann mapping theorem immediately and one does not need the intersection inequality. The square root trick is not available for Theorem 2 because one does not know in advance that the simply connected domain is realizable conformally as a subset of $\mathbb{C}$.

## 9. Weyl's lemma

The following result is called Weyl's lemma,

LEMMA 11. *Suppose $q$ is a complex valued function defined on a plane domain $D$ and $q$ is in $L^1(D)$. Suppose further that for every $C^1$ function $h$ with compact support in $D$*

$$\int\int h_{\overline{z}} q(z) dxdy = 0. \tag{21}$$

*Then there is an analytic function $\tilde{q}$ defined on $D$ such that*

$$q(z) = \tilde{q}(z)$$

*for almost all $z \in D$.*

PROOF. In the $z = x + iy$-chart, let $m$ be the Lebesgue measure, so that

$$m(S) = \int\int_S |dx \wedge dy|,$$

for every measurable set $S$. Put $J_\epsilon(z) = (1/2\pi\epsilon^2)$ in $\{|z| < \epsilon\}$ and equal to 0 elsewhere. $J_\epsilon$ is called an approximate identity because for every continuous function $f$ defined on a domain in $D$, the convolution

$$(J_\epsilon * f)(z) = \int\int J_\epsilon(w - z) f(w) dm(w)$$

approaches $f(z)$ as $\epsilon \to 0$.

Note that $J_\epsilon(z)$ is not continuous, but the convolution $j_\epsilon = J_\epsilon * J_\epsilon$ is and for any function $q$ in $L^1(D)$, $j_\epsilon * q$ has continuous first partial derivatives. Thus,

$$(q * j_\epsilon)_{\overline{z}}(z) = \int\int_D q(w)(j_\epsilon(w - z))_{\overline{z}} dm(w) = \int\int_D q(w)(j_\epsilon(w - z))_{\overline{w}} dm(w).$$

Substituting $h(w) = j_\epsilon(w - z)$ into the hypothesis (21) we see that $(j_\epsilon * q)_{\overline{z}}(z) = 0$ and so $j_\epsilon * q$ is holomorphic.

Note that $||q * j_\epsilon||_1 \leq ||q||_1$ and $||q||_1 = Dir(|du|)$, which is bounded. By normal families and the next lemma we know that there is a normal limit $\tilde{q}$ holomorphic in

$D$ such that $q * j_\epsilon \to \tilde{q}$ uniformly on compact subsets of $D$ and also $q * j_\epsilon$ converges to $q$ in $L^1$. This implies that $q = \tilde{q}$ almost everywhere. □

LEMMA 12. *Suppose $u_n$ is a sequence of harmonic functions defined in a bounded plane domain $U$ and*

$$\int\int_U |u_n| dxdy \leq B.$$

*Then there is a uniform convergent subsequence and the limit of any such convergent subsequence is harmonic.*

PROOF. Because of the mean value property for any disc $D_{r_0}(p_0)$ of radius $r_0$ centered at $p_0$ and contained in $U$,

$$|u_n(p_0)| \leq \frac{1}{\pi r_0^2} \int\int_D |u_n| dxdy \leq B/(\pi r_0^2). \tag{22}$$

Since the circumference of any disc with center $p$ and radius $r$ contained in $U$ can be covered by a finite number of discs contained in $U$, there is a constant $M$ such that $|u_n(z)| \leq M$ for all $z$ with $|z - p| = r$. By (22) the sequence $u_n$ is equicontinuous on this circle and must have a subsequence that converges uniformly. Since each of the functions $u_n$ satisfies the mean value property,

$$u_n(p) = \frac{1}{2\pi} \int_0^{2\pi} u(p + re^{i\theta}) d\theta,$$

the uniform limit $u$ also satisfies this property and is therefore harmonic. □

## 10. The long ray, the long line and the long cylinder

An immediate consequence of uniformization is Rado's theorem which says that any Riemann surface $X$ is separable, that is, it has countable basis for its topology. The proof is simple: $X$ must be conformal to either the Riemann sphere or the complex plane or the unit disc or a quotient space of one of the two latter mentioned surfaces by the action of a fixed point free, discontinuous group. Consequently, in the middle of any proof of uniformization there should appear a reason why any attempt to put a complex structure on any nonseparable surface will necessarily fail. In this section we construct a *long cylinder* which is a connected surface with differentiable structure and show how our proof of uniformization eliminates the possibility that it could have a subordinate complex structure and at the same time be "*long*."

Consider an uncountable well-ordered set $\omega_1$ with the order topology and the half closed unit interval $[0, 1)$ and form the *long ray*

$$L = \omega_1 \times [0, 1).$$

The *long line* is obtained by patching together two copies of the long ray but since we are concerned only with what at first appears to be a paradox concerning Rado's theorem we only need the long ray.

From the long ray we can construct the *long cyclinder* by putting

$$X = L \times \mathbb{S}^1.$$

Points of $X$ have coordinates $(\beta, x, \theta)$ where $\beta \in \omega_1, x \in [0, 1)$ and $\theta \in \mathbb{S}^1$. The standard Euclidean structure on the product of the unit circle $\mathbb{S}^1$ with the open unit interval $(0,1)$ induces a complex structure at all points of $X$ except those with coordinates $(\beta, 0, \theta)$ in $X$.

We may attempt to extend this to a complex structure on all of the long cylinder $X$, that is, including those points with coordinates $(\beta, 0, \theta)$ in the following way. Any element $\beta$ of the well-ordered set $\omega_1$ has an immediate successor, namely, the smallest element of the set of elements of $\omega_1$ stricty larger than $\beta$. We denote this element by $\beta+1$. Now we join together the two half-open annuli $\beta \times [0,1) \times \mathbb{S}^1$ and $(\beta+1) \times [0,1) \times \mathbb{S}^1$ by putting the first cylinder on the left and the second on the right and by identifying the point $(\beta, 1, \theta)$ on the left with the point $(\beta+1, 0, \theta)$ on the right.

If $\beta_0$ is the smallest element of $\omega_1$ we can delete from $X$ its initial circular boundary, namely, the copy of the unit circle which consists of the points with coordinates $(\beta_0, 0, \theta)$ to obtain a surface $Y$. At the interior points of the cyclindrical components

$$\beta \times [0,1) \times \mathbb{S}^1$$

of $Y$ there is the standard complex structure. Moreover, at the points $(\beta, 1) \times \mathbb{S}^1$ attached to the points $(\beta+1, 0) \times \mathbb{S}^1$ there is also a standard complex structure.

Since there must be some points $\alpha \in \omega_1$ for which there is no $\beta$ with the property that $\beta + 1 = \alpha$, the definitions given so far do not introduce a complex coordinate structure at every point of $Y$. In fact, because of Rado's theorem there cannot be an extension of this complex structure or in fact any complex structure to all of $Y$.

DEFINITION 17. We say that two points $P$ and $Q$ in $X$ are joined by a chain of parametric discs $D_1, \ldots, D_n$ if $P \in D_1$ and $Q \in D_n$ and each successive intersection $D_j \cap D_{j+1}$, $1 \leq j \leq n-1$ is non-empty.

PROPOSITION 1. *No matter what complex structure is introduced on the component cylinders of the long ray, there are points $\alpha$ and $\beta$ in $\omega_1$ with $\alpha < \beta$ such the cylinder joining $\alpha \times \mathbb{S}^1 \times [0,1]$ and $\beta \times \mathbb{S}^1 \times [0,1]$ in $Y$ cannot have finite modulus.*

PROOF. Let $\mathcal{F} = \mathcal{F}(\alpha, \beta)$ be the family of elements of the long ray that lie between $\alpha$ and $\beta$. The union component cylinders starting at $\alpha$ and coming before $\beta$ form a long component. For each $\gamma$ in $\mathcal{F}$ let $M(\gamma)$ be the modulus of the component cylinder which has coordinate $\gamma$ in this component must have positive modulus. From the comparison inequality

$$\frac{1}{M(A)} \geq \sum_{\gamma \in \mathcal{F}} \frac{1}{M(\gamma)},$$

this implies $M(A) = 0$ where $A$ is the cylinder starting at $\alpha$, and so it cannot be attached to the component cylinder corresponding to $\beta$. □

However, we see that in this case it is possible to introduce a nodal Riemann surface structure on $Y$. This of course is not a contradiction to Rado's theorem. A nodal complex structure on $Y$ may have uncountably many nodes.

## References


[1] W. Abikoff, *The real analytic theory of Teichmüller space*, Lecture Notes in Mathematics, vol. 820, Springer, Berlin, 1980. MR590044

[2] L. V. Ahlfors, *Conformal invariants: topics in geometric function theory*, McGraw-Hill Book Co., New York-Düsseldorf-Johannesburg, 1973. McGraw-Hill Series in Higher Mathematics. MR0357743

[3] L. V. Ahlfors, *Lectures on quasiconformal mappings*, 2nd ed., University Lecture Series, vol. 38, American Mathematical Society, Providence, RI, 2006. With supplemental chapters by C. J. Earle, I. Kra, M. Shishikura and J. H. Hubbard. MR2241787

[4] L. V. Ahlfors and L. Sario, *Riemann surfaces*, Princeton Mathematical Series, No. 26, Princeton University Press, Princeton, N.J., 1960. MR0114911

[5] L. Bers, *Riemann Surfaces, Lecture Notes.* Inst. Math. Sci. Lect. Notes, NYU, New York, 1957-58.

[6] L. Bers, *On Hilbert's 22nd problem*, Mathematical developments arising from Hilbert problems (Proc. Sympos. Pure. Math., Northern Illinois Univ., DeKalb, Ill., 1974), Amer. Math. Soc., Providence, R. I., 1976, pp. 559–609. Proc. Sympos. Pure Math., Vol. XXVIII. MR0427623

[7] H. P. de Saint-Gervais, *Uniformisation des surfaces de Riemann* (French), ENS Éditions, Lyon, 2010. Retour sur un théorème centenaire. [A look back at a 100-year-old theorem]; The name of Henri Paul de Saint-Gervais covers a group composed of fifteen mathematicians : Aurélien Alvarez, Christophe Bavard, François Béguin, Nicolas Bergeron, Maxime Bourrigan, Bertrand Deroin, Sorin Dumitrescu, Charles Frances, Étienne Ghys, Antonin Guilloux, Frank Loray, Patrick Popescu-Pampu, Pierre Py, Bruno Sévennec, and Jean-Claude Sikorav. MR2768303

[8] S. Donaldson, *Riemann surfaces*, Oxford Graduate Texts in Mathematics, vol. 22, Oxford University Press, Oxford, 2011. MR2856237

[9] H. M. Farkas and I. Kra, *Riemann surfaces*, Graduate Texts in Mathematics, vol. 71, Springer-Verlag, New York-Berlin, 1980. MR583745

[10] A. Fathi, F. Laudenbach, and V. Poénaru, *Travaux de Thurston sur les Surfaces*, volume 66-67 of *Asterisque*. Soc. Math. de France, Seminaire Orsay, 1979.

[11] F. P. Gardiner, *Measured foliations and the minimal norm property for quadratic differentials*, Acta Math. **152** (1984), no. 1-2, 57–76, DOI 10.1007/BF02392191. MR736212

[12] F. P. Gardiner, *Teichmüller theory and quadratic differentials*, Pure and Applied Mathematics (New York), John Wiley & Sons, Inc., New York, 1987. A Wiley-Interscience Publication. MR903027

[13] F. P. Gardiner and H. Masur, *Extremal length geometry of Teichmüller space*, Complex Variables Theory Appl. **16** (1991), no. 2-3, 209–237. MR1099913

[14] H. Grötszch, Über einige Extremalprobleme der konformen Abbildung I, II. *Acad. Wiss.*, 80:367–376, 497–502, 1918.

[15] Y. Imayoshi and M. Taniguchi, *An introduction to Teichmüller spaces*, Springer-Verlag, Tokyo, 1992. Translated and revised from the Japanese by the authors. MR1215481

[16] P. Koebe, *Über die Uniformisierung der algebraischen Kurven. I* (German), Math. Ann. **67** (1909), no. 2, 145–224, DOI 10.1007/BF01450180. MR1511526

[17] H. Kwan, Uniformization of Riemann Surfaces. *Ph.D. Thesis, Harvard University*, pages 1–29, 2004.

[18] A. Marden and K. Strebel, *The heights theorem for quadratic differentials on Riemann surfaces*, Acta Math. **153** (1984), no. 3-4, 153–211, DOI 10.1007/BF02392377. MR766263

[19] C. T. McMullen. Notes. *Harvard University*, pages 1–17, 2005.

[20] C. D. Minda, *The Hahn metric on Riemann surfaces*, Kodai Math. J. **6** (1983), no. 1, 57–69, DOI 10.2996/kmj/1138036663. MR698327

[21] Y. N. Minsky, *Harmonic maps, length, and energy in Teichmüller space*, J. Differential Geom. **35** (1992), no. 1, 151–217. MR1152229

[22] Ch. Pommerenke, *Boundary behaviour of conformal maps*, Grundlehren der Mathematischen Wissenschaften [Fundamental Principles of Mathematical Sciences], vol. 299, Springer-Verlag, Berlin, 1992. MR1217706

[23] T. Rado, Über den Begriff der Riemannschen Fläche. *Acta Szeged*, 2:101–121, 1925.

[24] M. Schiffer and D. C. Spencer, *Functionals of finite Riemann surfaces*, Princeton University Press, Princeton, N. J., 1954. MR0065652

[25] G. Springer, *Introduction to Riemann surfaces*, Addison-Wesley Publishing Company, Inc., Reading, Mass., 1957. MR0092855

[26] K. Strebel, *Quadratic differentials*, Ergebnisse der Mathematik und ihrer Grenzgebiete (3) [Results in Mathematics and Related Areas (3)], vol. 5, Springer-Verlag, Berlin, 1984. MR743423

[27] W. Thurston. *The Geometry and Topology of Three Manifolds*, volume I. Princeton Lecture notes, 1979.

[28] M. Tsuji, *Potential theory in modern function theory*, Maruzen Co., Ltd., Tokyo, 1959. MR0114894

[29] H. Weyl, *Die Idee der Riemannschen Fläche* (German), B. G. Teubner Verlagsgesellschaft, Stuttgart, 1955. 3te Aufl. MR0069903

Department of Pure Mathematics, Graduate School and University Center of CUNY, 365 Fifth Avenue, New York, New York 10016

*E-mail address*: frederick.gardiner@gmail.com

Contemporary Mathematics
Volume **696**, 2017
http://dx.doi.org/10.1090/conm/696/14019

# Winding and unwinding and essential intersections in $\mathbb{H}^3$

Jane Gilman and Linda Keen

ABSTRACT. Let $\mathbb{G} = \langle \mathbb{A}, \mathbb{B} \rangle$ be a non-elementary two generator subgroup of the isometry group of $\mathbb{H}^2$. If $G$ is discrete, free and generated by two hyperbolic elements with disjoint axes, its quotient is a pair of pants and in **[3]** we produced a recursive formula for the number of essential self intersections (ESIs) of any primitive geodesic on the quotient. An alternative non-recursive proof of the formula was given in **[10]**. We used the formula to give a geometric description of how the primitive geodesic winds around the "cuffs" of the pants.

On a pair of pants, an ESI is a point of a "seam" where a primitive geodesic has a self-intersection. Here we generalize the definition of ESI's to non-elementary purely loxodromic two generator groups $G \subset \text{Isom}(\mathbb{H}^3)$ which are discrete and free. We also associate an open geodesic to each generalized ESI we call a *Connector*. In the quotient manifold the connectors correspond to "opened up self-intersections". We then define a class of groups we call *Winding groups*; the connectors for these groups give us a way generalize the concept of winding of primitive geodesics in the quotient manifold to this context. For winding groups, the ESI's and connectors satisfy the same formula as they do in the two-dimensional case. Our techniques involve using the *stopping generators* we defined in **[3]** for the group $\mathbb{G}$ to make it a model. We then use the model to obtain an algebraic rather than geometric definition for ESI's and generalize from the algebraic definition.

## 1. Introduction

In **[3]** we considered some geometric implications of the Gilman-Maskit discreteness algorithm, **[6]**, for non-elementary two generator subgroups of $\text{Isom}(\mathbb{H}^2)$ whose generators were hyperbolic with disjoint axes and whose product was also hyperbolic. We showed that when the algorithm determines that the group is discrete and free, the quotient is a three holed sphere and the algorithm has determined discreteness by finding a special pair of *stopping generators* that correspond to the only three simple geodesics on the quotient surface, those around the waist and the cuffs. Every other geodesic on the quotient winds around the waist and cuffs and has self-intersections. We concentrate our attention on the *primitive geodesics* on

2010 *Mathematics Subject Classification.* Primary 30F10, 30F35, 30F40, 30F45; Secondary 14H30, 11F99, 22E40.

*Key words and phrases.* Fuchsian groups, Schottky groups, discrete groups, hyperbolic geometry, intersections, geodesics.

Some of this work was carried out while the first author was a supported visitor at ICERM. The second author was supported by a CUNY Collaborative grant and grants from PSC-CUNY.

the quotient. A primitive element of the fundamental group of the surface is one that together with one other element generates the fundamental group. A *primitive curve* is the geodesic in the free homotopy class of such a primitive element. A pair of such geodesics coming from a pair of elements that generate the fundamental group is a *primitive pair*. These conjugacy classes of primitive geodesics can be identified with the set of rational numbers, [**5**, **7**].

In discreteness considerations, it is standard to consider a group generated by three involutions in which the two generator group sits as a subgroup of index two. This can be done for two generator groups both in $\mathrm{Isom}(\mathbb{H}^2)$ and $\mathrm{Isom}(\mathbb{H}^3)$. For the subgroups of $\mathbb{H}^2$ under consideration here, the fixed points of these involutions project to geodesics on the surface we call *seams*. Some of the self-intersections of the primitive geodesics necessarily lie on these seams. In [**3**] we called these *Essential Self-Intersections* to differentiate them from other self-intersections. We produced a recursive formula to count the essential self intersections of the primitive geodesics. We showed that in the quotient, one could view the primitive geodesics as curves that *wind around* the simple geodesics and that the winding reflects the rational identified with the geodesic. A non-recursive version of the formula for essential self-intersections was given in [**9**] along with a proof of the recursive version.

In this paper we generalize the concept of Essential Self-Intersections (ESI's) of primitive geodesics to non-elementary two generator free discrete subgroups of $\mathrm{Isom}(\mathbb{H}^3)$. The hyperbolic quotient manifold is now a genus two handlebody and the group elements correspond to non-trivial free homotopy classes of curves in the manifold. Each class has a unique geodesic and to simplify the exposition, we do not distinguish between the group element and the geodesic in the free homotopy class it represents. Also, for the sake of brevity and clarity of the exposition we restrict our discussion to purely loxodromic groups. The results can be extended to groups containing parabolic elements as our two dimensional treatment did.

We note that in [**9**] the original definition of ESI's and the recursive formulas were extended to discrete groups that included elliptic elements and it is likely that these results can also be extended to the three dimensional case.

Generically, the projections of the axes of primitive elements of the group are simple closed curves in the quotient manifold so the there are no self-intersections. Therefore, in order to generalize the ESI's we need to understand what happens to the intersection points when the subgroup of $\mathrm{Isom}(\mathbb{H}^2)$ is deformed into a subgroup of $\mathrm{Isom}(\mathbb{H}^3)$; that is, to understand how the geodesic "opens up" under the deformation. To this end we associate open geodesic arcs we call *connectors* to the generalized ESI's we define for these more general groups. The connectors are perpendicular to a primitive geodesic in the quotient manifold and represent the opened up intersection point.

In [**2**] we showed that the involutions of the degree two extension of the group induce a *hyperelliptic* involution of the quotient manifold that generalizes the hyperelliptic involution of its boundary. The fixed set of this involution consists of three bi-infinite open geodesic arcs and is a generalization of the seams. We prove

THEOREM (Theorem A). *The connectors associated to the generalized ESI's are arcs joining points on the seams and they are invariant under the generalized hyperelliptic involution; they satisfy the same two formulas as ESI's satisfy in the two dimensional case.*

We then define a certain class of these non-elementary rank two discrete free discrete subgroups of $\mathrm{Isom}(\mathbb{H}^3)$ that we call *Winding groups.* For these groups we can define an analogue of the stopping generators and a geometric analogue of winding. For this class of groups we prove

THEOREM (Theorem B). *For every primitive geodesic in the quotient manifold, there is an analogue of the winding we found in the two dimensional case. This winding reflects the rational number identified with the geodesic.*

The paper is organized as follows. In Section 2 we give some background, define the terms we use and the enumeration scheme. In Section 3 we discuss the model group $\mathbb{G} = \langle \mathbb{A}, \mathbb{B} \rangle \subset \mathbb{H}^2$ and review Essential Self Intersections for this fixed group as an example of a non-elementary discrete free rank two purely hyperbolic group in $\mathrm{Isom}(\mathbb{H}^2)$ whose generators have disjoint axes. In Section 4 we use the natural isomorphism from the model group to the discrete free purely loxodromic subgroup $G$ of $\mathrm{Isom}(\mathbb{H}^3)$ to label elements of $G$ and its degree two extension $\tilde{G}$. We then use the labelling to define the generalized ESI's and the connectors and we prove Theorem A for non-elementary discrete, free, purely loxodromic groups. In Section 5 we define a class of groups for which we prove Theorem B. We call these *winding groups*; they have the properties that the quotient manifold has a rational pleating locus (defined in that section) and any pair of pleating curves are primitive geodesics that can be represented by a pair of generators of the group. We expect that there are other classes of groups for which Theorem B holds but we leave that discussion for the future.

## 2. Background and Terminology

We use the following terminology. We consider $\mathbb{H}^2 \subset \mathbb{H}^3$ so all of the following makes sense for both the two and three dimensional settings. If $M$ is a geodesic in $\mathbb{H}^2$ or $\mathbb{H}^3$ we let $H_M$ denote the half-turn about $M$. If $M$ is in $\mathbb{H}^2$, the half-turn $H_M$ is a reflection across $M$ and so is an orientation reversing isometry. If $M$ is in $\mathbb{H}^3$, $H_M$ is an involution whose full set of fixed points is the geodesic $M$; it is an orientation preserving isometry. Note that there are non-orientation preserving isometries of $\mathbb{H}^3$ of order two whose fixed point set is a hyperbolic plane in $\mathbb{H}^3$; these are not half-turns in our terminology.

If $X$ is an orientation preserving isometry we let $Ax_X$ denote the invariant geodesic joining its fixed points; these may be on the boundary of $\mathbb{H}^3$. This geodesic is called its axis. If $X$ has only one fixed point it is parabolic and the axis is the fixed point. When we restrict to $\mathbb{H}^2$, the axis of an elliptic is a point in $\mathbb{H}^2$. For brevity and clarity of exposition, in this paper, we will always assume that the isometries we encounter are not parabolic and those that are elliptic have order two. In $\mathbb{H}^3$ these are half-turns and their axes are fixed pointwise.

If $M$ is any geodesic orthogonal to $Ax_X$, there is another geodesic $M'$ orthogonal to $Ax_X$ such that $X$ is the product of the two half-turns, $X = H_{L_M} H_{L_{M'}}$. Note that any two geodesics in $\mathbb{H}^3$ have a unique common perpendicular. Any element of $\mathrm{Isom}(\mathbb{H}^3)$ can be factored in an infinite number of ways as a the product of half-turns about a pair of geodesics perpendicular to its axis; if $X$ is loxodromic, the hyperbolic distance between the pair along the axis is half the translation length while if $X$ is elliptic they intersect the axis at the same point and if $X$ is parabolic they intersect at a point on the boundary of the hyperbolic space.

We are interested in discrete free groups $G$ of rank two; we always assume the groups are non-elementary and, for the sake of brevity, give no further mention of that assumption. We will be concerned with *primitive elements of $G$*. For a free group of rank $n$, an element is *primitive* if it is part of a minimal generating set [**8**]. We call a geodesic *primitive* if it is in the free homotopy class of the fundamental group of the quotient represented by a primitive element. In this paper, a minimal generating set is a *primitive pair*.

If $G$ is generated by $A$ and $B$, the axes of $A$ and $B$ have a common perpendicular $L$ and there are a unique pair of geodesics $L_A$ and $L_B$ such that $A = H_L \circ H_{L_A}$ and $B = H_L \circ H_{L_B}$.

We denote by $\tilde{G}$ the group generated by the three half-turns

$$\tilde{G} = \langle H_L, H_{L_A}, H_{L_B} \rangle.$$

It is clear that $G \subset \tilde{G}$ and that the index of $G$ in $\tilde{G}$ is 2. It is a classical result that a group and a normal subgroup of finite index are simultaneously discrete and or non-discrete.

We have $A^{-1}B = H_{L_A} \circ H_{L_B}$ and its axis is the common perpendicular to $L_A$ and $L_B$; also $AB = H_L \circ H_{L_A} \circ H_L \circ H_{L_B}$ and its axis is the common perpendicular to $H_L \circ H_{L_A} \circ H_L$ and $H_{L_B}$. The cyclically ordered set of lines

$$Ax_A, L_A, Ax_{A^{-1}B}, L_B, Ax_B, L$$

consists of six geodesics with the property that each is orthogonal to both of its neighbors. They form a *right-angled hexagon* as defined by Fenchel in [**1**]. We denote this hexagon by $\mathcal{H}(A, B)$; it has three *axis sides* and three *half-turn sides.*

Note that in our discussion of subgroups of $\text{Isom}(\mathbb{H}^2)$ below, we always assume the axes of the given generators of our group are disjoint. The case where the axes intersect is subsumed in the general discussion of subgroups of $\text{Isom}(\mathbb{H}^3)$ above. Again for the sake of brevity, we leave a more detailed discussion of results about essential intersections for groups with intersecting axes for another time.

If $G$ acts invariantly on $\mathbb{H}^2$ and is discrete and free, and if the axes of the generators are disjoint, the quotient $S = \mathbb{H}^2/G$ is a three-holed sphere or pair of pants. A primitive curve on $S$ is the image of the axis of a primitive element in the group. The quotient $S$ is a two-fold covering of the quotient $\tilde{S} = \mathbb{H}^2/\tilde{G}$ which is an orbifold; we can visualize it as identifying the front and back sides of a pair of pants. The orbifold points are the projections of the half-turn lines; we call these *the seams* of the pants. The involution that identifies the front and back is induced by any one of the half-turns in $\tilde{G}$.

**2.1. Enumeration of Primitive Elements.** As we indicated in the introduction there are various ways to enumerate the conjugacy classes of primitive words. Each is an inductive scheme that associates a representative of the conjugacy class of primitive words to a rational number $p/q$. The scheme lists a word but not its inverse. The generators $A, B$ are assigned to $0/1$ and $1/0$ respectively and the representative word is built up from a sequence of steps using the Farey tessellation of the upper half plane. In [**5**] we denoted the representative words by $E_{p/q}$ and called the enumeration scheme produced there the *palindromic enumeration scheme.* In that scheme, if $p+q$ is odd ($pq$ is even), then the word $E_{p/q}$ is a palindrome in $A$ and $B$ while if $p+q$ is even ($pq$ is odd), the word $E_{p/q}$ is a product of the two words in the previous two steps of the scheme; these last two words are palindromes and the product is chosen so that the next word in the scheme is again a palindrome. This scheme reflects the $H_L$ symmetry of the group because the axes of the palindromic words in this scheme are all perpendicular to the half-turn line $L$.

We will use the palindromic scheme here and assume $p/q > 0$. Suppose first that $E_{p/q}$ is a palindrome. Then $p + q = 2n + 1$ and $E_{p/q}$ has the form $WU\overline{W}$ where $U \in \{A, B\}$ and $W = X_1X_2 \dots X_n$ is a word of length $n = (p + q - 1)/2$, $X_j \in \{A, B\}$ and $\overline{W} = X_n \dots X_1$. If $E_{p/q}$ is not a palindrome so that $p+q = 2n$ is even, we write $E_{p/q} = X_1 \dots X_{2n}$. There are a unique pair of rationals $l/m, r/s$ with $|ls - mr| = 1$ such that $\frac{p}{q} = \frac{l}{m} + \frac{r}{s}$. We set $E_{p/q} = E_{l/m}E_{r/s}$ or $E_{p/q} = E_{r/s}E_{l/m}$ where the choice is made so that the next words in the scheme, $E_{\frac{p+l}{q+m}}$ and $E_{\frac{p+r}{q+s}}$ are palindromes. We omit the details here. The interested reader can find them in [**3**, **5**].

The forms that the words take can be made more explicit, but we will not need them here. The only property of these words that we will need is

REMARK 2.1. In the identification of a rational and a primitive word, each time a generator appears in a given word, its exponent has the same sign. That is, in the palindromic enumeration scheme, if $p/q > 0$ all of the exponents of $A$ and $B$ are positive whereas if $p/q < 0$ the exponents of $A$ are all negative and the exponents of $B$ are all positive. Since $B^{-1}$ never appears, no word is an inverse of another.

## 3. $\mathbb{H}^2$: The Model Group

In this section we assume that we are given a fixed group $\mathbb{G} = \langle \mathbb{A}, \mathbb{B} \rangle \subset \mathrm{Isom}(\mathbb{H}^2)$ where the axes of $\mathbb{A}$ and $\mathbb{AB}$ are disjoint and that $\mathbb{A}, \mathbb{B}$ are stopping generators for the Gilman-Maskit algorithm. By results of [**3**], this means that their axes and the axis of $\mathbb{A}^{-1}\mathbb{B}$ project to simple curves on the quotient surface $S$ which is a three-holed sphere. All other axes of elements of $\mathbb{G}$ project to curves with self-intersections. We also consider the degree two extension $\tilde{\mathbb{G}} = \langle H_{\mathbb{L}}, H_{\mathbb{L}_{\mathbb{A}}}, H_{\mathbb{L}_{\mathbb{B}}} \rangle$ of $\mathbb{G}$.

The condition that $\mathbb{A}, \mathbb{B}$ are stopping generators also means that the hexagon $\mathcal{H}(\mathbb{A}, \mathbb{B})$ is convex in $\mathbb{H}^2$ and moreover, that

$$\mathcal{F} = \mathcal{H}(\mathbb{A}, \mathbb{B}) \cup H_{\mathbb{L}}(\mathcal{H}(\mathbb{A}, \mathbb{B}))$$

is a fundamental domain for $\mathbb{G}$. It has the property that it is symmetric about the line $\mathbb{L}$ where the symmetry is the half-turn $H_{\mathbb{L}}$. Note that we use an overline to denote the image under $H_{\mathbb{L}}$ so that $\overline{\mathcal{H}(\mathbb{A}, \mathbb{B})} = H_{\mathbb{L}}(\mathcal{H}(\mathbb{A}, \mathbb{B}))$ is also a convex hexagon with three axis sides, the axes of $\mathbb{A}$, $\mathbb{B}$ and $\mathbb{BA}^{-1}$, and three half-turn sides $\mathbb{L}$, $\mathbb{L}_{\bar{\mathbb{A}}} = H_{\mathbb{L}}(\mathbb{L}_{\mathbb{A}})$ and $\mathbb{L}_{\bar{\mathbb{B}}} = H_{\mathbb{L}}(\mathbb{L}_{\mathbb{B}})$. The hexagons $\mathcal{H}(\mathbb{A}, \mathbb{B})$ and $\overline{\mathcal{H}(\mathbb{A}, \mathbb{B})}$ are each fundamental domains for $\tilde{G}$.

In this model group the axes of the primitive elements $\mathbb{E}p/q$ have geometric properties we want to look at. We assume throughout that $p/q > 0$. To obtain the primitive words in the group in which $\mathbb{A}$ appears with negative exponent and $-1 < p/q < 0$, we need to replace $\mathcal{F}$ by the fundamental domain for $\mathbb{G}$ given by

$$\mathcal{F}_{\mathbb{B}} = \mathcal{H}(\mathbb{A}, \mathbb{B}) \cup H_{\mathbb{L}_{\mathbb{B}}}(\mathcal{H}(\mathbb{A}, \mathbb{B})$$

and in the rest of the discussion, to replace $\mathbb{L}$ by $\mathbb{L}_{\mathbb{B}}$. For $p/q < -1$, we interchange the roles of $\mathbb{A}$ and $\mathbb{B}$; that is we replace $\mathcal{F}$ by the fundamental domain for $\mathbb{G}$ given by

$$\mathcal{F}_{\mathbb{A}} = \mathcal{H}(\mathbb{A}, \mathbb{B}) \cup H_{\mathbb{L}_{\mathbb{A}}}(\mathcal{H}(\mathbb{A}, \mathbb{B})$$

and in the rest of the discussion replace $\mathbb{L}$ by $\mathbb{L}_{\mathbb{A}}$. We will not carry this out but leave it to the reader.

Some of the propositions below and their proofs appear in one or more of [**5**, **7**, **9**], but because of their importance to the new definitions in $\mathbb{H}^3$, we state the results using the current notation and give detailed proofs.

PROPOSITION 3.1. *The axis of* $\mathbb{E}_{p/q}$ *intersects* $\partial\mathcal{F}$ *in half-turn sides.*

PROOF. One of the intervals on the circle containing the limit set of $\mathbb{G}$ bounded by the fixed points of $\mathbb{A}$ is an interval of discontinuity. Similarly, one of the intervals bounded by the fixed points of $\mathbb{B}$ is, as is one of the intervals bounded by the axes of $\mathbb{A}^{-1}\mathbb{B}$ and $\mathbb{B}\mathbb{A}^{-1}$. Since these axes are mutually disjoint and the intervals of discontinuity are mutually disjoint, they are separated by the axes. The half-turn lines on $\partial\mathcal{F}$ are perpendicular to these axes so each has endpoints inside the intervals of discontinuity. No axis of a group element can enter the half planes bounded by these intervals of discontinuity so it cannot intersect any of the axis sides of $\mathcal{F}$. □

Note that the same argument shows that the axis any element of $\mathbb{G}$ cannot intersect any copy of $\mathcal{F}$ in an axis side.

PROPOSITION 3.2. *The axis of* $\mathbb{E}_{p/q}$ *intersects* $\mathbb{L}$ *inside* $\mathcal{F}$.

PROOF. By Proposition 3.1 the axis of $\mathbb{E}_{p/q}$ enters and leaves $\mathcal{F}$ through half-turn sides. Write $\mathbb{E}_{p/q}$ as a product of half-turns. If, for example, it enters through $\mathbb{L}_{\mathbb{A}}$ and leaves through the sides $\mathbb{L}_{\mathbb{B}}$ the half-turn expression for $\mathbb{E}_{p/q}$ would contain an adjacent pair $H_{\mathbb{L}_{\mathbb{A}}}H_{\mathbb{L}_{\mathbb{B}}}$ and hence the expression in terms of $\mathbb{A}$ and $\mathbb{B}$ would contain $\mathbb{A}^{-1}\mathbb{B}$. This cannot happen if $p/q > 0$ by Remark 2.1. The same contradiction would appear in every instance where the axis did not intersect $\mathbb{L}$. □

Our enumeration scheme reflects the symmetry of the domain $\mathcal{F}$ about $\mathbb{L}$. We want to exploit this to label the set of copies of $\mathcal{F}$ that a fundamental segment of the axis of $\mathbb{E}_{p/q}$ passes through. Suppose first that $p + q = 2n + 1$ and $\mathbb{E}_{p/q}$ is a palindrome so that its axis is symmetric about $\mathbb{L}$. This means that if $Ax_{\mathbb{E}_{p/q}}$ leaves $\mathcal{H}(\mathbb{A}, \mathbb{B})$ through $\mathbb{L}_{\mathbb{A}}$, it leaves $\overline{\mathcal{H}(\mathbb{A}, \mathbb{B})}$ through $\mathbb{L}_{\bar{\mathbb{A}}}$ and similarly for $\mathbb{L}_{\mathbb{B}}$. In addition to passing through $\mathcal{F}$, a symmetric fundamental segment of the axis passes through $n$ copies of $\mathcal{F} = \mathcal{F}_0$ on one side of $\mathbb{L}$ and $n$ copies on the other.

We label the sequence of $n$ copies of $\mathcal{F} = \mathcal{F}_0$, starting with the one adjacent to $\mathcal{H}(\mathbb{A}, \mathbb{B})$ by $\mathcal{F}_j, j = 1, \ldots, n$ and the sequence of $n$ copies of $\mathcal{F} = \mathcal{F}_0$, starting with the one adjacent to $\overline{\mathcal{H}(\mathbb{A}, \mathbb{B})}$ by $\mathcal{F}_{-j}, j = 1, \ldots, n,$. The axis of $\mathbb{E}_{p/q}$ leaves $\mathcal{F}_0$ through the half-turn sides $L_1^{p/q}$ and $L_{-1}^{p/q}$ that as it enters $\mathcal{F}_1$ and $F_{-1}$ respectively. It leaves each $\mathcal{F}_j$, $j = -n, \ldots, n$ through a half-turn side that is the image of one of the four half-turn sides of $\mathcal{F}$ by an element of $\mathbb{G}$. It also intersects a conjugate of $\mathbb{L}$ in each copy of $\mathcal{F}$. We could use the technique of cutting sequences to use group elements to label these sides but the notation is clumsy and we will not need it here (see [**7**, **9**]). Instead, we label the copy of $\mathbb{L}$ in $\mathcal{F}_j$ by $\mathbb{L}_{2j}^{p/q}$ and the boundary side through which it leaves $F_j$ by $L_{2j+1}^{p/q}$. We label the intersection point of $Ax_{\mathbb{E}_{p/q}}$ with each $\mathbb{L}_j^{p/q}$ by $Q_j^{p/q}$, $j = -2n - 1, \ldots, 2n + 1$.

Remember that for each $j$, $j = -2n-1, \dots, 2n+1$ there is a unique word $\mathbb{W}_j^{p/q}$ such that $\mathbb{L}_j^{p/q} = \mathbb{W}_j^{p/q}(\mathbb{Y})$, $\mathbb{Y} \in \{\mathbb{L}, \mathbb{L}_{\mathbb{A}}, \mathbb{L}_{\mathbb{B}}\}$ so we could use these words as our labels. The only information a labelling by group elements instead of indices would give us is that it would be obvious that $\mathbb{E}_{p/q}(\mathbb{L}_{-2n-1}^{p/q}) = \mathbb{L}_{2n+1}^{p/q}$ and $\mathbb{E}_{p/q}(Q_{-2n-1}^{p/q}) = Q_{2n+1}^{p/q}$. We take this for granted here. It then follows that $\mathbb{E}_{p/q} = H_{\mathbb{L}_{-2n-1}^{p/q}} H_{\mathbb{L}}$ so that the axis is perpendicular to $\mathbb{L}_{-2n-1}^{p/q}$ and $\mathbb{L}_{2n+1}^{p/q}$ at $Q_{-2n-1}^{p/q}$ and at $Q_{2n+1}^{p/q}$ respectively.

We also have

PROPOSITION 3.3. *The angles at intersection points $Q_j^{p/q}$, $0 < j < 2n+1$ are not right.*

PROOF. Suppose that for some $0 < |j| \leq 2n$ the angle between $L_j^{p/q}$ and $Ax_{\mathbb{E}_{p/q}}$ were right. Then the product $H_{L_j} H_{\mathbb{L}}$ would be an element sharing fixed points with $\mathbb{E}_{p/q}$. Since the group is discrete, this would imply that there is an integer $m > 1$ such that $(H_{L_j} H_{\mathbb{L}})^m = \mathbb{E}_{p/q}$. This cannot be since $\mathbb{E}_{p/q}$ is primitive. □

In [**5**] we noted that if $P_1$ and $P_2$ were palindromes in the scheme, then the common perpendicular to $P_1P_2$ and $P_2P_1$ would intersect the common orthogonal $\mathbb{L}$ to $P_1$ and $P_2$ and thus determine a point on it. We need an even more detailed analysis here.

That is we now suppose $p + q = 2n$ is even so that $\mathbb{E}_{p/q} = \mathbb{E}_{l/m}\mathbb{E}_{r/s}$ where $\frac{p}{q} = \frac{l}{m} + \frac{r}{s}$ with $|ls - mr| = 1$. In this case a fundamental segment of $Ax_{\mathbb{E}_{p/q}}$ goes through an even number of copies of $\mathcal{F}$.

We showed in [**5**] that $\mathbb{E}_{p/q}$ is a palindrome in $A$ and $AB$ or $B$ and $AB$ so that its axis is perpendicular to conjugates of both of the half-turn sides $\mathbb{L}_{\mathbb{A}}$ and $\mathbb{L}_{\mathbb{B}}$ of $\mathcal{F}$ but it is not perpendicular to $\mathbb{L}$. Now we also consider

$$(\tilde{\mathbb{E}}_{p/q})^{-1} = H_{\mathbb{L}}\mathbb{E}_{p/q}H_{\mathbb{L}} = H_{\mathbb{L}}\mathbb{E}_{l/m}\mathbb{E}_{r/s}H_{\mathbb{L}} = (\mathbb{E}_{l/m})^{-1}(\mathbb{E}_{r/s})^{-1} = (\mathbb{E}_{r/s}\mathbb{E}_{l/m})^{-1}$$

which gives us the other product, $\tilde{\mathbb{E}}_{p/q} = \mathbb{E}_{r/s}\mathbb{E}_{l/m}$. Note that this is a conjugate of $\mathbb{E}_{p/q}$,

$$\mathbb{E}_{l/m}\mathbb{E}_{r/s} = \mathbb{E}_{r/s}^{-1}\mathbb{E}_{r/s}\mathbb{E}_{l/m}\mathbb{E}_{r/s}.$$

In this case we follow the axis of $\mathbb{E}_{l/m}\mathbb{E}_{r/s}$ as it leaves $\mathcal{H}(\mathbb{A}, \mathbb{B})$ and mark off the $n-1$ adjacent fundamental domains it passes through. If it is not perpendicular to either of the sides through which it leaves $\mathcal{F}$, we go in both directions until we reach copies of $\mathcal{F}$ where the axis is perpendicular to the half-turn side it lands on. Starting with the furthest copy on either side, we label these $\mathcal{F}_j, j = 1, \dots, n$ and label conjugates of the half-turn side $\mathbb{L}$ in each copy of $\mathcal{F}$ by $L_{2j}^{p/q}$ and the half-turn sides through which it leaves $\mathcal{F}_j$ by $L_{2j+1}^{p/q}$. Note that $\mathcal{F} = \mathcal{F}_j$ for one of the $j$'s, say $j^*$.

Next we follow the axis of $\mathbb{E}_{r/s}\mathbb{E}_{l/m}$ as it leaves $\overline{\mathcal{H}(\mathbb{A}, \mathbb{B})}$ and mark off the $n-1$ adjacent fundamental domains it passes through until we reach copies of $\mathcal{F}$ where the axis is perpendicular to the half-turn side it lands on and, starting with the furthest copy either side, label them by $\mathcal{F}_{-j}, j = 1, \dots, n$; we label conjugates of the half-turn side $\mathbb{L}$ in each copy of $\mathcal{F}$ by $L_{-j}^{p/q}$ and the half-turn sides through which it leaves $\mathcal{F}_{-j}$ by $L_{-2j-1}^{p/q}$. As above we label the intersection points of the axes and the half-turn lines respectively by $Q_j^{p/q}$, $j = -2n-1, \ldots, -1, 1, \ldots 2n+1$. Note that by the symmetry under $H_{\mathbb{L}}$ we have $\mathcal{F}_{j*} = \mathcal{F}_{-j*}$

Now each axis passes through $n-1$ disjoint fundamental domains and each has a segment in $\mathcal{F} = \mathcal{F}_{j*} = \mathcal{F}_{-j^*}$. Again, if we were to use group elements to label the sides we would see that the union of the segments along the axis of $\mathbb{E}_{p/q}$ and the conjugate of the union of the segments along $\tilde{\mathbb{E}}_{p/q}$ by $\mathbb{E}_{r/s}$ give us a fundamental segment for $\mathbb{E}_{p/q}$.

Note that Proposition 3.3 holds for both these axes as well. We also have

PROPOSITION 3.4. *For $p+q$ even, the axes of $\mathbb{E}_{p/q}$ and $\tilde{\mathbb{E}}_{p/q}$ intersect $\mathbb{L}$ inside $\mathcal{F}$ in the same point.*

PROOF. A half-turn about a geodesic in the plane $\mathbb{H}^2$ is a reflection in the geodesic. It follows that since $Ax_{\mathbb{E}_{p/q}}$ intersects $\mathbb{L}$, its reflection also does and since the points of $\mathbb{L}$ are fixed, the intersection with $\mathbb{L}$ is a fixed point of the reflection. Because these axes are not orthogonal to $\mathbb{L}$ the intersection is not orthogonal. $\square$

This proposition implies that the two points we counted in $\mathcal{F}$ where the axes pass through $\mathbb{L}$ are actually the same point.

**3.1. Definition of ESI's.** We have shown that corresponding to any primitive $\mathbb{E}_{p/q}$ of the palindromic enumeration there is a unique set of $4n+3$ half-turn geodesics. An ESI is a pair $(L_j^{p/q}, Ax_{\mathbb{E}_{p/q}})$. This intersection of the axis of $\mathbb{E}_{p/q}$ and $L_j^{p/q}$ determines a point in $\mathbb{H}^2$. If $\mathbb{X}$ is any primitive element of the group, then $\mathbb{X}$ is conjugate to $\mathbb{E}_{p/q}$ for some $p/q$. Let $\mathbb{X} = \mathbb{W}\mathbb{E}_{p/q}\mathbb{W}^{-1}$ for $\mathbb{W} \in \mathbb{G}$. Then in the quotient, the image of the point determined by the pair $(\mathbb{W}(L_j^{p/q}), Ax_{\mathbb{X}})$ is the same as the image of the point determined by the pair $(L_j^{p/q}, Ax_{\mathbb{E}_{p/q}})$.

We say $P \in \mathbb{H}^2$ is equivalent to $Q_j^{p/q}$ for some $p/q$ and $j$ if there is $\mathbb{W} \in \mathbb{G}$ such that $P = \mathbb{W}(Q_j^{p/q})$. We denote the equivalence class by square brackets:

$$[Q_j^{p/q}] = \{\mathbb{W}(Q_j^{p/q})\}_{\mathbb{W}\in\mathbb{G}}.$$

DEFINITION 3.5. Let

$$[ESI_{p/q}] = \bigsqcup_{j=-2n+1}^{2n-1} [Q_j^{p/q}]$$

where $n = [p+q-1]/2$ and $\bigsqcup_{j=2n-1}^{2n-1}$ means that if $(p+q)$ is odd we omit the term $j=0$ and if $p+q$ is even we omit the term $j = j*$. Thus we have removed those $j$ such that the angle between $Ax_{\mathbb{E}_{p/q}}$ and $L_j^{p/q}$ at $Q_j^{p/q}$ is right. These are the *Essential Intersection points or ESI set* of $\mathbb{E}_{p/q}$. The full *ESI set of* $\mathbb{G}$ is

$$ESI(\mathbb{G}) = \bigcup_{p/q} [ESI_{p/q}].$$

The following theorem was proved in [**3**] using recursion and again in [**9**] where the proof was non-recursive.

THEOREM 3.6. *For any $p/q > 0$, the number of ESI points of $\mathbb{E}_{p/q}$ is $2(p+q-1)$. The number of points counted on the quotient surfaces is $p+q-1$.*

The points $Q_j^{p/q}$ are uniquely defined by the pair of geodesics $(L_j^{p/q}, Ax_{\mathbb{E}_{p/q}})$. Thus if $Q_j^{p/q}$ is equivalent to $\mathbb{W}(Q_j^{p/q})$, for some $\mathbb{W} \in \mathbb{G}$, then $\mathbb{W}(Q_j^{p/q})$ is uniquely defined by the pair of geodesics $(\mathbb{W}(L_j^{p/q}), \mathbb{W}(Ax_{\mathbb{E}_{p/q}}))$.

This gives us an equivalence relation and we again denote the equivalence class by square brackets $[(L_j^{p/q}, Ax_{\mathbb{E}_{p/q}})]$.

Another characterization of the ESI set of $\mathbb{G}$ is then given by

DEFINITION 3.7. The set of pairs, $\bigsqcup_{j=-2n+1}^{2n-1}[(L_j^{p/q}, Ax_{\mathbb{E}_{p/q}})]$ , where $n = [p+q-1]/2$ and $\bigsqcup$ has the same meaning as in Definition 3.5 are the *Essential Intersection points or ESI set* of $\mathbb{E}_{p/q}$. The full set is

$$ESI(\mathbb{G}) = \bigcup_{p/q} \bigsqcup_{j=-2n+1}^{2n-1} [(L_j^{p/q}, Ax_{\mathbb{E}_{p/q}})]$$

The equivalence of these definitions is clear since the points $Q_j^{p/q}$ are just the intersection points of the pairs of geodesics. The advantage of the second definition, as we will see, is that we can generalize it to other groups besides the model group.

**3.2. Loops and Winding in $\mathbb{H}^2$.** Under the projection $\pi : \mathbb{H}^2 \to \mathbb{H}^2/\mathbb{G} = S$, the axes $Ax_{\mathbb{A}}$, $Ax_{\mathbb{B}}$ and $Ax_{\mathbb{A}^{-1}\mathbb{B}}$ project to three simple geodesics $\gamma_0 = \pi(Ax_{\mathbb{A}}), \gamma_\infty = \pi(Ax_{\mathbb{B}})$ and $\gamma_{-1} = \pi(Ax_{\mathbb{A}^{-1}\mathbb{B}})$ which we now label by their corresponding rational. These are the only simple geodesics on $S$. The half-turn lines $\mathbb{L}$, $\mathbb{L}_{\mathbb{A}}$ and $\mathbb{L}_{\mathbb{B}}$ project to bi-infinite geodesics $\pi(L) = \ell, \pi(\mathbb{L}_{\mathbb{A}}) = \ell_{\mathbb{A}}$ and $\pi(\mathbb{L}_{\mathbb{B}}) = \ell_{\mathbb{B}}$ such that $\ell$ intersects $\gamma_0$ and $\gamma_\infty$ orthogonally and is disjoint from $\gamma_{-1}$, $\ell_{\mathbb{A}}$ intersects $\gamma_0$ and $\gamma_{-1}$ orthogonally and is disjoint from $\gamma_\infty$ and $\ell_{\mathbb{B}}$ intersects $\gamma_\infty$ and $\gamma_{-1}$ orthogonally and is disjoint from $\gamma_0$. The half-turn $H_{\mathbb{L}}$ on $\mathbb{H}^2$ induces an involution $\mathcal{I}_{hyp}$ of $S$. The geodesics $\{\ell, \ell_{\mathbb{A}}, \ell_{\mathbb{B}}\}$ are the seams of $S$.

Thus we can write $\gamma_0 = \sigma_0 \cup \mathcal{I}_{hyp}(\sigma_0)$ where $\sigma_0$ is the segment from $\ell$ to $\ell_{\mathbb{A}}$ to see that it is made up of two segments that form a loop. Similarly $\gamma_\infty = \sigma_\infty \cup \mathcal{I}_{hyp}(\sigma_\infty)$ is another loop. Each of these loops intersects exactly two of the seams once.

Write $\gamma_{p/q} = \pi(Ax_{E_{p/q}})$.

LEMMA 3.8. *For every $p/q$, $\gamma_{p/q}$ can be written as a union of $p+q$ loops each intersecting $\ell$ and either $\ell_{\mathbb{A}}$ or $\ell_{\mathbb{B}}$.*

PROOF. Consider the segment of $Ax_{\mathbb{E}_{p/q}}$ from $Q_j^{p/q}$ to $Q_{j+1}^{p/q}$. One of its endpoints lies inside a fundamental domain $\mathcal{F}_j$ and the other on its boundary. Denote the projection of the segment by $\sigma_j^{p/q}$; it is simple and one of its endpoints is on $\ell$ and the other is on either $\ell_{\mathbb{A}}$ or $\ell_{\mathbb{B}}$. Suppose for argument's sake that the other end is $\ell_{\mathbb{A}}$. The union $\sigma_j(p/q) \cup \mathcal{I}_{hyp}(\sigma_j^{p/q})$ is a closed loop on $S$. Since it contains one segment going from $\ell$ to $\ell_A$ and another from $\ell_A$ to $\ell$, it is homotopic to $\gamma_0$ on $S$.

Since we can do this for every adjacent pair of points, we obtain $p+q$ loops, each homotopic to either $\gamma_0$ or $\gamma_\infty$. □

REMARK 3.9. We interpret this decomposition of the geodesic $\gamma_{p/q}$ into loops as showing that it *winds* a total of $p$ times around $\gamma_0$ and a total of $q$ times around $\gamma_\infty$. Note however, that the order in which it goes around each of the loops $\gamma_0$ and $\gamma_\infty$ is determined by the form of the primitive word $\mathbb{E}_{p/q}$. The integer entries in the continued fraction expansion for $p/q$ determine the number of consecutive times it winds about each curve before it winds around the other curve. There are words in

$\mathbb{G}$ in which $\mathbb{A}$ appears $p$ times and $\mathbb{B}$ appears $q$ times that are not primitive. Our discussion does not apply to them.

REMARK 3.10. Note that we obtain this geometric interpretation only because we started with stopping generators.

## 4. $\mathbb{H}^3$: Generalized ESI's

**4.1. Transversals in the quotient.** The ESI points for our model group $\mathbb{G}$ have a geometric interpretation on the quotient surface. Definition 3.7, however, makes sense for an arbitrary free discrete group $G = \langle A, B\rangle \subset \mathrm{Isom}(\mathbb{H}^3)$ although the geometric interpretation is not clear because, generically, the pairs of lines do not intersect.

In the quotient manifold $M = \mathbb{H}^3$ we can look at the projections $\gamma_{p/q}$ of the axes of the words $E_{p/q}$ and the projections of the half-turn lines $L, L_A, L_B$. Adapting the notation in [**2, 3**] we call them $\ell, \ell_A, \ell_B$. In [**2**] we proved that $\ell, \ell_A, \ell_B$ are three bi-infinite geodesics in $M$ and that the half-turn about any one of them projects to an involution $\mathcal{I}_{hyp}$ of $M$ whose fixed set is precisely these three geodesics. We also proved that for any primitive element $E_{p/q} \in G$, $\gamma_{p/q}$ is invariant (but not pointwise invariant) under the involution $\mathcal{I}_{hyp}$.

Suppose for some $\xi \in \gamma_{p/q}$ there is a geodesic $\omega$ in $M$ joining $\xi$ to $\mathcal{I}_{hyp}(\xi) \neq \xi$ and suppose $\omega$ meets $\gamma_{p/q}$ orthogonally at these points. We call such an $\omega$ a *transversal* since $\omega$ must also intersect one of $\ell, \ell_A, \ell_B$ orthogonally. For the sake of discussion, assume it is $\ell$. Then we can lift $\gamma_{p/q}$ to $Ax_{E_{p/q}}$ and find a lift of $\ell$, $W(L)$ for some $W \in G$ such that $\omega$ lifts to the common orthogonal $O$ to $Ax_{E_{p/q}}$ and $W(L)$.

Clearly $W$ is not uniquely determined and $O$ depends on the choice of $W$. If, however, $W(L) = L_j^{p/q}$ for some $j$, then $O$ can be interpreted as a generalization of the intersection point. In general, the number of transversals we can find for a given $\gamma_{p/q}$ is unrelated to $p/q$. Since our goal is to relate the generalized intersection points to $p/q$, at least for certain groups, we proceed below to define our generalization from the labelling of the model group transferred to the abstract group via the natural isomorphism rather than the geometry of $M$.

**4.2. Generalized ESI's for a group $G \in \mathbb{H}^3$.** We now consider the group $G = \langle A, B\rangle$ as an arbitrary free purely loxodromic discrete group in $\mathrm{Isom}(\mathbb{H}^3)$ and let its degree two extension be $\tilde{G} = \langle H_L, H_{L_A}, H_{L_B}\rangle$. As such, we can talk about axes and half-turn lines. We can transfer the labelling we have for the model group to the group $G$: words $E_{p/q}$ in $G$ are just the words $\mathbb{E}_{p/q}$ with $\mathbb{A}$ replaced by $A$ and $\mathbb{B}$ replaced by $B$. Since each half-turn line $L_j^{p/q}$ for $\mathbb{G}$ can be written as $\mathbb{W}(\mathbb{L}_{\mathbb{X}_j})$ with $\mathbb{L}_{\mathbb{X}_j} \in \{\mathbb{L}, \mathbb{L}_{\mathbb{A}}, \mathbb{L}_{\mathbb{B}}\}$, we label a subset of half-turn lines for $G$ again by $L_j^{p/q} = W(L_{X_j})$, where $X_j$ corresponds to $\mathbb{X}_j$ under the isomorphism.

The equivalence relation between pairs also transfers to $G$ in the obvious way: $(L_j^{p/q}, Ax_{E_{p/q}})$ is equivalent to $(L_X, Ax_Y)$ for some half-turn line $L_X$ of $\tilde{G}$ and some primitive element of $Y \in G$ if there is a $W \in G$ such that $W(L_j^{p/q}) = L_X$ and $W(Ax_{(E_{p/q})}) = Ax_{E_{p/q}}$. Again using square brackets for equivalence classes, setting $n = [p + q - 1]/2$ and defining $\bigsqcup$ as above we have

DEFINITION 4.1. Given $G = \langle A, B\rangle \subset \mathrm{Isom}(\mathbb{H}^3)$. The *Generalized ESI set* for $G$ is defined as the collection of equivalence classes of pairs of geodesics:

$$SESI(G) = \bigcup_{p/q\in\mathbb{Q}} \{[ESI_{E_{p/q}}]\} = \bigsqcup_{j=-2n-1}^{2n+1} [(L_j(p/q), Ax_{E_{p/q}})]$$

where the labeling and notation is taken from the model group $\mathbb{G} = \langle \mathbb{A}, \mathbb{B}\rangle$ with stopping generators $\mathbb{A}, \mathbb{B}$.

THEOREM 4.2. *The set $SESI(G)$ is well-defined for any free, purely loxodromic discrete group $G = \langle A, B\rangle \subset \mathrm{Isom}(\mathbb{H}^3)$.*

PROOF. The set $SESI(G)$ is defined by pairs of geodesics. These are defined by their endpoints in $\partial\mathbb{H}^3$. For the set to be well-defined, these endpoints need to be distinct. This will be true of the axes if the group is free and all its elements are loxodromic. The half-turn lines will have distinct endpoints if the group is discrete. □

**4.3. The Connectors.** Let $(L_j^{p/q}, Ax_{E_{p/q}})$ be a representative pair for an equivalence class in $SESI(G)$. In general the two geodesics in a pair will not intersect so we consider their common orthogonal, $O_j^{p/q}$. Its endpoints can be computed using cross ratios from the endpoints of the geodesics of the pair. Thus, if the pair $(L_j^{p/q}, Ax_{E_{p/q}})$ does intersect, the $O_j^{p/q}$ determined by the cross ratios is the common orthogonal through the intersection point to the plane they span. Thus for the model group $\mathbb{G}$, this orthogonal $O_j^{p/q}$ is a geodesic perpendicular to the plane left invariant by $\mathbb{G}$ and intersecting it at the point $Q_j^{p/q}$; it is therefore the appropriate generalization of the intersection point.

Continuing our use of bar notation, we set $\overline{E}^j_{p/q} = H_{L_j^{p/q}} E_{p/q} H_{L_j^{p/q}}$. Then since $G$ has index two in $\tilde{G}$,

$$(L_j^{p/q}, Ax_{\overline{E}^j_{p/q}}) \in [(L_j^{p/q}, Ax_{E_{p/q}})].$$

Since $O_j^{p/q}$ is perpendicular to $L_j^{p/q}$, $H_{L_j^{p/q}}(O_j^{p/q}) = O_j^{p/q}$. That is $O_j^{p/q}$ is also the common orthogonal for the pair $(L_j^{p/q}, Ax_{\overline{E}^j_{p/q}})$. This leads us to define

DEFINITION 4.3. For each representative pair of geodesics $(L_j^{p/q}, Ax_{E_{p/q}})$ in $SESI(G)$, the common orthogonal $O_j^{p/q}$ to the pair is a geodesic called a *connector*. Note that the connectors depend on the presentation of $G$ and the labelling induced by the isomorphism to the model group.

The axes of $E_{p/q}$ and $\overline{E}^j_{p/q}$ project to the same geodesic $\gamma_{p/q}$ in $M$ and the connector $O_j^{p/q}$ projects to a geodesic $\omega_j^{p/q}$ joining two points on $\gamma_{p/q}$ and meeting it orthogonally at both these points. The geodesic $\omega_j^{p/q}$ also intersects the projection of the half-turn line $L_j^{p/q}$ orthogonally. This is one of the geodesics $\ell, \ell_A, \ell_B$. Otherwise said, the connectors project to transversals in the quotient $M$.

Thus we have proved Theorem A. We state it in two parts.

THEOREM 4.4. *[Theorem A in $\mathbb{H}^3$] Let $G = \langle A, B\rangle \subset \mathrm{Isom}(\mathbb{H}^3)$ be discrete, free and purely loxodromic. For each rational $p/q$, there is a geodesic $\gamma_{p/q}$ in the* *quotient manifold* $\mathbb{H}^3/G$ *with* $2(p+q-1)$ *marked points joined in pairs by geodesics* $\omega_j^{p/q}$, $j = 2, \ldots, p+q$ *that meet* $\gamma_{p/q}$ *orthogonally.*

We also have

THEOREM 4.5. *For* $G$ *as above, the union of the curves* $\gamma_{p/q}$ *and* $\omega_j^{p/q}$, $j = 1, \ldots, p+q-1$ *naturally divides into a set consisting of* $p+q$ *non-homotopically trivial loops. The order in which the same loop occurs a consecutive number of times is determined by the entries on the continued fraction expansion of* $p/q$.

PROOF. Denote the endpoints of $\omega_j^{p/q}$ on $\gamma_{p/q}$ by $\xi_j, \xi_j'$ so that they are joined by the projection of the connector $\omega_j^{p/q}$. Then following $\gamma_{p/q}$ from $\xi_j$ to $\xi_{j+1}$, next following $\omega_{j+1}^{p/q}$ to $\xi_{j+1}'$, next following $\gamma_{p/q}$ to $\xi_j'$ and finally following $\omega_j^{p/q}$ from $\xi_j'$ to $\xi_j$ we obtain a closed loop. Note that since each loop passes once through two distinct geodesics in the invariant set of the involution, $\{\ell, \ell_A, \ell_B\}$, it cannot be homotopically trivial.

□

We have used the presentation of the group to define the connectors. Note that while each connector is a transversal, there are many transversals that are not connectors. For example, in the model group $\mathbb{G}$, $\gamma_0$ has no connectors but many transversals. One is the projection of the common orthgonal to $Ax_A$ and $L_B$.

## 5. Winding Groups

In this section we define a class of groups, more general than the class of the model group, where the loops formed by the connectors are a generalization of *winding* around the generators. Key properties for these groups are that Propositions 3.1 and 3.2 hold. To define these groups we need some preliminaries.

We denote the *convex hull* of the limit set of $G$ in $\mathbb{H}^3$ by $\mathcal{C}$. It is invariant under $G$ and the quotient $\mathcal{C}/G$ is the *convex core* of $G$. For the model group $\mathbb{G}$ with stopping generators, the convex hull of the limit set is formed by deleting all the half-planes bounded by the intervals of discontinuity of $\mathbb{G}$. These are determined by the endpoints of the axes of $\mathbb{A}, \mathbb{B}, \mathbb{A}^{-1}\mathbb{B}$ and all of their conjugates. The convex core is the surface $S = \mathbb{H}^2/\mathbb{G}$ truncated along the geodesics $\gamma_0, \gamma_1$ and $\gamma_\infty$.

The boundary of the convex core $\tilde{S} = \partial\mathcal{C}/G$ is a surface of genus two. It is what is known as a *pleated surface*. For the purposes of this paper, what this means is that if more than one geodesic passes through a point on $\tilde{S}$ infinitely many do; that is, the surface looks like a piece of hyperbolic plane at that point. Points through which only one geodesic passes form the *pleating locus* of $\tilde{S}$. This consists of infinite open geodesics and/or closed simple geodesics. These are also called pleating curves. If there is a closed geodesic in the pleating locus, it is the projection of the axis of an element of $G$ and this axis lies on the boundary of $\mathcal{C}$.

DEFINITION 5.1. The pleating locus of $\tilde{S}$ is called rational if it consists entirely of closed geodesics.

A special class of groups whose convex core boundary has rational pleating locus are what we term *Winding Groups*.

DEFINITION 5.2. Let $G = \langle A, B \rangle$ be a group and suppose $\tilde{S} = \partial\mathcal{C}/G$ has rational pleating locus where the pleating curves are projections of the axes of $A, B, A^{-1}B$. Then $G$ is called a *winding group*.

We have

PROPOSITION 5.3. *If $G$ is a winding group, then any palindromic word in the palindromic enumeration scheme $E_{p/q}$ ($p+q$ odd) has an axis that intersects the common perpendicular $L$ to the axes of $A$ and $B$ between its intersections with $A$ and $B$. If $p+q$ is even, the common perpendicular to $\tilde{E}_{p/q} = H_L E_{p/q} H_L$ and $E_{p/q}$ intersects the axes of $A$ and $B$ between its intersections with $A$ and $B$.*

PROOF. This follows from the definition of the convex hull. No axis can intersect the boundary of the convex hull transversally; if it intersects the boundary it lies entirely in the boundary. If $p+q$ is odd, the palindromic property implies the axis of $E_{p/q}$ intersects $L$ so it must do it inside the convex hull. If $p+q$ is even, the axes of $\tilde{E}_{p/q}$ and $E_{p/q}$ lie inside the convex hull. By the symmetry in the definition their common orthogonal intersects $L$ and by the convexity the intersection point is inside the convex hull. □

The hexagons $\mathcal{H}(A,B), \overline{\mathcal{H}(A,B)}$ are well defined for $G$ and so we can think of their union $\mathcal{F}$ as a piecewise geodesic polygon in $\mathbb{H}^3$, but not a fundamental domain. The lines $L_j^{p/q}$ are sides of translates of these hexagons. For a winding group, the axis sides of the hexagon project to pleating curves. The content of Proposition 5.3 is that Propositions 3.1 and 3.2 hold for winding groups with this definition of $\mathcal{F}$. The generators of the winding groups play the role of the stopping generators of the model group.

Thus we have a geometric interpretation of Theorems 3.6 and 4.5 for these groups. We summarize this as

THEOREM 5.4. *[Theorem B] Given a winding group $G$, for every primitive geodesic in the quotient manifold, there is an analogue of the winding we found in the two dimensional case that reflects the rational number identified with it.*

## References


[1] W. Fenchel, *Elementary geometry in hyperbolic space*, De Gruyter Studies in Mathematics, vol. 11, Walter de Gruyter & Co., Berlin, 1989. With an editorial by Heinz Bauer. MR1004006

[2] J. Gilman and L. Keen, *The geometry of two generator groups: hyperelliptic handlebodies*, Geom. Dedicata **110** (2005), 159–190, DOI 10.1007/s10711-004-6556-8. MR2136025

[3] J. Gilman and L. Keen, *Word sequences and intersection numbers*, Complex manifolds and hyperbolic geometry (Guanajuato, 2001), Contemp. Math., vol. 311, Amer. Math. Soc., Providence, RI, 2002, pp. 231–249, DOI 10.1090/conm/311/05455. MR1940172

[4] J. Gilman and L. Keen, *Discreteness criteria and the hyperbolic geometry of palindromes*, Conform. Geom. Dyn. **13** (2009), 76–90, DOI 10.1090/S1088-4173-09-00191-X. MR2476657

[5] J. Gilman and L. Keen, *Enumerating palindromes and primitives in rank two free groups*, J. Algebra **332** (2011), 1–13, DOI 10.1016/j.jalgebra.2011.02.010. MR2774675

[6] J. Gilman and B. Maskit, *An algorithm for 2-generator Fuchsian groups*, Michigan Math. J. **38** (1991), no. 1, 13–32, DOI 10.1307/mmj/1029004258. MR1091506

[7] L. Keen and C. Series, *Pleating Coordinates for Teichmüller Space*, Bull. AMS, vol. 26, 1, 1992, 141-146

[8] W. Magnus, A. Karrass, and D. Solitar, *Combinatorial group theory: Presentations of groups in terms of generators and relations*, Interscience Publishers [John Wiley & Sons, Inc.], New York-London-Sydney, 1966. MR0207802

[9] V. Mailik, *Curves Generated on Surfaces by the Gilman-Maskit Algorithm*, AMS-CONM, vol. 510 (2010), 209-239.

Department of Mathematics and Computer Science, Rutgers University, Newark, New Jersey 07079

*E-mail address*: gilman@rutgers.edu

Department of Mathematics, The Graduate Center and Lehman College, CUNY, New York, New York 10016

*E-mail address*: LKeen@gc.cuny.edu

Contemporary Mathematics
Volume **696**, 2017
http://dx.doi.org/10.1090/conm/696/14020

# The 1-2 model

Geoffrey R. Grimmett and Zhongyang Li

Abstract. The current paper is a short review of rigorous results for the 1-2 model. The 1-2 model on the hexagonal lattice is a model of statistical mechanics in which each vertex is constrained to have degree either 1 or 2. It was proposed in a study by Schwartz and Bruck of constrained coding systems, and is strongly connected to the dimer model on a decoration of the lattice, and to an enhanced Ising model and an associated polygon model on the graph derived from the hexagonal lattice by adding a further vertex in the middle of each edge.

The general 1-2 model possesses three parameters $a$, $b$, $c$. The fundamental technique is to represent probabilities of interest as ratios of counts of dimer coverings of certain associated graphs, and to apply the Pfaffian method of Kasteleyn, Fisher, and Temperley.

Of special interest is the existence (or not) of phase transitions. It turns out that all clusters of the infinite-volume limit are almost surely finite. On the other hand, the existence (with strictly positive probability) of infinite 'homogeneous' clusters, containing vertices of given type, depends on the values of the parameters.

A further type of phase transition emerges in the study of the two-edge correlation function, and in this case the critical surface may be found explicitly. For instance, when $a \geq b \geq c > 0$, the surface given by $\sqrt{a} = \sqrt{b} + \sqrt{c}$ is critical.

## 1. Origin of the 1-2 model

The 1-2 model originated in the work of computer scientists Schwartz and Bruck [**23**] on constrained coding systems. They studied an array of variables on the hexagonal lattice $\mathbb{H}$ subject to the 'not all equal' constraint. Of particular interest to them was the asymptotic behaviour of the number of acceptable configurations on large bounded regions $\mathbb{H}_n$ of the lattice, in the 'thermodynamic limit' as $\mathbb{H}_n \uparrow \mathbb{H}$. Using the method of so-called 'holographic reduction', they were able to map their counting problem to one of counting the number of perfect matchings (or 'dimer coverings') on a certain graph derived from the hexagonal lattice. This last problem may be solved using the Pfaffian representation of Kasteleyn [**15**], Fisher [**5**], and Temperley and Fisher [**24**].

When rephrased in the language of statistical mechanics, the work of Schwartz and Bruck amounts to the calculation of the partition function of the following

2010 *Mathematics Subject Classification.* Primary 82B20, 60K35, 05C70.

*Key words and phrases.* 1-2 model, dimers, polygon model, Ising model, perfect matching, Kasteleyn matrix, phase transition, percolation.

*1-2 model* of probability theory and mathematical physics. Let $\mathbb{H} = (\mathbb{V}, \mathbb{E})$ be the hexagonal lattice of Figure 1.1, and let $\Sigma = \{-1, 1\}^{\mathbb{E}}$ be the set of configurations of absent/present edges, where the local state $-1$ (respectively, 1) means *absent* (respectively, *present*). The sample space is the subset of $\Sigma$ containing all $\sigma \in \Sigma$ such that: every vertex of $\mathbb{H}$ is incident to either one or two present edges. Thus, a configuration comprises disjoint paths and cycles of present edges.

Let $\mathbb{H}_n$ be an $n \times n$ subgraph of $\mathbb{H}$ with periodic boundary conditions, and let $\mu_n$ be the uniform probability measure on the set of 1-2 configurations on $\mathbb{H}_n$. We ask for properties of $\mu_n$ in the limit as $n \to \infty$. In particular, does the limit measure $\mu_\infty := \lim_{n\to\infty} \mu_n$ exist, and, if so, what can be said about the long-range correlations of edge-states under $\mu_\infty$? It turns out that the connection to dimers may be exploited to answer such questions.

The above system is a lattice model whose partition function can be computed by the calculation of certain determinants using the holographic algorithm of Valiant [**25**]. By introducing an invertible $2 \times 2$ matrix on edges of a graph and conducting a base change, the partition function of a general vertex model on a graph $G$ is transformed into the partition function of perfect matchings on a certain decorated version of $G$. Valiant's original holographic algorithm can be generalized by assigning different bases to different edges (see [**16**]), and the ensuing algorithm can be used to compute partition functions of a larger class of vertex models in polynomial time.

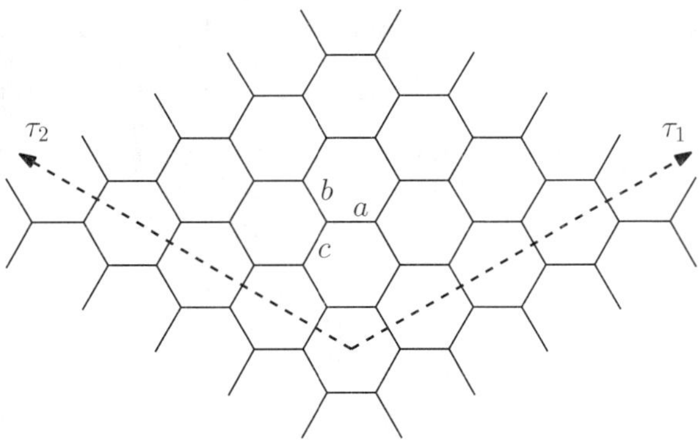


FIGURE 1.1. The graph $\mathbb{H}_n$ is an $n \times n$ lozenge wrapped onto a torus. A horizontal (respectively, NW/SE, NE/SW) edge is said to be of type $a$ (respectively, type $b$, $c$).

The holographic algorithm seems, however, not to be the most efficient way to solve the 1-2 model. In particular, the correspondence between the 1-2 partition function and the dimer partition function on a corresponding Fisher graph, via the base change method, is not measure-preserving; thus, the computation of local statistics and related probabilities becomes complicated, even if possible. An alternative measure-preserving correspondence was introduced in [**18**], and this permits a number of representations in closed form of probabilities associated with the 1-2 model. This method, and some of its consequences, will be described in the current review.

Certain properties of the underlying hexagonal lattice are utilized heavily in this work, such as trivalence, planarity, and support of a $\mathbb{Z}^2$ action. It may be possible to extend the results summarized here to certain other graphs with such properties, including the Archimedean $(3, 12^2)$ and $(4, 8^2)$ lattices, and also to certain periodic models.

The formal definition of the 1-2 model is presented in Section 2. The model has strong connections to the dimer and Ising models as well as to a certain polygon model, and these connections are laid out in Section 3. Two approaches to the issue of phase transition are outlined in Section 4, using the geometry and the correlation structure, respectively, and an exact formula for the critical surface is given in the second case.

## 2. Definition of the 1-2 model

Whereas the 1-2 model of [**23**] is uniform in that there is only one parameter, we present here the more general three-parameter model of [**16**].

Let $n \geq 1$, and let $\tau_1$, $\tau_2$ be the two shifts of $\mathbb{H}$ as in Figure 1.1. The pair $(\tau_1, \tau_2)$ generates a $\mathbb{Z}^2$ action on $\mathbb{H}$, and we write $\mathbb{H}_n = (V_n, E_n)$ for the (toroidal) quotient graph of $\mathbb{H}$ under the subgroup of $\mathbb{Z}^2$ generated by the powers $\tau_1^n$ and $\tau_2^n$. The configuration space $\Sigma_n$ is the set of all $\sigma \in \{-1, 1\}^{E_n}$ such that every $v \in V_n$ is incident to either 1 or 2 edges $e$ with $\sigma(e) = 1$. Note that $\sigma \in \Sigma_n$ if and only if $-\sigma \in \Sigma_n$. It is sometimes convenient to work with the vector $\sigma'$ given by $\sigma'(e) = \frac{1}{2}(1 + \sigma(e))$.

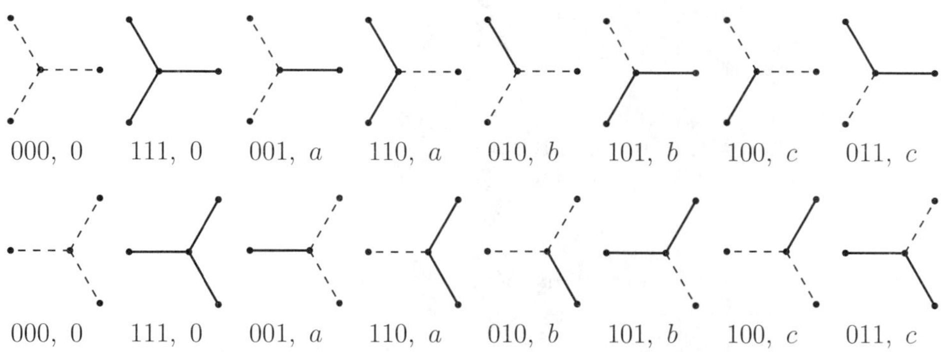


FIGURE 2.1. Each vertex has eight possible local configurations, six of which have nontrivial weights, $a$, $b$, or $c$ as above. A vertex $v$ is said to be of *type* $s$ in configuration $\sigma$ if $\sigma|_v$ has weight $s \in \{a, b, c\}$. Whereas the type of an edge is *deterministic* (see Figure 1.1), the type of a vertex is *random*.

A vertex $v \in V_n$ is incident to three edges $e_1, e_2, e_3$ of $\mathbb{H}_n$, which are in the respective orientations: horizontal, NW/SE, and NE/SW. Let $\sigma \in \Sigma_n$, and let the *signature* at $v$ be the triple $\sigma|_v := \sigma'(e_3)\sigma'(e_2)\sigma'(e_1) \in \{0, 1\}^3$ considered as a word with three letters in the alphabet with two letters. Let $a, b, c \in [0, \infty)$ be such that $(a, b, c) \neq (0, 0, 0)$. To the signature $\sigma|_v$ is allocated the weight $w(\sigma|_v) \in \{0, a, b, c\}$ given in Figure 2.1, and the weight function $w : \Sigma_n \to [0, \infty)$ is defined by

$$w(\sigma) = \prod_{v \in V_n} w(\sigma|_v). \tag{2.1}$$

This gives rise to the probability measure $\mu_n$ given by

$$\mu_n(\sigma) = \frac{1}{Z_n} w(\sigma), \tag{2.2}$$

where

$$Z_n = Z_n(a, b, c) = \sum_{\sigma \in \Sigma_n} w(\sigma) \tag{2.3}$$

is the *partition function*. A sample drawn (approximately) from $\mu_{10}$ is depicted in Figure 2.2.

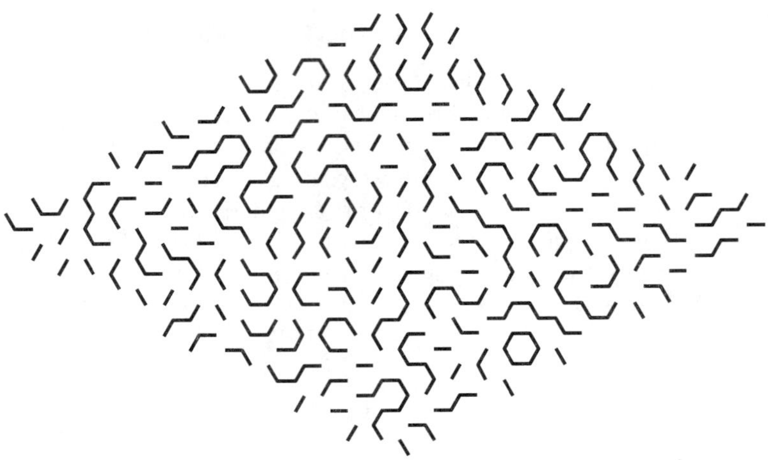

FIGURE 2.2. A realization of the (uniform) 1-2 model on $\mathbb{H}_{10}$ with $a = b = c = 1$, drawn by MCMC methods. Each vertex has degree either 1 or 2, and therefore all components are either paths or cycles.

It turns out that the weak limit of the sequence $(\mu_n : n \geq 1)$ exists. However, no simple correlation inequality is known, and the proof of existence of the limit follows a different route using the relationship to dimer configurations outlined in Section 3.1.

THEOREM 2.1 ([**10**, Thm 6.2]). *The weak limit*

$$\mu_\infty := \lim_{n \to \infty} \mu_n$$

*exists and is translation invariant.*

The infinite-volume limit $\mu_\infty$ is of course a Gibbs state in the sense that it satisfies the relevant Dobrushin–Lanford–Ruelle (DLR) condition (see [**7**, Sect. 4.4]). On the other hand, the structure of the space of such Gibbs measures is unknown. Neither is it known for which parameter-values $\mu_\infty$ is ergodic (it is not ergodic under the conditions of [**18**, Thm 4.9] and Theorem 4.3(b) of the current paper).

REMARK 2.2. *For edges $e$, $f$ of $\mathbb{H}$ and for sufficiently large $n$, we write $\langle \sigma_e \sigma_f \rangle_n$ for the* two-edge correlation function *of the measure $\mu_n$. By Theorem 2.1, the limit*

$$\langle \sigma_e \sigma_f \rangle = \lim_{n \to \infty} \langle \sigma_e \sigma_f \rangle_n$$

*exists. In Section 4.3, we shall consider the limit* $\lim_{|e-f|\to\infty}\langle\sigma_e\sigma_f\rangle$ *as an order parameter that is indicative of phase transition.*

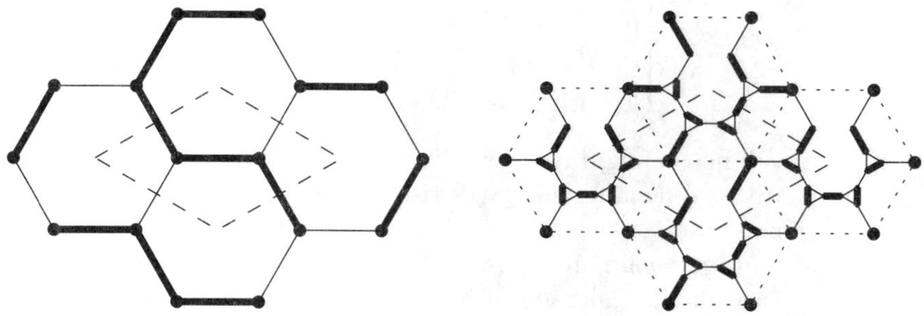

FIGURE 2.3. Part of the hexagonal lattice $\mathbb{H}$ is drawn on the left, together with a 1-2 configuration. The graph $\mathbb{H}_\Delta$ on the right is obtained by replacing each face of $\mathbb{H}$ by a certain 'gadget'. The left-hand 1-2 configuration gives rise to a dimer configuration on the decorated graph on the right, as described in the text. The fundamental domain is outlined as a lozenge, and expanded in Figure 3.1.

## 3. The dimer, Ising, and polygon models

The relationship between the 1-2 model and the dimer model is pivotal to the study of the former. Dimers are relevant also to the study of Ising models in two dimensions (see, for example, [**17**]), and the Ising model gives rise in turn to a 'high temperature' polygon model (see for example, [**1**, p. 75] and [**9**, **21**, **26**]). Therefore, the 1-2 model is connected firmly to the Ising and polygon models. These connections play roles in the theory of the 1-2 model, and are summarized in this section.

**3.1. The dimer model.** Let $\mathbb{H} = (\mathbb{V}, \mathbb{E})$ be the hexagonal lattice, and let $\mathbb{H}_\Delta = (V_\Delta, E_\Delta)$ be the 'decorated graph' drawn on the right side of Figure 2.3. The graph in the figure is obtained by replacing each face $F$ of $\mathbb{H}$ by a certain 'gadget' comprising a path which is joined to the vertices of $F$ in the manner drawn in the figure.

A 1-2 configuration of the left side of Figure 2.3 gives rise to a dimer configuration on the decorated graph on the right, as follows. Note first that $\mathbb{V} \subseteq V_\Delta$. Each $v \in \mathbb{V}$ has three incident edges in $\mathbb{H}_\Delta$, and these edges in $\mathbb{H}_\Delta$ may be regarded as the *bisector edges* of the three angles in $\mathbb{H}$ at $v$.

Let $\sigma \in \Sigma$. An edge $e \in E_\Delta$, incident to a vertex $v \in \mathbb{V}$, is designated *present* if and only if the two sides of the corresponding angle of $\mathbb{H}$ have the same states, that is, either both are present or both are absent. Once we have determined the states of the bisector edges of $\mathbb{H}_\Delta$, there is a unique extension to a dimer configuration on $\mathbb{H}_\Delta$. See Figure 2.3. Note that the two 1-2 configurations $\sigma, -\sigma \in \Sigma$ give rise to the same dimer configuration, and thus the above correspondence is two-to-one.

Consider now the toroidal graph $\mathbb{H}_n$ and the corresponding decorated graph $\mathbb{H}_{n,\Delta}$. The edges of $\mathbb{H}_{n,\Delta}$ are weighted, with the edge $e = \langle i, j\rangle$ having weight

$$w_{i,j} = \begin{cases} a & \text{if } e \text{ is a horizontal bisector edge,} \\ b & \text{if } e \text{ is a NW/SE bisector edge,} \\ c & \text{if } e \text{ is a NE/SW bisector edge,} \\ 1 & \text{otherwise.} \end{cases} \tag{3.1}$$

The weight of a dimer configuration is defined as the product of the weights of the edges that are present, and this gives rise (as in (2.2)) to a probability measure $\delta_{n,\Delta}$ on dimer configurations. Now, $\delta_{n,\Delta}$-probabilities may be represented as weighted counts of dimer configurations, and such quantities on planar graphs may be computed by the Pfaffian method of Kasteleyn [**15**], Fisher [**5**], and Temperley and Fisher [**24**]. This leads to the following limit theorem.

THEOREM 3.1 ([**18**, Prop. 3.3]). *Let $a, b, c > 0$. The limit measure*

$$\delta_\Delta := \lim_{n\to\infty} \delta_{n,\Delta}$$

*exists and is translation-invariant and ergodic.*

Theorem 2.1 follows by Theorem 3.1 and the above correspondence between 1-2 model configurations and dimer configurations. Note that the weak limit for 1-2 measures need not be ergodic.

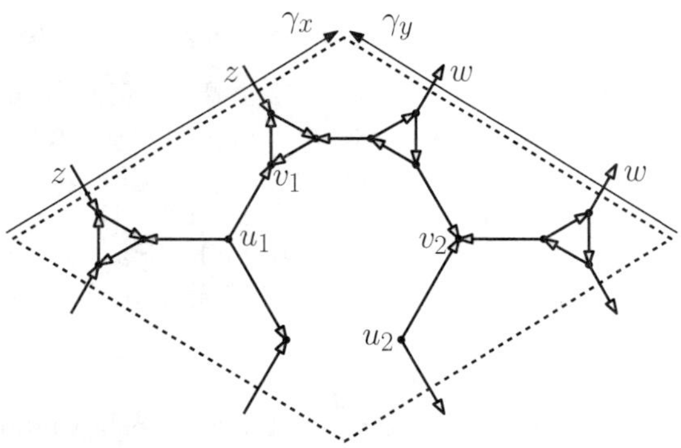


FIGURE 3.1. A single fundamental domain of the decorated graph $\mathbb{H}_{n,\Delta}$ obtained from the central lozenge of Figure 2.3. This illustrates the relationship between the fundamental domain and the original hexagonal lattice $\mathbb{H}$. Note the homology cycles $\gamma_x$, $\gamma_y$ of the torus.

A $\delta_{n,\Delta}$-probability may be expressed in terms of weighted counts of dimers, and these are studied via the Pfaffian representation. As explained in [**10**] and the references therein, the asymptotics (as $n \to \infty$) of such Pfaffians depend on the so-called *characteristic polynomial* of the model. We do not define the characteristic polynomial here beyond saying that it is the determinant $P(z, w)$ of the weighted adjacency matrix of the fundamental domain of $\mathbb{H}_\Delta$, oriented in a 'clockwise odd' manner, and illustrated in Figure 3.1. It is a function of the parameters $a$, $b$, $c$, and of two complex variables $z$, $w$. It is shown in [**18**] that

$$P(z,w) = a^4+b^4+c^4+6a^2b^2+6a^2c^2+6b^2c^2-2ab\left(z+\frac{1}{z}\right)(a^2+b^2-c^2)$$
$$-2ac\left(w+\frac{1}{w}\right)(a^2+c^2-b^2)-2bc\left(\frac{z}{w}+\frac{w}{z}\right)(b^2+c^2-a^2).$$

The *spectral curve* is the zero locus of the characteristic polynomial, that is, the set of roots of the equation $P(z,w)=0$. As explained in [**18**], it is important to understand the intersection of the spectral curve with the unit torus

$$\mathbb{T}^2 = \{(z,w)\in\mathbb{C}^2 : |z|=|w|=1\}.$$

It turns out that the intersection is either empty or is a single real point $(1,1)$. Moreover, when $P(1,1)=0$, the zero $(1,1)$ has multiplicity 2. Evidently,

$$P(1,1) = (a^2+b^2+c^2-2ab-2bc-2ac)^2, \tag{3.2}$$

and therefore the spectral curve intersects $\mathbb{T}^2$ if and only if

$$\sqrt{a}\pm\sqrt{b}\pm\sqrt{c}=0. \tag{3.3}$$

We shall return to this equation in the study of phase transition in Section 4.3.

**3.2. The half-edge graph.** When considering correlations, it will be convenient to work on a graph derived from the hexagonal lattice by replacing each edge by two half-edges. Let $A\mathbb{H}_n = (AV_n, AE_n)$ be the graph derived from $\mathbb{H}_n = (V_n, E_n)$ by adding a vertex at the midpoint of each edge in $E_n$. Let $ME_n = \{Me : e\in E_n\}$ be the set of such midpoints, and $AV_n = V_n\cup ME_n$. The edges $AE_n$ are the half-edges of $E_n$, each being of the form $\langle v, Me\rangle$ for some $v\in V_n$ and incident edge $e\in E_n$.

The 1-2 model on $\mathbb{H}_n$ can be viewed as a spin-model on the set $ME_n$ of midpoints, as we explain next.

**3.3. The Ising model.** It turns out that the 1-2 model is the marginal of a certain Ising-type model on the half-edge graph $A\mathbb{H}_n$ of Section 3.2, that is reminiscent of the Edwards–Sokal coupling of the Potts and random-cluster measures (see [**7**, Sect. 1.4]). It is constructed via a weight function on configuration space, using weights that are permitted in general to be $\mathbb{C}$-valued.

Let $\Sigma^{\mathrm{e}} = \{-1,+1\}^{ME_n}$ and $\Sigma^{\mathrm{v}} = \{-1,+1\}^{V_n}$. An edge $e\in E_n$ is identified with the element of $ME_n$ at its centre. A spin-vector is a pair $(\sigma^{\mathrm{e}},\sigma^{\mathrm{v}})\in\Sigma^{\mathrm{e}}\times\Sigma^{\mathrm{v}}$ with $\sigma^{\mathrm{e}} = (\sigma_{v,s} : v\in V_n,\ s=a,b,c)$ and $\sigma^{\mathrm{v}} = (\sigma_v : v\in V_n)$, where $\sigma_{v,a}$, $\sigma_{v,b}$, $\sigma_{v,c}$ denote the spins on midpoints of the corresponding edges of given types incident to $v\in V_n$ (see Figure 1.1). We allocate the (possibly complex) weights

$$w(\sigma^{\mathrm{v}},\sigma^{\mathrm{e}}) = \prod_{v\in V_n}(1+\epsilon_a\sigma_v\sigma_{v,a})(1+\epsilon_b\sigma_v\sigma_{v,b})(1+\epsilon_c\sigma_v\sigma_{v,c}), \tag{3.4}$$

where $\epsilon_a,\epsilon_b,\epsilon_c\in\mathbb{C}$ are constants associated with horizontal, NW/SE, and NE/SW edges, respectively. In (3.4), each factor $1+\epsilon_s\sigma_v\sigma_{v,s}$ $(s=a,b,c)$ corresponds to a half-edge of $\mathbb{H}_n$.

When considering the relationship to the 1-2 model, it will be convenient to choose the constants $\epsilon_s$, $s=a,b,c$, as follows. Let $a,b,c>0$, and

$$A=\frac{a-b-c}{a+b+c},\quad B=\frac{b-a-c}{a+b+c},\quad C=\frac{c-a-b}{a+b+c}, \tag{3.5}$$

where we assume for simplicity that $ABC \neq 0$. The appropriate values of the $\epsilon_s$ are

$$\epsilon_a = \sqrt{\frac{BC}{A}}, \quad \epsilon_b = \sqrt{\frac{AC}{B}}, \quad \epsilon_c = \sqrt{\frac{AB}{C}}. \tag{3.6}$$

THEOREM 3.2 ([**10**, Sect. 4]).

(a) Marginal on $V_n$. *For* $\sigma^{\mathrm{v}} \in \Sigma^{\mathrm{v}}$, $w(\sigma^{\mathrm{v}}) := \sum_{\sigma^{\mathrm{e}} \in \Sigma^{\mathrm{e}}} w(\sigma^{\mathrm{v}}, \sigma^{\mathrm{e}})$ *satisfies*

$$w(\sigma^{\mathrm{v}}) = 2^{|E_n|} \prod_{g=\langle u,v\rangle \in E_n} \left(1 + \epsilon_g^2 \sigma_u \sigma_v\right). \tag{3.7}$$

*That is, the marginal weights on* $\Sigma^{\mathrm{v}}$ *are those of an Ising-type model on* $\mathbb{H}_n$ *with (possibly complex) edge-interactions. Here,* $\epsilon_g$ *denotes the parameter associated with edge* $g = \langle u, v\rangle$.

(b) Marginal on $ME_n$. *Let* $a, b, c > 0$ *and assume the* $\epsilon_s$ *satisfy* (3.5)–(3.6) *where* $ABC \neq 0$. *For* $\sigma^{\mathrm{e}} \in \Sigma^{\mathrm{e}}$, $w(\sigma^{\mathrm{e}}) := \sum_{\sigma^{\mathrm{v}} \in \Sigma^{\mathrm{v}}} w(\sigma^{\mathrm{v}}, \sigma^{\mathrm{e}})$ *satisfies*

$$\begin{aligned} w(\sigma^{\mathrm{e}}) &= \prod_{v \in V_n} \left(1 + A\sigma_{v,b}\sigma_{v,c} + B\sigma_{v,a}\sigma_{v,c} + C\sigma_{v,a}\sigma_{v,b}\right) \\ &\propto \mu_n(\sigma^{\mathrm{e}}). \end{aligned}$$

*That is, the marginal weights on* $\Sigma^{\mathrm{e}}$ *are proportional to those of the 1-2 model.*

(c) Two-edge correlation function. *Let* $e = \langle u, v\rangle, f = \langle x, y\rangle \in E_n$. *Subject to the notation of part* (b) *above, the two-edge correlation function of Remark* 2.2 *satisfies*

$$\langle \sigma_e \sigma_f \rangle_n = \sum_{\sigma^{\mathrm{v}} \in \Sigma^{\mathrm{v}}} D_{e,f}(\sigma^{\mathrm{v}}) w(\sigma^{\mathrm{v}}) \bigg/ \sum_{\sigma^{\mathrm{v}} \in \Sigma^{\mathrm{v}}} w(\sigma^{\mathrm{v}}), \tag{3.8}$$

*where*

$$D_{e,f}(\sigma^{\mathrm{v}}) = \frac{\epsilon_e(\sigma_u + \sigma_v)\epsilon_f(\sigma_x + \sigma_y)}{(1 + \epsilon_e^2)(1 + \epsilon_f^2)}.$$

If the weights $w(\sigma^{\mathrm{v}})$ of (3.7) are real and non-negative (which they are not in general), the ratio on the right side of (3.8) may be interpreted as an expectation. The weights $w(\sigma^{\mathrm{v}})$ correspond to a ferromagnetic Ising model if and only the edge-weights of (3.7) satisfy $0 < \epsilon_g^2 < 1$. Unfortunately, this never occurs with the $\epsilon_g$ derived from the 1-2 model as in (3.5)–(3.6). If, however, one assumes that the $a$, $b$, $c$ satisfy the *acute angle condition*

$$a^2 < b^2 + c^2, \quad b^2 < c^2 + a^2, \quad c^2 < a^2 + b^2,$$

then $-1 < \epsilon_g^2 < 0$, and the corresponding Ising model is antiferromagnetic. Since $\mathbb{H}$ is bipartite, this process may be transformed into a ferromagnetic system by changing the sign of every other vertex (see [**6**, p. 17]), and this transformation greatly facilitates its analysis.

The above Ising model may be regarded as a special case of the eight-vertex model of Lin and Wu [**20**].

**3.4. The hexagonal polygon model.** Let $\mathbb{H}_n = (V_n, E_n)$ as before, and let $\Pi_n = \{0,1\}^{E_n}$. The sample space of the polygon model on $\mathbb{H}_n$ is the subset $\Pi_n^{\text{poly}} \subseteq \Pi_n$ containing all $\pi = (\pi(e) : e \in E_n) \in \Pi_n$ such that

$$\sum_{e \ni v} \pi(e) \quad \text{is either 0 or 2,} \qquad v \in V_n. \tag{3.9}$$

Each $\pi \in \Pi_n^{\text{poly}}$ may be considered as a union of vertex-disjoint cycles of $\mathbb{H}_n$, together with isolated vertices. We identify $\pi \in \Pi_n$ with the set $\{e \in E_n : \pi(e) = 1\}$ of 'open' edges under $\pi$. Thus (3.9) requires that every vertex is incident to an even number of open edges.

Let $\epsilon_a, \epsilon_b, \epsilon_c \in \mathbb{C}$. To the configuration $\pi \in \Pi_n^{\text{poly}}$, we assign the (possibly complex) weight

$$w(\pi) = \epsilon_a^{2|\pi(a)|} \epsilon_b^{2|\pi(b)|} \epsilon_c^{2|\pi(c)|}, \tag{3.10}$$

where $\pi(s)$ is the set of open $s$-type edges. The weight function $w$ gives rise to the partition function

$$Z_n^{\text{poly}} = \sum_{\pi \in \Pi_n^{\text{poly}}} w(\pi).$$

Let $a, b, c > 0$. We now choose the constants $\epsilon_s$, $s = a, b, c$, according to (3.5)–(3.6), where it is assumed that $ABC \neq 0$. The corresponding polygon model is related to the high-temperature expansion of the Ising-type model of Section 3.3 (see, for example, [**8**, eqn 5.1] and [**9**, Thm 1.7]).

In considering correlation functions, it is convenient to view the polygon model on the half-edge graph $A\mathbb{H}_n$ of Section 3.2. A polygon configuration on $\mathbb{H}_n$ induces a polygon configuration on $A\mathbb{H}_n$, namely a subset of $AE_n$ with the property that every vertex in $AV_n$ has even degree. For an $a$-type edge $e \in E_n$, the two half-edges of $e$ have weight $\epsilon_a$ each (and similarly for $b$- and $c$-type edges). The weight function $w$ of (3.10) may now be expressed as

$$w(\pi) = \epsilon_a^{|\pi(a)|} \epsilon_b^{|\pi(b)|} \epsilon_c^{|\pi(c)|},$$

where $\pi(s)$ is the set of open half-edges of type $s$, as $\pi$ ranges over polygon configurations on $A\mathbb{H}_n$.

Let $e, f \in ME_n$ be distinct midpoints of $A\mathbb{H}_n$, and let $\Pi_{e,f}$ be the subset of all $\pi \in \{0,1\}^{AE_n}$ such that: (i) every $v \in AV_n$ with $v \neq e, f$ is incident to an even number of open half-edges, and (ii) the midpoints $e$ and $f$ are incident to exactly one open half-edge. Let

$$M_n(e, f) = \frac{Z_{n, e \leftrightarrow f}}{Z_n^{\text{poly}}}, \tag{3.11}$$

where

$$Z_{n, e \leftrightarrow f} := \sum_{\pi \in \Pi_{e,f}} \epsilon_a^{|\pi(a)|} \epsilon_b^{|\pi(b)|} \epsilon_c^{|\pi(c)|}. \tag{3.12}$$

THEOREM 3.3 ([**12**]). *Subject to* (3.6), *the two-edge correlation function* $\langle \sigma_e \sigma_f \rangle_n$ *of the 1-2 model on* $\mathbb{H}_n$ *satisfies* $\langle \sigma_e \sigma_f \rangle_n = M_n(e, f)$.

The polygon model with general parameters is studied in [**12**].

## 4. Phase transition

We discuss two forms of phase transition for the 1-2 model. Of these, the first concerns the existence (or not) of infinite 'homogeneous' clusters of $\mathbb{H}$ containing vertices of the same type, and the second considers as order parameter the limiting two-edge correlation function. Thus, the first studies the geometry of the model, and the second its correlation structure. There may exist other forms of phase transition, as yet unstudied.

**4.1. Occurrence of paths.** Every connected component ('cluster') in a realization of the 1-2 model is either a self-avoiding path or a cycle (see Figure 2.2). It turns out that all such clusters are $\mu_\infty$-a.s. finite, when $a, b, c > 0$. We formalize this statement in this subsection, and begin with an exact formula. Using the correspondence between 1-2 model configurations on $\mathbb{H}$ and dimer configurations on $\mathbb{H}_\Delta$, as described in Section 3.1, we have the following.

THEOREM 4.1 ([**18**, Thm 3.4]). *Let $a, b, c > 0$ and let $\mu_\infty$ be the limit 1-2 measure of Theorem 2.1. Let $\ell$ be a self-avoiding path of $\mathbb{H}_n$ containing $l+1$ edges, and write $E_\ell = \{e_k = \langle u_k, v_k\rangle : 1 \le k \le l\}$ for the set of bisector edges of $\mathbb{H}_\Delta$ encountered along $\ell$, as in Figure 2.3. Then*

$$\mu_\infty(\ell \text{ is present}) = \tfrac{1}{2}\left(\prod_{k=1}^{l} w_{e_k}\right)\left|\operatorname{Pf} K_\ell^{-1}\right|,$$

*where $w_e$ is the weight of the edge $e$ in $\mathbb{H}_\Delta$, $K_\ell^{-1}$ is the submatrix of the inverse of the weighted adjacency matrix of $\mathbb{H}_\Delta$ with rows and columns indexed by $u_1, v_1, \dots, u_l, v_l$, and $\operatorname{Pf} M$ is the Pfaffian of the matrix $M$.*

The mass transport principle, introduced in [**3, 13**], is a valuable tool in the study of interacting systems including percolation and self-avoiding walks on Cayley graphs, see [**2, 11, 14**]. It may also be applied to the 1-2 model, where it is used to prove the following.

THEOREM 4.2 ([**18**, Thm 2.4]). *If $a, b, c > 0$, we have that*

$$\mu_\infty\big(\text{there exists an infinite path}\big) = 0.$$

**4.2. Existence of infinite homogeneous clusters.** The concept of 'phase transition' hinges on the non-smoothness of some so-called 'order parameter'. For the Ising model, one may take as order parameter the magnetization at the origin in the infinite-volume measure with + boundary conditions. This corresponds in the universe of percolation and the random-cluster model (see [**7**]) to studying whether or not there there exists an infinite cluster. By Theorem 4.2, the 1-2 model possesses no infinite cluster for any values of $a, b, c > 0$. 'Clusters' may however be defined in another manner.

Let $\sigma \in \Sigma$. Each vertex of $\mathbb{H}$ has a random type, given in Figure 2.1. For $s \in \{a, b, c\}$, a *type-$s$* cluster is a maximal connected subgraph of $\mathbb{H}$ every vertex of which has type $s$. As illustrated in the figure, type $s$ comes in two sub-types; for example, a type-$a$ vertex has signature either 001 or 110. Thus, one may speak of a 001-cluster, etc. By examining the figure again, it is seen that a type-$a$ cluster is either a 001-cluster or a 110-cluster, but may not contain vertices of both types (and similarly for types $b$ and $c$). A *homogeneous cluster* of $\sigma \in \Sigma$ is a $w$-cluster for some $w \in \{0,1\}^3$, $w \neq 000, 111$. We concentrate now on the existence (or not) of an infinite homogeneous cluster.

THEOREM 4.3. *Let $a, b, c > 0$.*

(a) ([**19**, Thm 1.1]) *Let $w \in \{0,1\}^3$, $w \neq 000, 111$. The number $I_w$ of infinite $w$-clusters is $\mu_\infty$-a.s. no greater than* 1.

(b) ([**18**, Thm 4.4, Prop. 4.7]) *Fix $b, c > 0$. For sufficiently small $a > 0$, there exists $\mu_\infty$-a.s. no infinite type-a cluster. For sufficiently large $a$, the $\mu_\infty$-probability that the origin belongs to an infinite type-a cluster is strictly positive.*

Part (a) is proved in [**19**] using an adaptation of the method of Burton and Keane (see [**4**, **22**]) to the 1-2 model, subject to the complication that $\mu_\infty$ does not have the so-called 'finite energy property'. It is unknown whether infinite $w$-clusters and $w'$-clusters can coexist with $w \neq w'$.

Part (b) indicates the existence of a phase transition. It is not known if there exists a single critical point $a_c = a_c(b, c)$ for the given property. Furthermore, there is currently no indication of the exact value of such a point.

**4.3. Non-analyticity of the two-edge correlation function.** For distinct edges $e, f \in \mathbb{E}$, we write

$$\langle \sigma_e \sigma_f \rangle = \lim_{n \to \infty} \langle \sigma_e \sigma_f \rangle_n,$$

which exists by Theorem 2.1, (see [**10**, Thm 6.2]). We consider here the asymptotic behaviour of $\langle \sigma_e \sigma_f \rangle$ as $|e - f| \to \infty$. The behaviour of this limit is unknown in general, but a great deal is known if $e$ and $f$ are related in the 'diagonal' manner of the forthcoming assumption (4.1), as illustrated in Figure 4.1.

THEOREM 4.4 ([**10**, Thm 3.1]). *Let $a, b, c > 0$, and let $e, f \in \mathbb{E}$ be NW/SE edges such that:*

(4.1) *there exists a path $\ell = \ell(e, f)$ of $\mathbb{H}_n$ from $e$ to $f$ using only horizontal and NW/SE half-edges.*

(a) *Let $a \geq b > 0$. For $c > 0$ satisfying*

$$\text{either } \sqrt{a} > \sqrt{b} + \sqrt{c} \quad \text{or} \quad \sqrt{c} > \sqrt{a} + \sqrt{b}$$

*except possibly on some set of isolated points, the limit $\lim_{|e-f| \to \infty} \langle \sigma_e \sigma_f \rangle^2$ exists and is non-zero.*

(b) *If $a \geq b > 0$ and*

$$\sqrt{a} - \sqrt{b} < \sqrt{c} < \sqrt{a} + \sqrt{b},$$

*then $\langle \sigma_e \sigma_f \rangle \to 0$ as $|e - f| \to \infty$.*

A strengthened version of this theorem may be found in [**10**], allowing a partial relaxation of assumption (4.1), and including a proof of an exponential rate of convergence.

By Theorem 4.4, when $a \geq b \geq c > 0$, the phase transition occurs when $\sqrt{a} = \sqrt{b} + \sqrt{c}$. The proof is along the following lines. The square $\langle \sigma_e \sigma_f \rangle^2$ of the two-edge correlation may be expressed as the determinant of an explicit block Toeplitz matrix (this is where (4.1) is used); see the forthcoming Lemma 4.5. Its limit as $|e - f| \to \infty$ is given by Widom's theorem (see [**27**, **28**] and [**10**, Thm 8.7]) as the determinant of the limiting (infinite) block Toeplitz matrix. This determinant

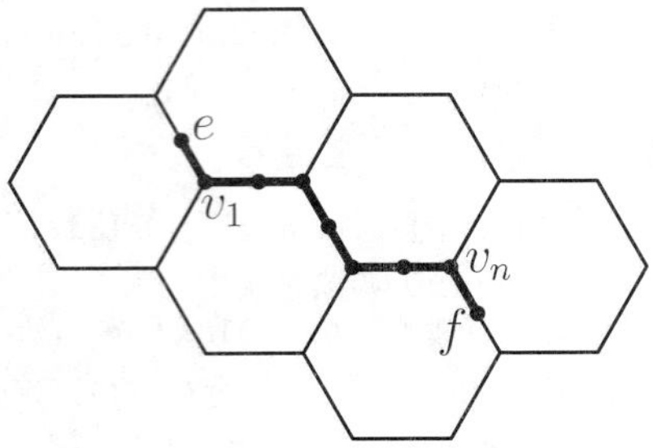


FIGURE 4.1. A path $\ell$ comprising horizontal and NW/SE mid-edges, connecting the midpoints of two NW/SE edges $e$ and $f$.

is complex analytic with respect to the parameters $a$, $b$, $c$ except when the spectral curve has a unique real zero on the unit torus. As remarked at the end of Section 3.1, the last occurs if and only if $\sqrt{a} \pm \sqrt{b} \pm \sqrt{c} = 0$. When $a \geq b \geq c$, this equation becomes $\sqrt{a} - \sqrt{b} - \sqrt{c} = 0$.

The 'isolated points' of part (b) arise through the use in the proof of the fact that, for an analytic function $\Lambda$, either $\Lambda \equiv 0$ or the zeros of $\Lambda$ are isolated.

It is easily seen from Figures 2.1 and 4.1 that

$$\langle \sigma_e \sigma_f \rangle = 1 \qquad \text{if either} \quad a, b > 0,\ c = 0, \quad \text{or} \quad a = b = 0,\ c > 0,$$

and part (b) of Theorem 4.4 follows by the analyticity. For part (a), one uses the representation of the 1-2 model as the Ising model of Section 3.3.

The key step in the proof of Theorem 4.4 is the following exact formula. Let $Y_1$ be the $2 \times 2$ matrix

$$Y_1 = \begin{pmatrix} 0 & 1 \\ -1 & 0 \end{pmatrix},$$

and let $Y_{2k}$ be the $4k \times 4k$ block diagonal matrix with diagonal $2 \times 2$ blocks equal to $Y_1$. That is,

$$Y_{2k} = \begin{pmatrix} Y_1 & 0 & \cdots & 0 \\ 0 & Y_1 & \cdots & 0 \\ \vdots & \vdots & \ddots & \vdots \\ 0 & 0 & \cdots & Y_1 \end{pmatrix}.$$

LEMMA 4.5 ([**10**, Lemma 8.4]). *Suppose the path $\ell = \ell(e, f)$ of (4.1) passes $2k$ bisector edges of $\mathbb{H}_\Delta$. We have that*

$$\langle \sigma_e \sigma_f \rangle = \operatorname{Pf}\,[Y_{2k} + 2cK_\ell^{-1}],$$

*where $K_\ell^{-1}$ is as in Theorem 4.1.*


## Acknowledgements

This work was supported in part by the Engineering and Physical Sciences Research Council under grant EP/I03372X/1. ZL acknowledges support from the Simons Foundation grant #351813 and the National Science Foundation grant DMS 1608896. The authors thank the anonymous referee for the careful reading and useful comments.

## References

[1] R. J. Baxter, *Exactly solved models in statistical mechanics*, Academic Press, Inc. [Harcourt Brace Jovanovich, Publishers], London, 1982. MR690578

[2] I. Benjamini, R. Lyons, Y. Peres, and O. Schramm, *Critical percolation on any nonamenable group has no infinite clusters*, Ann. Probab. **27** (1999), no. 3, 1347–1356, DOI 10.1214/aop/1022677450. MR1733151

[3] I. Benjamini, R. Lyons, Y. Peres, and O. Schramm, *Group-invariant percolation on graphs*, Geom. Funct. Anal. **9** (1999), no. 1, 29–66, DOI 10.1007/s000390050080. MR1675890

[4] R. M. Burton and M. Keane, *Density and uniqueness in percolation*, Comm. Math. Phys. **121** (1989), no. 3, 501–505. MR990777

[5] M. E. Fisher, *Statistical mechanics of dimers on a plane lattice*, Phys. Rev. (2) **124** (1961), 1664–1672. MR0136399

[6] H.-O. Georgii, O. Häggström, and C. Maes, *The random geometry of equilibrium phases*, Phase Transit. Crit. Phenom., vol. 18, Academic Press, San Diego, CA, 2001, pp. 1–142, DOI 10.1016/S1062-7901(01)80008-2. MR2014387

[7] G. R. Grimmett, *The Random-Cluster Model*, Springer, Berlin, 2006, available at `http://www.statslab.cam.ac.uk/~grg/books/rcm.html`.

[8] G. Grimmett, *Flows and ferromagnets*, Combinatorics, complexity, and chance, Oxford Lecture Ser. Math. Appl., vol. 34, Oxford Univ. Press, Oxford, 2007, pp. 130–143, DOI 10.1093/acprof:oso/9780198571278.003.0008. MR2314565

[9] G. Grimmett and S. Janson, *Random even graphs*, Electron. J. Combin. **16** (2009), no. 1, Research Paper, 46, 19. MR2491648

[10] G. R. Grimmett and Z. Li, *Critical surface of the 1-2 model*, Int. Math. Res. Notes (2017), DOI 10.1093/imrn/rnx066.

[11] G. R. Grimmett and Z. Li, *Connective constants and height functions of Cayley graphs*, Trans. Amer. Math. Soc. **369** (2017), 5961–5980.

[12] G. R. Grimmett and Z. Li, *Critical surface of the hexagonal polygon model*, J. Stat. Phys. **163** (2016), no. 4, 733–753, DOI 10.1007/s10955-016-1497-9. MR3488571

[13] O. Häggström, *Infinite clusters in dependent automorphism invariant percolation on trees*, Ann. Probab. **25** (1997), no. 3, 1423–1436, DOI 10.1214/aop/1024404518. MR1457624

[14] O. Häggström and Y. Peres, *Monotonicity of uniqueness for percolation on Cayley graphs: all infinite clusters are born simultaneously*, Probab. Theory Related Fields **113** (1999), no. 2, 273–285, DOI 10.1007/s004400050208. MR1676835

[15] P. W. Kasteleyn, *The statistics of dimers on a lattice, I. The number of dimer arrangements on a quadratic lattice*, Physica **27** (1961), 1209–1225.

[16] Z. Li, *Local statistics of realizable vertex models*, Comm. Math. Phys. **304** (2011), no. 3, 723–763, DOI 10.1007/s00220-011-1240-y. MR2794546

[17] Z. Li, *Critical temperature of periodic Ising models*, Comm. Math. Phys. **315** (2012), no. 2, 337–381, DOI 10.1007/s00220-012-1571-3. MR2971729

[18] Z. Li, *1-2 model, dimers, and clusters*, Electron. J. Probab. **19** (2014), no. 48, 28, DOI 10.1214/EJP.v19-2563. MR3217336

[19] Z. Li, *Uniqueness of the infinite homogeneous cluster in the 1-2 model*, Electron. Commun. Probab. **19** (2014), no. 23, 8, DOI 10.1214/ECP.v19-3105. MR3197119

[20] K. Y. Lin and F. Y. Wu, *General 8-vertex model on the honeycomb lattice: equivalence with an Ising model*, Modern Phys. Lett. B **4** (1990), no. 5, 311–316, DOI 10.1142/S0217984990000398. MR1044723

[21] B. McCoy and T. T. Wu, *The Two-Dimensional Ising Model*, Harvard University Press, Cambridge MA, 1973.

[22] C. M. Newman and L. S. Schulman, *Infinite clusters in percolation models*, J. Statist. Phys. **26** (1981), no. 3, 613–628, DOI 10.1007/BF01011437. MR648202

[23] M. Schwartz and J. Bruck, *Constrained codes as networks of relations*, IEEE Trans. Inform. Theory **54** (2008), no. 5, 2179–2195, DOI 10.1109/TIT.2008.920245. MR2450856

[24] H. N. V. Temperley and M. E. Fisher, *Dimer problem in statistical mechanics—an exact result*, Philos. Mag. (8) **6** (1961), 1061–1063. MR0136398

[25] L. G. Valiant, *Holographic algorithms*, SIAM J. Comput. **37** (2008), no. 5, 1565–1594, DOI 10.1137/070682575. MR2386281

[26] B. L. van der Waerden, *Die lange Reichweite der regelmässigen Atomanordnung in Mischkristallen*, Zeit. Physik **118** (1941), 473–488.
[27] H. Widom, *On the limit of block Toeplitz determinants*, Proc. Amer. Math. Soc. **50** (1975), 167–173, DOI 10.2307/2040534. MR0370254
[28] H. Widom, *Asymptotic behavior of block Toeplitz matrices and determinants. II*, Advances in Math. **21** (1976), no. 1, 1–29, DOI 10.1016/0001-8708(76)90113-4. MR0409512

Statistical Laboratory, Centre for Mathematical Sciences, Cambridge University, Wilberforce Road, Cambridge CB3 0WB, United Kingdom
*E-mail address*: g.r.grimmett@statslab.cam.ac.uk
*URL*: http://www.statslab.cam.ac.uk/~grg/

Department of Mathematics, University of Connecticut, Storrs, Connecticut 06269-3009
*E-mail address*: zhongyang.li@uconn.edu
*URL*: http://www.math.uconn.edu/~zhongyang/

Contemporary Mathematics
Volume **696**, 2017
http://dx.doi.org/10.1090/conm/696/14021

# Meromorphic quadratic differentials with complex residues and spiralling foliations

Subhojoy Gupta and Michael Wolf

ABSTRACT. A meromorphic quadratic differential with poles of order two, on a compact Riemann surface, induces a measured foliation on the surface, with a spiralling structure at any pole that is determined by the complex residue of the differential at the pole. We introduce the space of such measured foliations, and prove that for a fixed Riemann surface, any such foliation is realized by a quadratic differential with second order poles at marked points. Furthermore, such a differential is uniquely determined if one prescribes complex residues at the poles that are compatible with the transverse measures around them. This generalizes a theorem of Hubbard and Masur concerning holomorphic quadratic differentials on closed surfaces, as well as a theorem of Strebel for the case when the foliation has only closed leaves. The proof involves taking a compact exhaustion of the surface, and considering a sequence of equivariant harmonic maps to real trees that do not have a uniform bound on total energy.

## 1. Introduction

A holomorphic quadratic differential on a Riemann surface induces a singular foliation equipped with a transverse measure. For a closed Riemann surface $\Sigma$, the Hubbard-Masur theorem ([**HM79**]) asserts that this correspondence provides a homeomorphism between the space of holomorphic quadratic differentials and the space of equivalence classes of measured foliations on $\Sigma$. Both provide useful descriptions of the space of infinitesimal directions in Teichmüller space at the point $\Sigma$, and the theorem established an important bridge between the complex-analytic and topological aspects of the theory. The purpose of this article is to provide a generalization of this result to quadratic differentials with poles of order two.

For this, we shall introduce a space $\mathcal{MF}_2$ of the corresponding topological objects on the surface, namely equivalence classes of measured foliations with *centers* (disks foliated by closed circles or spirals - see Figure 3) around finitely many marked points. The data for such a foliation includes the transverse measure of a loop linking each distinguished point, in addition to the transverse measures of a system of arcs on the surface that provide coordinates on the usual space of measured foliations introduced by Thurston (see [**FLP12**]).

---

2010 *Mathematics Subject Classification.* Primary 30F30, 30F60.

On the complex-analytic side, at a pole $p$ of order two, a quadratic differential is locally of the form :

$$\left(\frac{a^2}{z^2}\right) dz^2 \text{ where } a \in \mathbb{C}^* \tag{1.1}$$

with respect to a choice of coordinate chart around the pole.

DEFINITION 1.1 (Residue). The constant $a$ in (1.1) shall be called the *(complex) residue* of the differential at the pole. The ambiguity of sign of the square root is resolved by choosing the one that makes the real part of $a$ positive, or in case it is zero, the sign such that the imaginary part of $a$ is positive. We shall denote the residue at a pole $p$ by $\text{Res}(p)$.

The residue is in fact a coordinate-independent quantity. We also note that the coefficient $a^2$ is often called the residue (or *quadratic* residue) in the literature; the present definition makes subsequent notation less cumbersome. For example, the transverse measure of a loop around the pole, for the induced measured foliation, is a constant multiple of the real part of the residue as defined above (see (2.3) in §2.2).

We shall prove:

THEOREM 1.2. *Let $\Sigma$ be a closed Riemann surface of genus $g$, with marked points $P = \{p_j\}_{1\leq j\leq n}$, such that $2g-2+n>0$. Let $F \in \mathcal{MF}_2(\Sigma, P)$ be a measured foliation on $\Sigma$ with centers at points of $P$. Then for any tuple $(c_1, c_2, \dots c_n) \in \mathbb{R}^n_+$ there is a unique meromorphic quadratic differential with*

- *a pole of order two at each $p_j$,*
- *an induced measured foliation that is equivalent to $F$, and*
- *a residue at $p_j$ having a prescribed imaginary part, namely,*

$$\Im(Res(p_j)) = \begin{cases} c_j & \text{if } \Re(Res(p_j)) = 0 \\ \ln c_j & \text{if } \Re(Res(p_j)) > 0 \end{cases}$$

*for each $1 \leq j \leq n$.*

*Remarks.* (i) In the case when the real part of the residue is non-zero (and hence positive, by our sign convention), a neighborhood of the pole is foliated by leaves that spiral into it. The spiralling nature is *not* captured by the parameters (namely, transverse measures of loops) determining the measured foliation $F$; indeed, different spiralling behaviour at the pole are isotopic to each other (see §4). It is only the total transverse measure around the pole which is fixed by the choice of $F$, and this determines the real part of the residue. The free parameter is therefore the imaginary part of the residue; this then determines the spiralling behaviour, and its sign represents the "handedness" of the spiralling, that is, clockwise or counter-clockwise with respect to the orientation on the surface.

(ii) The condition of prescribing the imaginary part of the residue can be replaced with the geometric requirement of

(1) prescribing the circumference of the cylinder surrounding the pole $p_j$ in the induced singular-flat metric (see §2.2) and
(2) prescribing the "handedness" of the spiralling mentioned in (i) above.

Note that the circumference in (1) must not be less than the total transverse measure around the pole.

(iii) A special case is when the foliation $F$ has the global property that all leaves are closed (excepting a connected critical graph) and foliate punctured disks around each pole. In this case, the transverse measure of any loop around a pole vanishes, since each linking loop can be chosen to be a closed leaf of $F$ (which has no transverse intersection with the foliation). From the relation between the residue and transverse measure, it follows that the real part of the residue is zero. Theorem 1.2 then asserts that an additional positive real parameter (the imaginary part of the residue) uniquely determines a differential realizing $F$. This recovers a well-known theorem of Strebel (see Theorem 23.5 of [**Str84**] or [**Liu08**]).

The present paper deals with the case of poles of order *exactly* two. The case of order-1 poles, that we elide, falls under the purview of classical theory: the corresponding quadratic differentials are then integrable, with the induced foliation having a "fold" at the puncture. The Hubbard-Masur theorem can be applied after taking a double cover of the surface branched at such poles.

The proof of Theorem 1.2 involves the technique of harmonic maps to $\mathbb{R}$-trees used by one of us ([**Wol96**]) in an alternative proof of the Hubbard-Masur theorem. Such an $\mathbb{R}$-tree forms the leaf-space of the measured foliation when lifted to the universal cover $\widetilde{\Sigma}$, and the required quadratic differential is then recovered as (a multiple of) the Hopf differential of the equivariant harmonic map from $\widetilde{\Sigma}$ to the tree. The construction of the $\mathbb{R}$-tree as the leaf-space ignores the differences in foliations caused by Whitehead moves; thus the analysis avoids issues relating to the presence of higher order zeroes of the differential, and their possible decay under deformation to multiple lower order zeroes, that was a difficulty in the approach of Hubbard and Masur.

In the present paper the main difficulty lies in the fact that surfaces have punctures and the corresponding harmonic maps have infinite energy. The strategy for the proof of existence is to establish uniform energy bounds for harmonic maps defined on a compact exhaustion of the surface. In the conclusion of the proof, some additional work is needed to verify that the Hopf differential of the limiting map has the desired complex residue.

A similar strategy was used in ([**GW16**]) for a generalization of Strebel's theorem (mentioned above) to the case of quadratic differentials with poles of higher order. There, the global "half-plane structure" of the foliation allowed us to consider harmonic maps from the punctured Riemann surface $(\Sigma \setminus P)$ to a $k$-pronged tree. In this paper we consider arbitrary foliations, which necessitates working in the universal cover and considering a target space that is a more general $\mathbb{R}$-tree. Furthermore, a difficulty unique to the case of double order poles is the "spiralling" nature of the foliations at a (generic) such pole.

There are two novel technical elements in meeting the difficulties of the infinite energy in this setting of harmonic maps for such spiralling foliations. First, we use a version of the length-area method that is adapted to the collapsing maps for such foliations; the resulting estimate (Proposition 4.6) takes into account both horizontal and vertical stretches. Second, we need to control the supporting annulus of the spirals or twists for a map in the approximating sequence: this is done (Lemma 5.3) by using transverse measures (that determine distances in the target $\mathbb{R}$-tree) to bound the size of the subsurface to which the twisting can retreat.

Harmonic maps provides a convenient analytical tool in this paper; apart from the advantage over the approach of Hubbard-Masur in avoiding issues related to Whitehead moves mentioned earlier, a crucial fact used in the proof of the uniform energy bound (Lemma 5.6) is that for a non-positively curved target, such a map is unique and energy-minimizing.

In the forthcoming work [**GW**] we use ideas in these papers to prove a full generalization of the Hubbard-Masur theorem to quadratic differentials with higher-order poles.

**Acknowledgements.** The authors thank the anonymous referee for very helpful comments, and gratefully acknowledge support for collaborative travel by NSF grants DMS-1107452, 1107263, 1107367 "RNMS: GEometric structures And Representation varieties" (the GEAR network). The first author was supported by the Danish Research Foundation Center of Excellence grant and the Centre for Quantum Geometry of Moduli Spaces (QGM), as well as the hospitality of MSRI (Berkeley). He would like to thank Daniele Allessandrini and Maxime Fortier-Bourque for helpful discussions. The second author gratefully acknowledges support from the U.S. National Science Foundation (NSF) through grants DMS-1007383 and DMS-1564374.

## 2. Preliminaries

**2.1. Quadratic differentials, measured foliations.** Let $\Sigma$ be a closed Riemann surface of genus $g \geq 2$.

DEFINITION 2.1. A *meromorphic quadratic differential* on $\Sigma$ is a $(2,0)$-tensor that is locally of the form $q(z)dz^2$ where $q(z)$ is meromorphic. Alternatively, it is a meromorphic section of $K^{\otimes 2}$, where $K$ is the canonical line bundle on $\Sigma$.

By Riemann-Roch, the complex vector space of meromorphic quadratic differentials with poles at points $p_1, p_2, \dots p_n$ of orders at most $k_1, k_2, \dots k_n$ respectively, has (complex) dimension $3g - 3 + \sum\limits_{i=1}^{n} k_i$.

DEFINITION 2.2. The *horizontal foliation* induced by a quadratic differential $q$ is a (singular) foliation on the surface whose leaves are the integral curves of the line field along which the quadratic differential is real and positive (that is, $\pm v \in T\Sigma$ such that $q(v \otimes v) \in \mathbb{R}_+$). The singularities are at the zeros of $q$, where there is a "prong"-type singularity (see Figure 1), and at the poles (for poles of order two, as in this paper, see Figure 3).

DEFINITION 2.3. The *singular-flat metric* induced by a quadratic differential $q$ is a singular metric $ds_q^2$ which is locally of the form $ds_q^2 = |q(z)||dz^2|$. The singularities of the metric are at the zeros and first order poles of $q$, where there is a cone-angle $\pi(k+2)$ where $k$ is the order of the zero or pole. The metric is complete in the neighborhood of a second order pole; we can choose such a neighborhood to be isometric to a portion of a half-infinite cylinder (see §2.2).

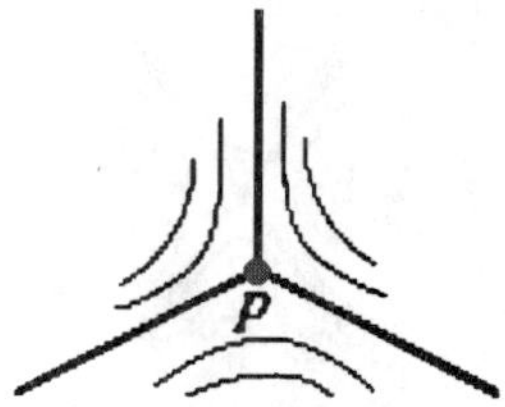


FIGURE 1. The induced (horizontal) foliation at a simple zero has a 3-prong singularity with a cone angle of $3\pi$.

The *horizontal* and *vertical* lengths of an arc $\alpha$ are defined to be the absolute values of the real and imaginary parts, respectively:

$$l_h(\alpha) = |\Re \int_\alpha \sqrt{q(z)}dz|, \; l_v(\alpha) = |\Im \int_\alpha \sqrt{q(z)}dz| \tag{2.1}$$

where we note that the length of a geodesic arc $\alpha$, which avoids singularities, in the induced metric is $l(\alpha) = \left(l_h(\alpha)^2 + l_v(\alpha)^2\right)^{1/2}$.

The horizontal foliation $\mathcal{F}(q)$ induced by a quadratic differential $q$ (Defn. 2.2) can be equipped with a *transverse measure*, namely the vertical length of any transverse arc. Thus, the foliation $\mathcal{F}$ is *measured* in the following sense:

DEFINITION 2.4 (Measured foliation). A measured foliation on a smooth surface $S$ of genus $g \geq 2$ is a smooth foliation with finitely many "prong" singularities, equipped with a measure on transverse arcs that is invariant under transverse homotopy. A pair of measured foliations are said to be *equivalent*, if they differ by an isotopy of leaves and Whitehead moves (see Figure 2).

The space $\mathcal{MF}$ of equivalence classes of measured foliations is homeomorphic to $\mathbb{R}^{6g-6}$. (See for example, Exposé 6 of [**FLP12**].)

One way to obtain this parameterization is to consider a pants decomposition of the surface, and use the transverse measures of each of the $3g-3$ pants curves, together with the *twisting number* of each. Here, the twisting number of a curve can be defined to be the nonnegative real number that is the transverse measure across a maximal annulus on the surface with a core curve homotopic to the given one. This is known as the Dehn-Thurston parametrization (see [**Thu**] and [**PH92**] for details).

The same definition of a measured foliation as above holds for a surface with boundary; the leaves of the foliation may be parallel or transverse at the boundary components, and prong singularities may also lie on the boundary.

*Notation.* In this paper the *induced measured foliation* for a (meromorphic) quadratic differential $q$ shall mean its horizontal foliation, denoted $\mathcal{F}(q)$.

As mentioned in the introduction, this article is motivated by the following correspondence between measured foliations and *holomorphic* quadratic differentials.

THEOREM (Hubbard-Masur [**HM79**]). *Given a closed Riemann surface $\Sigma$ of genus $g \geq 2$, and a measured foliation $F \in \mathcal{MF}$, there is a unique holomorphic quadratic differential whose induced measured foliation is equivalent to $F$.*

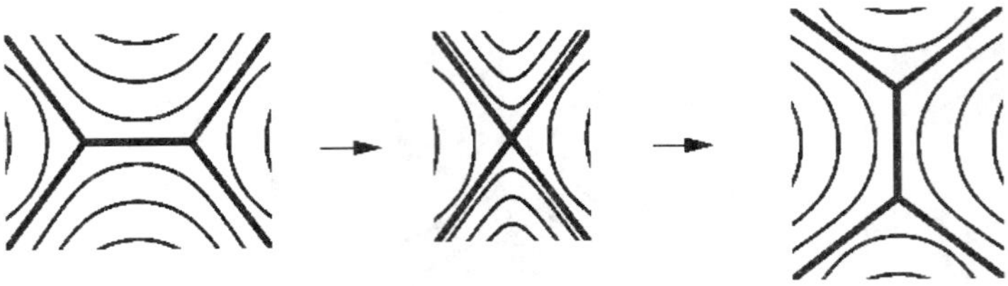

FIGURE 2. A Whitehead move preserves the equivalence class of a measured foliation.

**2.2. Structure at a double order pole.** In this article, we shall be concerned with quadratic differentials with poles of order two, that is, when the local expression of the differential at the pole is:

$$\left(\frac{a^2}{z^2} + \frac{c_1}{z} + c_0 + \cdots\right) dz^2 \tag{2.2}$$

where $a \in \mathbb{C}^*, c_0, c_1 \in \mathbb{C}$. Note that it is always possible to choose a coordinate $z$ where this expression is of the form (1.1) - see §7.2 of [**Str84**]. Here the complex constant $a$, with a choice of sign as defined in Definition 1.1, is the *residue* at the pole, and is easily seen to be independent of choice of coordinate chart.

**Induced measured foliation.** There is always a neighborhood $U$ of an order two pole such that any horizontal trajectory in $U$ continues to the pole in one direction, unless the trajectories close up to form closed leaves foliating the neighborhood. We shall refer to such a neighborhood as a *sink neighborhood* of the foliation. See Chapter III, §7.2 of [**Str84**] for details.

There are three possibilities for what the induced foliation looks like in a sink neighborhood of an order two pole, depending on the residue, as defined in Definition 1.1:

- $\Re(a) = 0$: The foliation is by concentric circles around the pole. In particular, each leaf in $U$ is closed.
- $\Im(a) = 0$: The foliation is by radial lines from the pole.
- $\Re(a) \neq 0$ and $\Im(a) \neq 0$: Each horizontal leaf in $U$ spirals into the pole.

See Figure 3 and section 4 of [**MP98**] for examples.

The transverse measure of a loop $\gamma$ linking the pole is computed by an integral as in (2.1). We obtain:

$$\tau(\gamma) = |\Im\left(\int_\gamma \sqrt{q}\right)| = |\Im(2\pi i a)| = 2\pi\Re(a) \tag{2.3}$$

and is thus determined by the real part of the complex residue. (Recall from Definition 1.1 that we have chosen the sign of the residue such that the real part is non-negative.)

**Metric.** By a similar computation using (2.1), one can check that in the induced singular-flat metric, a sink neighborhood of the pole is a half-infinite Euclidean cylinder of circumference $2\pi|a|$, where $a \in \mathbb{C}^*$ is the residue.

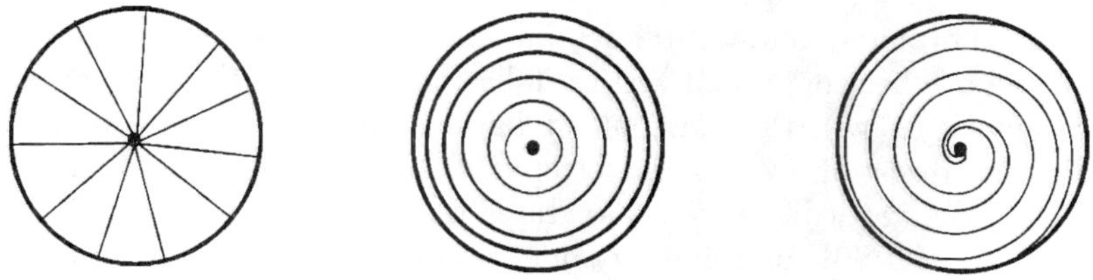

FIGURE 3. The induced foliation at a double order pole.

**2.3. Harmonic maps to $\mathbb{R}$-trees.** Let $F$ be a measured foliation on a smooth surface $S$ (see Defn. 2.4). The leaf space of its lift $\tilde{F}$ to the universal cover, metrized by the projection along the leaves of its transverse measure, defines an $\mathbb{R}$-tree: namely, a simply connected metric graph $T$ such that any two points of $T$ has a unique geodesic (isometric embedding of a segment) between them. For definitions and details we refer to [**Wol95**]; see also [**Kap09**] for more background on the construction and uses of $\mathbb{R}$-trees, and [**Sun03**] for some useful analytical background, including restrictions on the shape of the image of a harmonic map. See [**Sch**] for an insightful treatment of the following notion of Hopf differentials, and [**DW07**] for a very nice survey of both results and techniques in the uses of harmonic maps in Teichmüller theory. Jost [**Jos91**] provides detailed proofs of some of the results mentioned in the [**DW07**] survey, together with a tasteful choice of useful techniques and perspectives.

DEFINITION 2.5 (Harmonic map, Hopf differential). Let $X$ be a (possibly non-compact) Riemann surface. A map $h : X \to T$ that has locally square-integrable derivatives is *harmonic* if it is a critical point of the energy functional

$$\mathcal{E}(f) = \int\limits_X \left(|h_z|^2 + |h_{\bar{z}}|^2\right) dz d\bar{z} \tag{2.4}$$

in the case when this integral takes a finite value.

When this energy is infinite, as in a number of situations in this paper, we adapt the definition to say a map is harmonic when the above holds for the restriction of the energy functional to any relatively compact set $U$ in $X$. Namely, the derivative of the above energy functional, *restricted to* $U$, vanishes for any variation of the map with compact support in $U$.

*Notation.* In later sections, when we define equivariant maps on the universal cover of a surface, $\bar{\mathcal{E}}(\widetilde{f})$ shall denote the *equivariant* energy, that is, the energy of the restriction of $\widetilde{f}$ to a fundamental domain of the action. (In particular, if the surface is compact, then the equivariant energy is finite.)

For a harmonic map $h$ with locally square-integrable derivatives, we define the *Hopf differential* as:

$$\text{Hopf}(h) = -4(h_z)^2 dz^2.$$

This quadratic differential is then locally in $L^1$ on the Riemann surface, and Schoen [**Sch84**] points out the stationary nature of the map $h$ implies that this differential is weakly holomorphic, hence strongly holomorphic by Weyl's Lemma. Thus it is a classical holomorphic quadratic differential with isolated zeroes (if not identically zero) and smooth horizontal and vertical foliations. Now, in the convention chosen (see the remark below), the horizontal foliation relates to the harmonic map as integrating the directions of minimal stretch of the differential $dh$ of the harmonic map $h$, while the vertical foliation integrates the directions of maximal stretch of $dh$. (See e.g. [**Wol89**] for the easy computations which justify these geometric statements.) Thus, when the harmonic map $h$ is a projection to a tree, the minimal stretch direction lies along the kernel of the differential map $dh$, and thus the horizontal leaves are the level sets of points in the tree. We see then that, away from the isolated zeroes of the Hopf differential, the harmonic map takes disks to geodesic segments in the $\mathbb{R}$-tree $T$.

*Remark.* The constant in the definition above is chosen such that the Hopf-differential of the map $z \mapsto \Im z$ is $dz^2$, that is, the Hopf-differential realizes the horizontal foliation. In particular, the sign convention differs from the one used in, say [**Wol96**], and in other places in the harmonic maps literature.

In particular, for a holomorphic quadratic differential $q$ on $\Sigma$, let $c = c(q) : \tilde{\Sigma} \to T$ be the collapsing map to the leaf space of the lift of the measured foliation $\mathcal{F}(q)$, to the universal cover. Namely, the map takes each leaf to a unique point in $T$. Recall that the $\mathbb{R}$-tree $T$ is metrized: distance is given by the pushforward of the transverse measure. The collapsing map is harmonic in the sense above: locally, it is the harmonic function $z \mapsto \Im(z)$, and the Hopf differential recovers the quadratic differential $q$, as a the reader can easily verify. The energy (2.4) equals half the area of the domain in the singular-flat metric induced by the differential. As in the previous paragraph, geometrically the horizontal foliation is along the directions of least stretch for the harmonic map, and we shall refer to this as the *collapsing foliation*.

The above observation has been used by one of us ([**Wol96**]) in a proof of the Hubbard-Masur Theorem (stated in §2.1). Namely, for a Riemann surface $\Sigma$ and measured foliation $F$, let $T$ be the leaf space of the lift of the foliation to the universal cover, metrized by the transverse measure, and consider the Hopf differential of the energy-minimizing map $h : \tilde{\Sigma} \to T$ that is equivariant with respect to the surface group action on both spaces. To prove such an energy-minimizing harmonic map exists, one crucially needs the surface is compact, which provides a global energy bound. This Hopf differential of the energy-minimizer then descends to $\Sigma$ to give the required holomorphic quadratic differential. In this paper, we adapt the strategy above for the corresponding measured foliations, and their harmonic collapsing maps of *infinite energy*; more precisely, we study limits of sequences $h_n$ of harmonic maps from a compact exhaustion of a punctured surface. There will be no *a priori* bound on the total energies of the elements of these sequences, so we will need a finer analysis to control the shapes of the maps.

## 3. Measured foliations with centers

Let $S$ be a smooth surface of genus $g$ with a set $P$ of marked points such that $|P| = n$. We shall henceforth assume that the Euler characteristic of the punctured surface is negative, namely $\chi(S \setminus p) = 2 - 2g - n < 0$, so the surface admits a complete hyperbolic metric.

We refer to the previous section for the usual notion of measured foliations.

DEFINITION 3.1 (Center). A *center* is a foliation on the punctured unit disk $\mathbb{D}^*$ that is isotopic, relative to the boundary, to one of the two standard foliations:

(1) (Radial center) A foliation by radial lines to the puncture, defined by the kernel of the smooth 1-form $\Im(z^{-1}dz)$,
(2) (Closed center) A foliation by concentric circles around the puncture, defined by the kernel of the smooth 1-form $\Re(z^{-1}dz)$.

*Remark.* As discussed in §2.2, a pole of order two as in (1.1) induces a foliation on $\mathbb{D}^*$ that generically has leaves that spiral into the puncture. Any such spiralling foliation is isotopic in the punctured disk, to a radial center: namely, the smoothly varying family of foliations induced by

$$\frac{(\Re a + (1-t)\Im a)^2}{z^2} dz^2 \text{ for } 0 \leq t \leq 1$$

has the initial foliation at $t = 0$, and a radial center at $t = 1$.

DEFINITION 3.2. A *measured foliation on $S$ with centers at $P$* is a smooth foliation on $S$ away from finitely many singularities such that

- a neighborhood of each point of $P$ is a center,
- all other singularities are of prong type, and
- the foliation is equipped with a finite positive measure on compact arcs on $S \setminus P$ transverse to the foliation, that is invariant under transverse homotopy.

*Remark.* Equivalently, a measured foliation as in the definition above can be described as follows. Consider an atlas of charts $\{(U_i, \phi_i)\}$ away from $S \setminus (P \cup Z)$, where $\phi_i$ are closed smooth 1-forms such that $\phi = \pm\phi_j$ on $U_i \cap U_j$. The foliation away from $P$ is locally defined to be the integral curves of the line field associated with the kernel of the 1-forms. Moreover, in a neighborhood of the the points of $Z$, the foliation is given by the kernel of the 1-form $\Im(z^{k/2}dz)$ (see Figure 1 for the case $k = 1$) and of the points of $P$, that the kernel of $\Im(z^{-1}dz)$ or $\Re(z^{-1}dz)$ (see Definition 3.1). The line element $|\phi_i|$ provides a well-defined transverse measure, that is invariant under transverse homotopy because the forms are closed.
In this description, the horizontal measured foliation induced by a meromorphic quadratic differential $q$ (*cf.* Definition 2.2) is obtained by taking the 1-forms to be locally $\pm\Im(\sqrt{q})$.

The main theorem of the paper can then be interpreted as a Hodge theorem for such smooth objects, namely, that any measured foliation with centers as above can be realized (up to equivalence) as one induced by a meromorphic quadratic differential.

In analogy with the space $\mathcal{MF}$ on a closed surface (*cf.* Definition 2.4) we have:

DEFINITION 3.3. Let $\mathcal{MF}_2(S, P)$ denote the space of measured foliations with centers, up to the equivalence relations of isotopy and Whitehead moves.

(We shall drop the reference to the surface $S$ and set of centers $P$ when it is obvious from the context.)

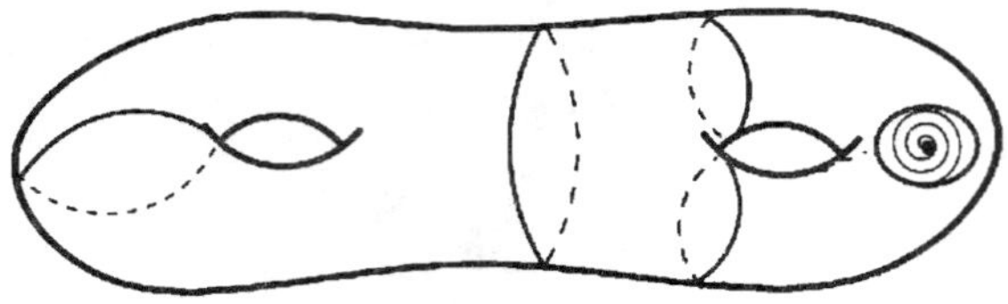

FIGURE 4. The space of measured foliations with a single center on a genus 2 surface has dimension 9: each interior pants curve contributes two real parameters, and the closed curve around the center contributes one (Proposition 3.4)

Namely, the space $\mathcal{MF}_2(S,P)$ is parametrized by an adaptation of the Dehn-Thurston coordinates for measured foliations that we mentioned in Definition 2.4. We refer to Proposition 3.9 of [**ALPS16**] for details (see also §3 of that paper), and provide a sketch of the argument.

PROPOSITION 3.4. *Let $S$ be a surface of genus $g \geq 2$ and $P$ be a set of $n$ marked points, as above. The space $\mathcal{MF}_2$ of measured foliations on $S$ with centers at $P$ is homeomorphic to $\mathbb{R}^{6g-6+3n}$.*

SKETCH OF THE PROOF. Given a measured foliation $F$ in $\mathcal{MF}_2$, we can excise the centers from $S$ and consider the measured foliation on the surface with boundary. It suffices to parameterize the space of such foliations, as $F$ is recovered (up to isotopy) by re-attaching a *radial center* in the case the boundary has a positive transverse measure, and by a *closed center* when the transverse measure vanishes. (See Definition 3.1 for these notions.) In the former case, by an isotopy relative to the boundary, one can arrange that the leaves of $F$ on the surface-with-boundary are orthogonal to the boundary component, and in the latter case, parallel to it.

In either case, doubling the surface across the boundaries by an orientation-reversing reflection results in a doubled surface with a measured foliation. Such measured foliations on the doubled surface can then be parametrized by coordinates provided by the transverse measures and twisting numbers for a system of pants curves that that have an involutive symmetry, and include the curves obtained from doubling the boundary components (*cf.* §3.4 of [**ALPS16**]).

The two-fold symmetry implies that the parameters for pairs of "interior" pants curves (that is, other than ones obtained from the boundary curves) will agree.

There are a total of $3g-3+n$ interior pants curves (on the excised surface before doubling) and $n$ boundary curves. Each boundary curve contributes one parameter, namely the transverse measure, to the foliation on the doubled surface. Each interior pants curve contributes two (the length and twist parameters). Hence the total number of parameters determining $F$ is $6g-6+3n$. (See Figure 4 for an example.) Note that the boundary curve having transverse measure zero corresponds to the case of a closed center.

Conversely, by reversing the steps, any such collection of $6g-6+3n$ parameters determines a symmetric foliation on the doubled surface, and hence a unique foliation $F$ in $\mathcal{MF}_2$. Thus the space $\mathcal{MF}_2$ is homeomorphic to $\mathbb{R}^{6g-6+3n}$. □

*Remark.* Measured foliations with 1-prong singularities (associated with poles of order one) arise in the context of classical Teichmüller theory for a surface with punctures. For these, the usual parametrization for closed surfaces extends by associating two real parameters for each such singularity. Namely, the corresponding space $\mathcal{MF}_1$ of foliations is of dimension $6g-6+2n$ (see Exposé XI of [**FLP12**]). For $\mathcal{MF}_2$ as in the proposition above, the additional real parameter is the transverse measure of a loop around each center.

## 4. Model maps and cylinder ends

As discussed in §2, a meromorphic quadratic differential induces a measured foliation and a harmonic collapsing map to its metrized leaf-space. Recall that a quadratic differential with a double order pole has a neighborhood that is a "center" (see Definition 3.1). The leaf space of the foliation on a radial center is a circle of circumference given by the transverse measure around the pole, and in the case of a closed center, a copy of $\mathbb{R}_{\geq 0}$. (See Figure 5.) In this section we describe a model family of harmonic maps from a neighborhood of a pole to $S^1$ (or $\mathbb{R}_{\geq 0}$), parametrized by the complex residue at the pole. Up to bounded distance, these form all the possibilities for the restriction of the harmonic collapsing map to that neighborhood.

*Notation.* Throughout this paper, $S^1_R$ shall denote a circle of circumference $R$, and $\mathcal{C}_R$ the half-infinite Euclidean cylinder of circumference $R$:

$$\mathcal{C}_R = \{(x,\theta) | x \in [0,\infty), \theta \in S^1_R\}.$$

A finite subcylinder $(0 \leq x \leq L)$ of $\mathcal{C}_R$ is denoted by $\mathcal{C}_R(L)$.

**A basic example.** The differential $\frac{1}{z^2}dz^2$ on the punctured disk $\mathbb{D}^*$ yields a half-infinite Euclidean cylinder $\mathcal{C}_{2\pi}$ in the induced singular-flat metric. In fact, in the coordinates $(x,\theta)$ on this flat cylinder, the differential is $d\omega^2$ for the complex coordinate $\omega = x + i\theta$. The induced foliation then comprises the "horizontal rays" ($\theta = const.$) along the length of the cylinder, and the collapsing map is

$$h : \mathcal{C}_{2\pi} \to S^1_{2\pi} \text{ where } h(x,\theta) = \theta$$

that is clearly harmonic.

Other model maps are then obtained by introducing "twists" to this basic example above, together with scaling the circumference. This amounts to multiplying the quadratic differential $dw^2$ on $\mathcal{C}_{2\pi}$ by a non-zero complex number (*cf.* Lemma 4.3). The entire family of model maps also includes the "limiting" case of a closed center, which is obtained as the "twists" tend to infinity, or equivalently, the angle of the leaves with the longitudinal direction tends to $\pm\pi/2$.

**4.1. Definitions.** For $R \in \mathbb{R}_+$ and $\alpha \in (-\pi/2, \pi/2)$, let $\mathcal{C}_R$ be a half-infinite Euclidean cylinder of radius $R$ as above, and define $m_{R,\alpha} : \mathcal{C}_R \to S^1_{R\cos\alpha}$ to be the map

$$m_{R,\alpha}(x,\theta) = \theta\cos\alpha + x\sin\alpha \tag{4.1}$$

where the right-hand side is considered modulo $R\cos\alpha$. This map is harmonic: in local Euclidean coordinates on the cylinder, it is a linear function and hence its Laplacian vanishes.

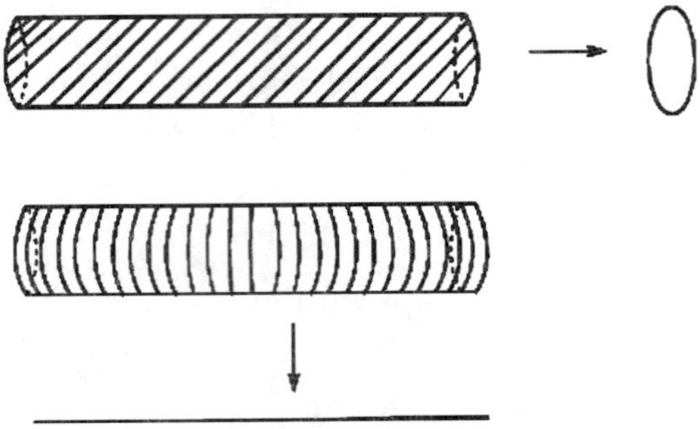

FIGURE 5. The collapsing map for the induced foliation in a neighborhood of an order two pole yields a harmonic map to $S^1$ or $\mathbb{R}$.

Here the angle $\alpha$ shall be referred to as the foliation angle: this denotes the angle from the longitudinal direction, of the leaves of the collapsing foliation for the model map.

For $\alpha = \pm\pi/2$, we define $m_{R,\alpha} : \mathcal{C}_R \to \mathbb{R}_{\geq 0}$ by $(x, \theta) \mapsto x$. This is the case corresponding to when the collapsing foliation "degenerates" to a foliation of $\mathcal{C}_R$ by closed circles.

Together, we obtain

DEFINITION 4.1 (Model maps). The expression (4.1) defines the family

$$\mathsf{M} = \{m_{R,\alpha} | R \in \mathbb{R}_+, \alpha \in S^1_\pi\}$$

of harmonic maps from $\mathcal{C}_R$ to a target that is either $S^1$ (of circumference $R\cos\alpha$, when $\cos\alpha \neq 0$) or $\mathbb{R}$ (when $\cos\alpha = 0$).

Note that the interval $[-\pi/2, \pi/2]$ that is the range of values for $\alpha$ has been identified with a circle $S^1_\pi$, where $\pm\pi/2$ defines a single point.

As in the basic example at the beginning of the section, these model maps $m_{R,\alpha}$ are the collapsing maps for a quadratic differential on $\mathbb{D}^*$ of the standard form (1.1), and are parameterized by its complex residue $a$. (For the relation between the parameters $R, \alpha$ with $a$ see Lemma 4.3.)

DEFINITION 4.2 (Model end). For a choice of parameters $R \in \mathbb{R}_+$ and $\alpha \in S^1_\pi$, we shall refer to the half-infinite cylinder $\mathcal{C}_R$ with the foliation at angle $\alpha$ described above, as the *model cylindrical end* determined by those parameters.

Since the metric on the leaf-space of a measured foliation is determined by transverse measures, the transverse measure of an arc on a model cylindrical end is equal to the length of the image segment under the collapsing map (4.1) to the leaf-space. In particular, the transverse measure of a meridional circle is $R\cos\alpha$, and the transverse measure of a longitudinal segment of length $L$ (such as across a truncated cylinder $\mathcal{C}_R(L)$, is $L\sin\alpha$. (The latter quantity is considered modulo $R\cos\alpha$, which is the circumference of the image circle, but is exactly $L\sin\alpha$ when we consider the lift to the universal cover in the next subsection.)

**4.2. Lifts to the universal cover.** We fix the parameters $R, \alpha$ for the rest of this section.

Let $\mathbb{H} = \{x, \theta | x \in \mathbb{R}_+, \theta \in \mathbb{R}\}$ be the universal cover of the interior of $\mathcal{C}_R$.

A model map $m \in \mathsf{M}$ lifts to a map

$$\widetilde{m} : \mathbb{H} \to \mathbb{R}$$

which is equivariant with respect to the $\mathbb{Z}$-action that acts

(a) on the target by translation by $R\cos\alpha$ when $\alpha \neq \pi/2$, and trivially when $\alpha = \pm\pi/2$, and

(b) on the domain by a translation in the $\theta$-coordinate by $R$.

Any quadratic differential on $\mathbb{D}^*$ with a pole of order two at the puncture also has an induced foliation that lifts to the universal cover $\mathbb{H}$. (The universal covering map $\pi : \mathbb{H} \to \mathbb{D}^*$ is the quotient by the translation $z \mapsto z + R$.) The following lemma asserts that the corresponding collapsing map is bounded distance from *some* model map in $\mathsf{M}$.

LEMMA 4.3. *Let $q$ be a meromorphic quadratic differential on $\mathbb{D}^*$ with an order two pole of residue $a \in \mathbb{C}^*$. Then the collapsing map of the induced foliation on $\mathbb{H}$ is bounded distance from the lift of the model map $m_{R,\alpha}$ where $R = 2\pi|a|$ and $\alpha = Arg(a)$.*

PROOF. For a pole of the form (1.1) the collapsing map is identical to $m_{R,\alpha}$, using the fact that the "spiralling" ray

$$r(t) = (t\cos\alpha, t\sin\alpha)$$

is a horizontal trajectory on the cylinder $\mathcal{C}_R$ with the usual $(x,\theta)$-coordinates, as one can check by a straightforward calculation.

In general, the remaining terms in the Laurent expansion (2.2) contribute to a bounded perturbation of the collapsing map, since by (2.1) the additional transverse measure is bounded by

$$\left|\int_0^r \left(\frac{c_1}{z} + c_0 + f(z)\right)^{1/2} dz\right| = O\left(\int_0^r z^{-1/2}dz\right) = O(1)$$

for any $0 < r < 1$, where $f$ is a holomorphic function on $\mathbb{D}$ that vanishes at the origin.

This transverse measure determines distances in the leaf-space that is the image of the collapsing map. Hence the above observation implies that up to a bounded error, the image is determined by the leading order $(a^2/z^2)$ term in (2.2), which as we previously noted, is identical to $m_{R,\alpha}$.

(Alternatively, note that with respect to the coordinate $\omega = x + i\theta$ on the half-infinite flat cylinder, the expression (2.2) is $(a^2 + c_1e^{-\omega})d\omega^2$: the exponentially decaying term contributes a bounded amount to the associated collapsing map.) □

The following Lemma now asserts that the model map in Lemma 4.3 at a bounded distance from the collapsing map, is uniquely determined, if one also knows the parameter $R$.

LEMMA 4.4. *Two model maps $m_1, m_2 \in \mathsf{M}$ lift to maps*

$$\widetilde{m_1}, \widetilde{m_2} : \mathbb{H} \to \mathbb{R}$$

*that are a bounded distance apart if and only if their foliation angles $\alpha_1$ and $\alpha_2$ are equal.*

PROOF. The 'if' direction is straightforward. The model maps $\widetilde{m_i}$ are determined by the two parameters of the radius $R_i$ and the angle $\alpha_i$. If the foliation angles $\alpha_1$ and $\alpha_2$ and the radii $R_1$ and $R_2$ are equal, then by the expression (4.1) the two lifts are in fact identical, up to a translation of base point. Further, if the angles $\alpha_1$ and $\alpha_2$ are identical, but the maps differ in their parameter $R$, then the lifts are still the same (and it is only the translation action of $\mathbb{Z}$ on $\mathbb{H}$ and $\mathbb{R}$ that differs).

For the other direction, consider the case when neither foliation angle equals $\pm\pi/2$ (which is a single point in the parameter space for angles, $S^1_\pi$, that corresponds to a closed center). Then the expression (4.1) also represents the lift of the model map to $\mathbb{R}$, where the latter is thought of as the universal cover of $S^1_{R\cos\alpha}$. From (4.1), we see that two such maps with different values of $\alpha$, say $\alpha_1$ and $\alpha_2$, are an unbounded distance apart: consider, for example, the restriction to a $\theta = const.$ vertical line. Then the distance between the images equals $x|\sin\alpha_1 - \sin\alpha_2| \to \infty$ as $x \to \infty$.

If one of the foliation angles is $\pm\pi/2$, then the image of an entire horizontal line is bounded (it maps to a point in the target $\mathbb{R}$) which, again by (4.1), is not the case for $\alpha \neq \pm\pi/2$. □

We shall use the previous two lemmas in the §5.2.4, the endgame of the proof, where we need to verify that the harmonic map we produce, at a bounded distance from a given model map near the puncture, indeed has the required residue.

**4.3. Least-energy property.** We begin by noting the energy of a model map, that we will need in the next section.

LEMMA 4.5. *The restriction of the model map $m_{R,\alpha} : \mathcal{C}_R \to S^1_{R\cos\alpha}$ to a subcylinder $\mathcal{C} = \mathcal{C}_R(L)$ has energy equal to $\frac{1}{2}RL = \frac{1}{2}R^2 \cdot Mod(\mathcal{C})$.*

PROOF. The expression for energy (2.4) can be converted to Euclidean $(x,\theta)$-coordinates on $\mathcal{C}$ to obtain:

$$\mathcal{E}(f) = \frac{1}{2}\int\limits_{\mathcal{C}} \left(|h_x|^2 + |h_\theta|^2\right) dx d\theta$$

and then using the expression (4.1) of the map, we see that the integrand is 1. Hence the energy equals half the Euclidean area of the cylinder, that is, $\frac{1}{2}RL$.

Note that the modulus of the cylinder $\text{Mod}(\mathcal{C}) = L/R$. □

What we shall now prove is a crucial property of a model map, namely its restriction to subcylinders solves a certain least-energy problem for maps to an $\mathbb{R}$-tree target.

In what follows, let $A$ be an annulus of modulus $0 < M < \infty$ that we uniformize to the flat cylinder $\mathcal{C}_R(L)$, that is, of circumference $R$ and length $L$, such that $L/R = M$. As usual, we equip this cylinder with coordinates $x, \theta$.

Let $\widetilde{A} = [0, L] \times \mathbb{R}$ be the universal cover of $A$, and $T$ be an $\mathbb{R}$-tree with a $\mathbb{Z}$-action (that is, in fact, part of a larger surface-group action, which we shall not need).

Consider a map

$$\tilde{u} : \widetilde{A} \to T$$

which is $\mathbb{Z}$-equivariant, where the action on the domain is by translation by $R$ on the $\mathbb{R}$-factor. Let $D = [0, L] \times [0, R] \subset \widetilde{A}$ be a fundamental domain of the action.

PROPOSITION 4.6. *Let*

$$u : [0, L] \times [0, R] \to T$$

*be a map to an $\mathbb{R}$-tree, defined on a fundamental domain in the universal cover of an annulus of modulus $M := L/R$ as defined above. Suppose that there exist constants $C, \tau > 0$ such that $u$ satisfies:*

(1) *each vertical ($\theta$-) arc in $D$ maps to a segment of length $C$ in $T$, and*
(2) *each horizontal ($x$-) arc in $D$ maps to a segment of length $\tau$ in $T$.*

*Then the energy of the map satisfies the lower bound*

$$\mathcal{E}(u) \geq \frac{1}{2} M \cdot \left( C^2 + \frac{\tau^2}{M^2} \right). \tag{4.2}$$

PROOF. We adapt the classical length-area argument.

Since every vertical arc in $D = [0, L] \times [0, R]$ maps to an arc of length $C$, we have

$$\int_0^R u_\theta d\theta \geq C$$

Integrating over the longitudinal ($x$-) direction we obtain:

$$\int_0^R \int_0^L |u_\theta| dx d\theta \geq LC.$$

By the Cauchy-Schwarz inequality,

$$\int_0^R \int_0^L |u_\theta|^2 dx d\theta \int_0^R \int_0^L 1^2 dx d\theta \geq L^2 C^2$$

which implies

$$\int_0^R \int_0^L |u_\theta|^2 dx d\theta \geq \frac{L^2 C^2}{LR} = C^2 M. \tag{4.3}$$

On the other hand, for each horizontal arc we have:

$$\int_0^L u_x dx \geq \tau$$

and by the same calculation as above we obtain

$$\int_0^R \int_0^L |u_x|^2 dx d\theta \geq \frac{\tau^2}{M}. \tag{4.4}$$

Combining (4.3) and (4.4) we find:

$$\mathcal{E}(u) = \frac{1}{2} \int_0^R \int_0^L \left( |u_x|^2 + |u_\theta|^2 \right) dx d\theta \geq \frac{1}{2} M \cdot \left( C^2 + \frac{\tau^2}{M^2} \right)$$

as desired. □

*Remark.* The restriction of a model map $m = m_{R,\alpha} : \mathcal{C}_R \to S^1_{R\cos\alpha}$ to a subcylinder $\mathcal{C} = \mathcal{C}_R(L)$ lifts to a $\mathbb{Z}$-equivariant map

$$\tilde{m} : [0, L] \times \mathbb{R} \to \mathbb{R}$$

that satisfies (1) and (2) with $\tau = L\sin\alpha$ and $C = R\cos\alpha$, which are the transverse measures of the foliation across the subcylinder (see Definition 4.2).

A simple substitution in (4.2) then yields $\mathcal{E}(\tilde{m}|_{\widetilde{\mathcal{C}}}) \geq \frac{1}{2}R^2 \cdot \mathrm{Mod}(\mathcal{C})$. Lemma 4.5 then confirms that, in fact, equality holds, and so the lift of the model map realizes the least energy for all $\mathbb{Z}$-equivariant maps from $\widetilde{\mathcal{C}}$ to $\mathbb{R}$ satisfying (1) and (2).

## 5. Proof of Theorem 1.2

We fix throughout the Riemann surface $\Sigma$ with a single marked point that we denote by $P$. Note that in the case of several marked points, we need add only an additional subscript, and the entire argument holds *mutatis mutandis.*

We also fix a measured foliation $F \in \mathcal{MF}_2(\Sigma, P)$ with center $U$ around $P$. Let $T$ be an $\mathbb{R}$-tree that is the leaf-space of the lifted foliation $\widetilde{F}$ on the universal cover. Recall that this acquires a metric from the projection of the transverse measures of the foliation. The collapsing map to the leaf space is then a map from $\widetilde{\Sigma \setminus P}$ to $T$.

The proof of Theorem 1.2 involves showing there is a *harmonic* map

$$\widetilde{h} : \widetilde{\Sigma \setminus P} \to T$$

whose collapsing foliation is $F$, with a Hopf differential with a prescribed double order pole at $P$. We begin with the (easier) proof of the uniqueness of such a map in §5.1, and then show existence in §5.2.

**5.1. Uniqueness.** Let $q_1$ and $q_2$ be meromorphic quadratic differentials realizing the same measured foliation $F$, and with the same imaginary part of the residue at $P$ (up to sign).

These determine harmonic maps $\widetilde{h_1}, \widetilde{h_2} : \widetilde{\Sigma \setminus P} \to T$.

Recall from §1 that the real part of the residue at the pole is determined by $F$, namely, it is the transverse measure of a loop around the pole. Our convention is that this real part is always taken to be non-negative.

Since we have assumed that $q_1$ and $q_2$ have the same imaginary parts, the (complex) residues for $q_1$ and $q_2$ at the pole $P$ are equal.

By Lemma 4.3, this residue then determines the (lift of the) model map $m_{R,\alpha}$ to which $\widetilde{h_1}$ and $\widetilde{h_2}$ are asymptotic. (In particular, $R$ is the circumference of the half-infinite cylinder in the induced metric, that is, equals $2\pi|\mathrm{Res}(P)|$.) That is, both $\widetilde{h_1}$ and $\widetilde{h_2}$ are bounded distance from the lift $\widetilde{m_{R,\alpha}}$. Hence the distance function between $\widetilde{h_1}$ and $\widetilde{h_2}$ is bounded. The distance function is invariant under the action of a deck transformation, hence descends to the (original) Riemann surface. However it is well-known (this is an application of the chain-rule - see [**Jos91**] for Riemannian targets and [**KS93**] or [**Wol96**] for tree-targets) that such a distance function is subharmonic, and since a punctured Riemann surface is parabolic in the potential-theoretic sense, the distance must be constant. Moreover, the constant is zero because of a local analysis around a prong singularity, as in the proof of Proposition 3.1 in [**DDW00**] (see also section 4 of [**Wol96**]).

**5.2. Existence.** In this section, we show that given

- the topological data of a measured foliation $F \in \mathcal{MF}_2$ with a center at $P$, and
- a complex residue $a$ at the pole $P$ that is compatible with the transverse measure of the linking loop around it, namely, having the corresponding real part,

there exists a quadratic differential on the punctured surface $\Sigma \setminus P$ with a double order pole at $P$ of that residue (or its negative), realizing $F$.

Recall that $T$ is the leaf-space of the lift of the foliation $\mathcal{F}$ to the universal cover. The required quadratic differential shall be obtained as a Hopf differential of a harmonic map $\widetilde{h} : \widetilde{\Sigma \setminus P} \to T$, which in turn is obtained as a limiting of a sequence of harmonic maps $\widetilde{h_n}$ defined on (lifts) of a compact exhaustion of the punctured surface.

The following is an outline of the proof:

- *Step 1.* We start with $\Sigma \setminus P$ equipped with $F$, such that on the neighborhood $U$ of the puncture, the foliation $F$ restricts to that of a model cylindrical end for the parameters $R$, $\alpha$. Consider a compact exhaustion $\Sigma_n$ of $\Sigma \setminus P$. For each $n$, by a holomorphic doubling across the boundary $\partial\Sigma_n$, we use the Hubbard-Masur Theorem to produce a holomorphic quadratic differential on $\Sigma_n$ that realizes the restricted foliation $F|_{\Sigma_n}$. Lifting to the universal covers, we obtain harmonic maps $\widetilde{h_n} : \widetilde{\Sigma_n} \to T$. What is crucial is that they are the energy-minimizing maps for their Dirichlet boundary conditions.

- *Step 2.* Consider the annular region $A_n := \Sigma_n \setminus \Sigma_0$. We show that the energies of the restrictions of $\widetilde{h_n}$ to the lifts of $A_n$ have a lower bound proportional to the modulus of $A_n$. Together with their energy-minimizing property, an argument adapted from [**Wol91**] then provides a uniform energy bound on the compact subsurface $\Sigma_0$. Standard arguments then apply, and show the uniform convergence of $\widetilde{h_n}$ to a harmonic map $\widetilde{h} : \widetilde{\Sigma \setminus P} \to T$.

- *Step 3.* We verify that the Hopf differential of $\tilde{h}$ yields a holomorphic quadratic differential on the quotient surface $\Sigma \setminus P$ with a double order pole at $P$ having the required residue, and realizing the desired foliation $F$.

We shall carry out the above steps in the following sections.

5.2.1. *Step 1: The approximating sequence* $\widetilde{h_n}$. Let $a \in \mathbb{C}^*$ be the desired residue at $P$. Associated with this complex number are the parameters of circumference $R$ and "foliation angle" $\alpha$ (see Lemma 4.3) and the model cylindrical end for $F$ around $P$. In particular, the parameter $R$ equals $2\pi$ times the absolute value of the residue.

Choose a coordinate disk $U$ around $P$. After an isotopy if necessary, we can assume the foliation $F$ restricts to the foliation on a model cylindrical end on $U$, for the parameters $R$ and $\alpha$, as in Definition 4.2.

Let $\Sigma_0$ be the Riemann surface with boundary obtained by excising $U$.

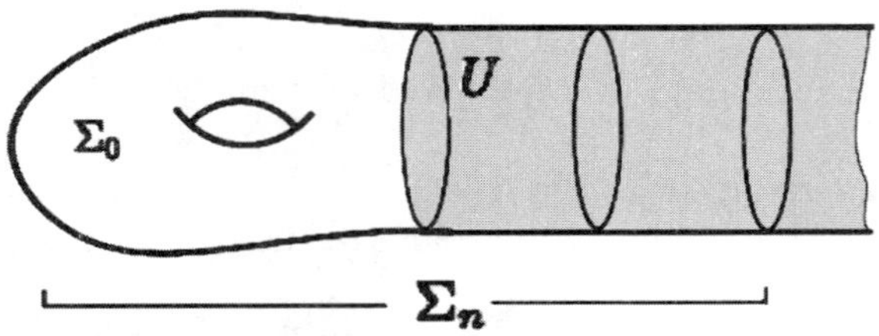


FIGURE 6. Exhaustion of the punctured surface $\Sigma \setminus P$.

For $n \geq 1$ consider a compact exhaustion $\Sigma_n$ of $\Sigma$ obtained by excising a succession of increasingly smaller disks in $U$ around $P$. (See Figure 6.)

As noted in the outline, we set $A_n = \Sigma_n \setminus \Sigma_0$ to be the intermediate annulus.

Let $F_n$ be the restriction of the measured foliation $F$ to the compact surface-with-boundary $\Sigma_n$. By construction, on the annulus $A_n$, this foliation is the restriction of that on a model cylindrical end.

**Doubling.** Let $n \geq 1$. Take two identical copies of $\Sigma_n$, and perform a holomorphic doubling across $\partial\Sigma_n$: namely, take two identical copies of the resulting surface-with-boundary $S_0$, and glue corresponding boundary components by an orientation-reversing homeomorphism of the circle, with exactly two fixed points (see Figure 7). The resulting doubled surface $S_d$ has an orientation-preserving involution $\mathcal{I} : S_d \to S_d$.

Using this involution, one can push forward a measured foliation on $S_0$ to one on $S_d$ that is symmetric under $\mathcal{I}$.

[Note that this construction differs from the usual "doubling-by-reflection" in the proof of Proposition 3.4 that fixes each boundary component and has an orientation-reversing involution. The advantage of the present construction is to ensure that the foliation on $S_0$, that can be isotoped to be at a constant angle $\alpha$ at the boundary, extends smoothly to the doubled surface $S_d$ maintaining the twists about the boundary curves.]

We obtain a compact Riemann surface

$$\widehat{\Sigma_n} = \Sigma_n \sqcup_\partial \Sigma_n \tag{5.1}$$

with a holomorphic involution $\mathcal{I} : \widehat{\Sigma_n} \to \widehat{\Sigma_n}$ that takes one copy of $\Sigma_n$ to the other.

We define the corresponding "doubled" measured foliation $\widehat{F_n}$ on $\widehat{\Sigma_n}$. Note that by the construction of $F$ on $U$, the foliation $F_n$ is incident at a constant angle $\alpha$ at the boundary, and hence the foliation $\widehat{F_n}$ is smooth across the doubled boundary. (See Figure 9.)

Moreover, the foliation $\widehat{F_n}$ is defined by the same parameters of twist and intersection-number on the pants curves in the interior of $\Sigma_n$, and its restriction to the central cylinder $\hat{A}_n$ obtained by doubling $A_n$ is equivalent to the foliation on a model cylindrical end for parameters $R, \alpha$.

That is, for $n \geq 1$, the restriction of $\widehat{F_n}$ to this annulus is the foliation of the finite subcylinder $\mathcal{C}_R(2L_n)$ of a model cylindrical end for the parameters $R$ and $\alpha$, where

$$\operatorname{mod}(A_n) = L_n/R. \tag{5.2}$$

is the modulus of the subcylinder.

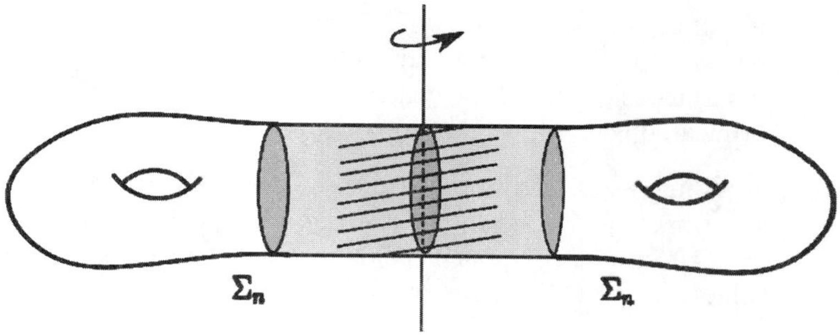


FIGURE 7. The foliation $\hat{F}_n$ is invariant under the holomorphic involution $\mathcal{I}$, indicated here as a $\pi$-rotation about a central axis.

The transverse measure of a meridional curve around the cylinder is $L_n \cos\alpha$, and that of a longitudinal arc across it is $L_n \sin\alpha$ (see Definition 4.2). Note that $L_n \to \infty$ as $n \to \infty$.

The involution $\mathcal{I}$ preserves the foliation $\widehat{F_n}$, since the foliation $\widehat{F_n}$ was defined on $\hat{\Sigma}_n \setminus \Sigma_n$ to agree with $\mathcal{I}(\widehat{F_n|_{\Sigma_n}})$, and $\mathcal{I}$ squares to be the identity.

Now, by the Hubbard-Masur Theorem (see §2.1), there exists a holomorphic quadratic differential $\hat{q_n}$ that realizes the measured foliation $\widehat{F_n}$ on the doubled surface $\widehat{\Sigma}_n$.

Since the involution $\mathcal{I}$ preserves the foliation $\widehat{F_n}$ of $\hat{q_n}$, by the uniqueness part of the Hubbard-Masur theorem, the holomorphic involution $\mathcal{I}$ takes $\hat{q_n}$ to itself, namely $\mathcal{I}_*\hat{q_n} = \hat{q_n}$.

Therefore taking the quotient by the holomorphic involution $\mathcal{I}$ on $\widehat{\Sigma}_n \setminus \partial\Sigma_n$, we obtain a (well-defined) holomorphic quadratic differential $q_n$ on $\Sigma_n$ (that in fact equals the restriction of $\hat{q_n}$ on a copy of $\Sigma_n \subset \widehat{\Sigma}_n$). Its induced foliation $\mathcal{F}(q_n)$ realizes the (well-defined quotient) measured foliation $F_n$ on the surface.

We now identify a foliated annular subsurface which converges to the spiralling end in the limiting construction in the final sections of the paper:

LEMMA 5.1. *There is an embedded annulus $A(q_n)$ on $\Sigma_n$ such that the foliation $\mathcal{F}(q_n)$ restricted to $A(q_n)$ is identical, via a leaf-preserving diffeomorphism, to a truncation of a model cylindrical end with parameters $R$ and $\alpha$, and that has modulus mod$(A_n) = L_n/R$.*

PROOF. On the doubled surface $\hat{\Sigma}_n$, the (doubled) measured foliation $\hat{F}_n$ has a foliated cylinder $\mathcal{C}_R(2L_n)$ that is a truncation of a model cylindrical end, coming from doubling the restriction of $F_n$ to $A_n$. Now, the foliation $\mathcal{F}(\hat{q}_n)$ is measure-equivalent to $\hat{F}_n$, and differs from it by at worst isotopy and Whitehead moves. Thus, the foliation $\mathcal{F}(\hat{q}_n)$ still contains an annulus, denoted by $A(\hat{q}_n)$, which is measure equivalent to $\mathcal{C}_R(2L_n)$; that is, there is a leaf-preserving diffeomorphism of the foliated annulus $A(\hat{q}_n)$ to $\mathcal{C}_R(2L_n)$, such that the transverse measure around the annulus is $R\cos\alpha$, and any arc between the boundary components has transverse measure $2L_n \sin\alpha$; measurements that agree with those in a truncation of a model end (see Definition 4.2). By the two-fold symmetry of the foliation under the involution $\mathcal{I}$, the foliated annulus $A(\hat{q}_n)$ descends to an annulus $A(q_n)$ on $\Sigma_n$ that is measure equivalent to the truncation $\mathcal{C}_R(L_n)$ of a model cylindrical end. □

DEFINITION 5.2 (Truncated center). The foliated cylinder $A(q_n)$ of Lemma 5.1 above will be referred to as the "truncated center" of the quadratic differential $q_n$ in the Riemann surface $\Sigma_n$. [When the context is clear, we shall often just refer to $A(q_n)$ as the "truncated center".]

The transverse measures, across the truncated center $A(q_n)$ or around it, are identical to those across or around $\mathcal{C}_R(L_n)$ of the model cylindrical end with parameters $R$ and $\alpha$. A crucial component of the proof of convergence is to control the position of the truncated center relative to the initial foliated annulus $A_n$ (see §5.2.2).

**The harmonic map $\widetilde{h_n}$.** By considering the lift $\widetilde{q_n}$ of the holomorphic quadratic differential $q_n$ to the universal cover $\widetilde{\Sigma}_n$ and the collapsing map of its induced measured foliation, one obtains a harmonic map $\widetilde{h_n}$ which maps to the leaf-space of the lift of $F_n$. That image leaf-space is a sub-$\mathbb{R}$-tree of the leaf space $T$ of $\widetilde{F}$, namely, the leaf-space of the restriction of the foliation to $\widetilde{\Sigma}_n$ that we henceforth denote by $T_n$. Thus we have:

$$\widetilde{h_n} : \widetilde{\Sigma}_n \to T_n \subset T.$$

By the harmonicity and the uniqueness of such (non-degenerate) harmonic maps to $\mathbb{R}$-trees (or more generally, non-positively curved spaces, see [**Mes02**]), this map solves the least energy (Dirichlet) problem for an equivariant map to the tree $T_n$ with the given boundary conditions.

5.2.2. *Step 2: Energy restricted to the annulus $A_n$.* The purpose of this subsection is to establish a lower bound on the energy of $\widetilde{h}_n$ when restricted to lifts of the annulus $A_n$.

Recall from Lemma 5.1 that on the doubled surface, a cylinder with a spiralling foliation persists and descends to a foliated embedded annulus $A(q_n)$ on $\Sigma_n$, with one boundary component $\partial\Sigma_n$. This is what we call the "truncated center".

Note that since the truncated center $A(q_n)$ is a subcylinder of a model cylindrical end with parameters $R$ and $\alpha$, the transverse measure of a longitude is $L_n \sin\alpha$, where $L_n$ is as in (5.2), and that of a meridional circle is $R\cos\alpha$ (see Definition 4.2).

Let $\gamma$ denote the boundary curve $\partial\Sigma_n$ that defines an essential closed curve on the doubled surface; thus $\gamma$ is also the core of the cylinder $\hat{A}_n$. (Topologically, this is the same curve for all $n$, and hence has no subscript "$n$".)

Similarly, let $\hat{\beta}$ denote a non-trivial curve on $\widehat{\Sigma_n}$, symmetric under the involution $\mathcal{I}$, that intersects $\gamma$ twice (see Figure 8) and let $\beta$ be the embedded arc in $\Sigma_n$ obtained under the quotient by the involution. In what follows, we shall use the transverse measures of these curves to control the position of the annulus $A(q_n)$ relative to $A_n$.

*Notation.* In what follows $\tau_n(\sigma)$ denote the transverse measure of a topologically non-trivial embedded closed curve or arc $\sigma$ on $\Sigma_n$, where the transverse measure is with respect to the measured foliation $F_n$.

LEMMA 5.3. *For each $n \geq 1$, we have the following estimates of transverse measures:*

(a) $\tau_n(\gamma) = R\cos\alpha$.

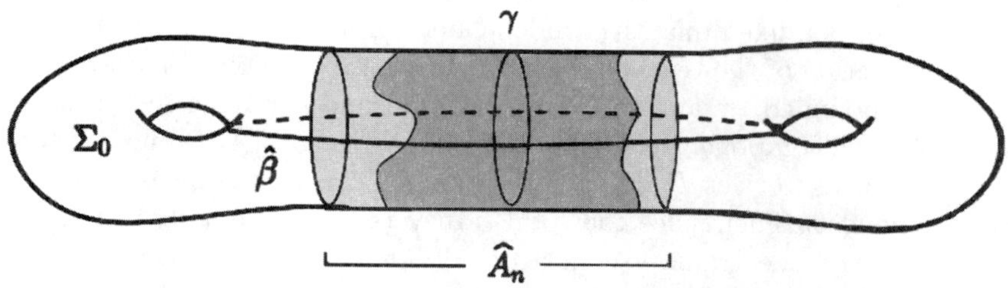


FIGURE 8. The position of the "truncated center" for $\mathcal{F}(\hat{q}_n)$ (shown shaded), relative to $\hat{A}_n$, is controlled by Lemma 5.3.

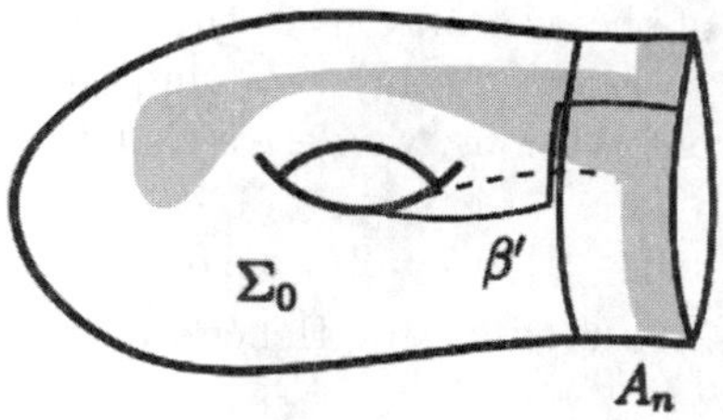


FIGURE 9. Construction of the arc $\beta'$ on $\Sigma_n$ avoiding a "tongue" of the truncated center (shown shaded).

(b) $\tau_n(\beta) \geq 2L_n \sin\alpha$, *where* $L_n$ *is given by equation* (5.2).
(c) $\tau_n(\beta \cap \Sigma_0) \leq T_0$, *where* $T_0$ *is independent of* $n$.
(d) $\tau_n(\beta) \leq 2L_n \sin\alpha + T_0$.
(e) $\tau_n(\beta \cap A_n) \geq 2L_n \sin\alpha - T_0$.

PROOF. Part (a) follows from the fact that the asserted transverse measure is that of $\partial\Sigma_n$ with respect to $F_n$. However recall that $F_n$ is the restriction of the foliation on a model cylindrical end with parameters $R$ and $\alpha$, and hence the boundary acquires the transverse measure $R\cos\alpha$ of the core curve.

Part (b) follows from the fact that by construction, the truncated center is an annulus with core curve $\gamma$, and the arc $\beta$ crosses the truncated center (see Defn. 5.1) twice, and hence the transverse measure is greater than twice the transverse measure across $A_n$. As in part (a), the latter computation is just that of the model cylindrical end for parameters $R, \alpha$, and gives $2L_n \sin\alpha$.

Part (c) is the most important estimate. First note that $\beta \cap \Sigma_0$ forms an arc on the fixed subsurface $\Sigma_0$ whose topological type is independent of $n$.

The transverse measure of $\beta \cap \Sigma_0$ depends on how the foliation $\mathcal{F}(q_n)$ (that is equivalent to $F_n$) and in particular, the truncated center $A(q_n)$ meets the subsurface $\Sigma_0$. (We need to check that the truncated center $A(q_n)$ does not move "far" from $A_n$.)

Note that the transverse measure with respect to $\mathcal{F}(q_n)$ of a fixed embedded arc in the *complement* of the truncated center is uniformly bounded.

Hence, it suffices to prove that the truncated center for $\mathcal{F}(q_n)$ cannot venture too "deep" in $\Sigma_0$. For this, we consider the following two cases:
(1) The intersection of the truncated center with $\Sigma_0$ is an annular region with $\partial\Sigma_0$ as one of the boundaries. In this case, the modulus of such a peripheral annulus that can embed in $\Sigma_0$ is bounded above by the reciprocal of the extremal length of

$\partial\Sigma_0$. The latter is a fixed quantity (independent of $n$).
(2) The intersection of the truncated center with $\Sigma_0$ is a union of simply-connected regions, each bounded by an arc along $\partial\Sigma_0$, and another lying in the interior of $\Sigma_0$. Consider one component of this type. Such a "tongue" $\Omega$ cannot be too large in transverse measure: let $D$ denote the maximal transverse measure of any arc in that component. Then since the total transverse measure across the truncated center is $L_n \sin\alpha$, the transverse measure of some longitudinal arc in the part of the truncated center $\mathcal{F}(q_n) \cap A_n$ across the annulus $A_n$ will be $L_n \sin\alpha - D$. One can then construct an arc $\beta'$ on $\Sigma_n$, homotopic to $\beta$ through a homotopy fixing the endpoints on $\partial\Sigma_n$, by joining a sub-arc in $\Sigma_0$ in the complement of the "tongue" $\Omega$, and two longitudinal arcs across $A_n$ as above, by two arcs lying in $\partial\Sigma_0$. The transverse measure of the arc away from the "tongue" is uniformly bounded (independent of $n$), and so is that of any arc lying on $\partial\Sigma_0$ (bounded, in fact, by $R\cos\alpha$). Adding these contributions, we derive that the transverse measure of $\beta'$ satisfies

$$2L_n \sin\alpha \leq \tau_n(\beta') = 2L_n \sin\alpha - 2D + O(1)$$

where the inequality was observed in (b). Hence we obain a uniform upper bound on $D$, as required.

Together, (1) and (2) show that only a uniformly bounded part of the truncated center can lie in $\Sigma_0$, in the sense that any arc contained in the intersection of $\Sigma_0$ and the truncated center, has a uniform upper bound on its transverse measure.

Thus the truncated center contributes only a uniformly bounded transverse measure to the arc $\beta \cap \Sigma_0$. This completes the proof of (c).

Part (d) now follows the previous observations when one decomposes $\beta$ into arcs across the truncated center, and its complement.

Finally, part (e) follows from (b) and (c). □

COROLLARY 5.4. *Let $\gamma$ be any arc across the cylinder $A_n$ on $\Sigma_n$. The transverse measure of $\gamma$ satisfies*

$$L_n \sin\alpha - T_0 - \frac{R\cos\alpha}{2} \leq \tau(\gamma) \leq L_n \sin\alpha + T_0/2$$

*where $T_0$ is the constant (independent of $n$) in Lemma 5.3.*

PROOF. The upper and lower bounds are obtained as follows:
(1) If the transverse measure of some arc is less than $L_n \sin\alpha - T_0 - R\cos\alpha/2$, then two oppositely-oriented copies of that arc together with an arc in $\Sigma_0 \cup \partial\Sigma_0$ of transverse measure bounded above (See Lemma 5.3(a),(c)) by $T_0 + R\cos\alpha$ will produce an arc homotopic to $\beta$ and of transverse measure less than $2L_n \sin\alpha - T_0$, contradicting Lemma 5.3 (e).
(2) If the transverse measure of some arc is more than $L_n \sin\alpha + T_0/2$, then two oppositely-oriented copies of the arc together with an arc in $\Sigma_0$ produces an arc transversely homotopic to $\beta$ and of transverse measure greater than $2L_n \sin\alpha + T_0$, contradicting Lemma 5.3 (d). □ □

We combine the above estimate with an energy estimate in §4 to provide the following crucial *lower* bound on energy.

LEMMA 5.5. *For all $n$, and any lift $\widetilde{A}_n$ of the annulus $A_n$, we have*

$$\bar{\mathcal{E}}(\widetilde{h}_n|_{\widetilde{A}_n}) \geq \frac{1}{2}R^2 M_n - K_0$$

*where $M_n$ is the modulus of $A_n$, and $K_0 > 0$ is independent of $n$.*

*(Recall that $\bar{\mathcal{E}}(\cdot)$ is the equivariant energy, or equivalently, the energy of the restriction to a fundamental domain.)*

PROOF. Choose a fundamental domain of the $\mathbb{Z}$-action on $\widetilde{A}_n$ that we call $A_n$, by abuse of notation.

From Corollary 5.4, we know that any longitudinal arc across $A_n$ must have transverse measure at least $L_n \sin\alpha - T_0 - R\cos\alpha$.

Moreover, note that any meridional circle in $A_n$ is homotopic to $\gamma$ and hence has transverse measure at least $R\cos\alpha$ by part (a) of Lemma 5.3.

Then Proposition 4.6 applies with $C = R\cos\alpha$ and $\tau = L_n\sin\alpha - T_0$, and provides the lower bound

$$\begin{aligned}\bar{\mathcal{E}}(\widetilde{h}_n|_{\widetilde{A}_n}) &\geq \frac{1}{2}M_n \cdot \left(R^2\cos^2\alpha + \frac{(L_n\sin\alpha - (T_0 - R\cos\alpha/2))^2}{L_n^2/R^2}\right)\\ &\geq \frac{1}{2}M_n \cdot \left(R^2\cos^2\alpha + \frac{L_n^2\sin^2\alpha - 2(T_0 - R\cos\alpha/2)\cdot L_n\sin\alpha}{L_n^2/R^2}\right)\\ &= \frac{1}{2}M_nR^2 - (T_0 - R\cos\alpha/2)R\sin\alpha\end{aligned}$$

which is of the required form. □

*Remark.* The upper bound in Corollary 5.4 is used the final section of the paper.

5.2.3. *Step 2, continued: Convergence of $\widetilde{h}_n$.* Recall that $\Sigma_n = \Sigma_0 \cup A_n$ where $A_n$ is a conformal annulus, obtained by excising subdisks of a chosen coordinate disk $U$ about the puncture $P$. (See the beginning of §5.2.1)

We shall show that the sequence of harmonic maps defined on their universal covers converges after passing to a subsequence.

For this, we shall need the following uniform energy estimate that shall use the work of the previous subsection:

LEMMA 5.6 (Uniform energy bound). *The equivariant energy $\bar{\mathcal{E}}\left(\widetilde{h}_n|_{\widetilde{\Sigma_0}}\right) \leq K_1$ where $K_1$ is independent of $n$.*

PROOF OF LEMMA 5.6. We use an adaptation of a technique from [**Wol91**]:

We begin with the obvious equality

$$\bar{\mathcal{E}}(\widetilde{h}_n|_{\widetilde{\Sigma}_0}) + \bar{\mathcal{E}}(\widetilde{h}_n|_{\widetilde{A}_n}) = \bar{\mathcal{E}}(\widetilde{h}_n).$$

We can equivariantly build a map $\widetilde{g}$ which is a candidate for the energy-minimizing problem that $\widetilde{h}_n$ solves (see the last part of §5.2.1), by defining $\widetilde{g}$ to equal $\widetilde{h}_0$ on $\widetilde{\Sigma}_0$, and to equal the model map $\widetilde{m}_{R,\alpha}$ on $\widetilde{A}_n$, for each lift of $\Sigma_0$ and $A_n$.

We thus obtain:

$$\bar{\mathcal{E}}(\widetilde{h}_n) \leq \bar{\mathcal{E}}(\widetilde{g}) = \bar{\mathcal{E}}(\widetilde{h}_0) + \bar{\mathcal{E}}(\widetilde{m}_{R,\alpha}|_{\widetilde{A}_n}) = \bar{\mathcal{E}}(\widetilde{h}_0) + \frac{1}{2}R^2\cdot \mathrm{mod}(A_n) \tag{5.3}$$

where the last equality follows from Lemma 4.5.

However from Lemma 5.5 we know

$$\frac{1}{2}R^2\cdot \mathrm{mod}(A_n) \leq \bar{\mathcal{E}}(\widetilde{h}_n|_{\widetilde{A}_n}) + K_0$$

where $K_0$ is independent of $n$.

By combining the three displayed inequalities, we obtain

$$\bar{\mathcal{E}}(\widetilde{h}_n|_{\Sigma_0}) \leq \bar{\mathcal{E}}(\widetilde{h}_0) + K_0$$

where the right hand side is independent of $n$, as required. □

*Remark.* The same argument applies for the energy of the restriction $\widetilde{h}_n|_{\widetilde{\Sigma}_m}$, for any subsurface $\Sigma_m$ for some fixed $m \geq 1$, and hence to any compact set $K \subset \Sigma \setminus p$ since $K$ is contained in some such subsurface.

Using a standard argument (see [**Wol96**] for details) we then have:

LEMMA 5.7. *The maps $\widetilde{h}_n$ uniformly converge to an equivariant harmonic map $\widetilde{h} : \widetilde{\Sigma \setminus P} \to T$.*

SKETCH OF THE PROOF. Each map $\widetilde{h}_m$ in the family $\{\widetilde{h}_m|_{\widetilde{\Sigma}_n} | m \geq n\}$ is harmonic, hence all are projections along the horizontal foliations of their Hopf differentials to an NPC space. Thus a disk in $\widetilde{\Sigma}_n$ gets taken by each such map into a finite subtree bounded by the image of the boundary of the disk. Here we notice and use that such balls contain but a finite number of zeroes, hence have image with but a finite number of vertices. Thus, the proof of the Courant-Lebesgue Lemma together with the uniform energy bound on the compact domain $\Sigma_n$ shows that these maps $\{\widetilde{h}_m|_{\widetilde{\Sigma}_n} | m \geq n\}$ are equicontinuous on $\widetilde{\Sigma}_n$. Then, we would like to use Arzela-Ascoli to provide for convergence of a subsequence of those maps on that domain. However, note that the target $T$ is, in general, a non-locally-compact $\mathbb{R}$-tree, so we first need to show that, for each $z \in \widetilde{\Sigma}_n$, we have that $\{\widetilde{h}_m(z) | m \geq n\}$ lies in a compact set.

We begin with a sketch of the argument in Lemma 3.3 of [**Wol96**] proving the images are uniformly bounded in distance; because of the failure of local compactness, this is insufficient on its own to provide what we need, but it is a first step.

Consider two simple closed curves $B$ and $C$ in $\Sigma_n$ that intersect, and are transverse to the foliation. Let $\gamma_B$ and $\gamma_C$ be the corresponding elements of the fundamental group that act on the universal cover and tree $T$, preserving bi-infinite axes for each. In particular, the axes intersect in the image of a fundamental domain $\mathcal{D}$. The key observation is that if $w \in T$, then using the fact that $T$ is a tree and the action of $\pi_1(\Sigma_n)$ on $T$ is by isometries we find

$$d_T(w, \gamma_B w) > 2 d_T(w, \text{axis of } \gamma_B)$$

(and similarly for $\gamma_C$). By the equivariance of the action of $\pi_1(\Sigma_n)$, we have $\widetilde{h_n}(\gamma_B z) = \gamma_B \widetilde{h_n}(z)$ and $\widetilde{h_n}(\gamma_C z) = \gamma_C \widetilde{h_n}(z)$. Using the equicontinuity bound and the fact that $\mathcal{D} \cup \gamma_B \mathcal{D} \cup \gamma_C \mathcal{D}$ is of bounded diameter, we then have

$$d_T(\widetilde{h_n}(z), \widetilde{h_n}(\gamma_B z)) \leq C_1 \text{ and } d_T(\widetilde{h_n}(z), \widetilde{h_n}(\gamma_C z)) \leq C_1$$

for any $z \in \mathcal{D}$, where $C_1$ is independent of $n$.

From these facts we can derive that both

$$d_T(\widetilde{h_n}(z), \text{axis of } \gamma_B) \leq C_1/2 \quad \text{and} \quad d_T(\widetilde{h_n}(z), \text{axis of } \gamma_C) \leq C_1/2$$

for any $z \in \mathcal{D}$. Since these axes $\gamma_B$ and $\gamma_C$ diverge in $T$, the image $\widetilde{h_n}(z)$ lies in a set of uniformly bounded diameter, as required.

Finally, in order to prove that for each $z \in \widetilde{\Sigma}_n$, we have that $\{\widetilde{h}_m(z)|m \geq n\}$ lies in a compact set, we adapt an argument from [**Wol96**] (see Lemma 3.4) to our present case. Indeed, we can cite that argument almost verbatim, so we sketch it here, and then indicate the only change at the very end of the discussion.

The idea is to first find any point $z_0 \in \widetilde{\Sigma}_n$ so that $\{\widetilde{h}_m(z_0)|m \geq n\}$ lies in a compact set; we then leverage the existence of that point to prove that limits of singularities of $\widetilde{h}_m$ have compact images, and then use that fact to show to that all points $z \in \widetilde{\Sigma}_n$ have that property.

The key observation is that the lift to $\widetilde{\Sigma}_n$ of a homotopically non-trivial closed curve $C$ on $\Sigma_n$ has an $\widetilde{h}_m$-image on the tree $T$ which must meet the axis on $T$ for the action of the cyclic subgroup of $\pi_1(\Sigma_n)$ generated by $[C]$. (Recall that the *axis* of an isometry of the tree $T$ is an isometrically embedded line on which the isometry acts as a translation.) Thus, by considering points along a compact closure $\bar{C}$ of a lift to a fundamental domain in $\widetilde{\Sigma}_n$ of $C$, we find a limit point $z_0 \in \bar{C}$ which has compact image in the axis corresponding to $[C]$.

From there, we consider homotopically non-trivial and distinct loops through $z_0$ that pass through singularities of $\widetilde{h}_m$; using the fact that we can find isometries of $T$ whose axes diverge at a vertex in the tree $T$, we show that those singularities have images limiting on such a vertex. Next we connect those singularities by arcs which must have images in segments between the image vertices, finding that the compact image property holds for points on those segments. Finally, we cut up the surface into simple connected domains bounded by those segments: by harmonicity of the map, those domains then have images that lie in the convex hull of the images of their boundaries, hence in a compact set by construction.

The only new feature we need to address in the setting of the current argument is the presence of a non-empty boundary $\partial\widetilde{\Sigma}_n$. Of course, the previous constructions have shown that points on lifts $\bar{C}$ of loops $C$ that pass through singularities of $\widetilde{h}_m$ and also link the boundary have images lying in compact sets. Thus the lift of any small neighborhood of a boundary point has images under $\widetilde{h}_m$ that lie only in an embedded line in $T$ (except in the case when the boundary has zero transverse measure, in which case the image is a point). Thus these points also have precompact image, as they are a bounded set in a real segment; in fact, we may use that property as part of the final step above when we cut up the surface into simply connected pieces, the boundaries of which have the desired property.

As a consequence of these constructions, Arzela-Ascoli applies, and we obtain a convergent subsequence of the family. A diagonal argument applied to the exhaustion $\{\Sigma_n\}$ of $\Sigma$ then finishes the argument. □

*Remark.* In a forthcoming paper [**GW**], we provide a different approach to this problem of the convergence of harmonic maps to trees where the energy grows without bound. That argument exploits the discreteness of the vertex set of the image tree to trap the images of the singularities of the Hopf differentials of the harmonic maps.

5.2.4. *Step 3: The Hopf differential has the required foliation.* Once we have the limiting harmonic map $\widetilde{h} : \widetilde{\Sigma \setminus P} \to T$, it remains to check that its Hopf differential is the desired one, namely, that it descends to $\Sigma \setminus P$ to yield a quadratic differential that has:

(a) poles of order two at the punctures,
(b) the required residues at the poles, and
(c) an induced measured foliation that is measure-equivalent to $F$.

We provide the proof of these in this subsection.

Recall that we have a sequence of equivariant harmonic maps $\widetilde{h_n} : \widetilde{\Sigma_n} \to T$ where $\widetilde{h_n} \to \widetilde{h}$ uniformly on compact sets.

*Proof of (a).* Since (see Lemma 5.5) the equivariant energies of the harmonic map $\bar{\mathcal{E}}(\widetilde{h_n}) \to \infty$ as $n \to \infty$ in the lifts of any neighborhood of each puncture of $\Sigma$, the Hopf differential of $h$ has poles of order greater or equal to two at any puncture.

So at any fixed puncture, the Hopf differential has the local form $z^{-k}Q(z)dz^2$ for some $k \geq 2$, where $Q(z)$ is a bounded holomorphic function.

Assume $k > 2$. Then if we uniformize $A_n$ to a round annulus of outer radius 1 and inner radius $0 < r < 1$, we have the following computation of energy:

$$\bar{\mathcal{E}}(\widetilde{h}_n|_{\widetilde{A}_n}) = \frac{1}{2}\int_0^{2\pi}\int_r^1 |z^{-k}Q(z)|rdrd\theta \asymp \frac{1}{r^{k-2}}.$$

Now note that $r^{-1} = e^{2\pi \mathrm{mod}(A_n)}$, so that $\bar{\mathcal{E}}(\widetilde{h}_n|_{\widetilde{A}_n}) \asymp e^{2\pi(k-2)\mathrm{mod}(A_n)}$.

Yet estimate (5.3) asserts that the equivariant energy of the restriction to $\widetilde{A}_n$ is at most linear in the modulus of $A_n$, so it is not possible that $k > 2$. (Indeed, for $k = 2$ the same calculation above yields an expression that is linear in $\ln r$.)

*Proof of (b).* We first refine the previous estimate and show that the distance of each approximating $\widetilde{h}_n$ from the lift of the model map $\widetilde{m}_{R,\alpha}$ is uniformly bounded on their restrictions on the strip $\widetilde{A}_n = [0, L_n] \times \mathbb{R}$ (a lift of the annulus $A_n$). Recall that $m_{R,\alpha}$ was the model map corresponding to the desired residue at the pole.

LEMMA 5.8. *The distance function $d_T(\widetilde{h}_n|_{\widetilde{A}_n}, \widetilde{m_{R,\alpha}}|_{\widetilde{A}_n}) < d_0$ where $d_0$ is independent of $n$.*

PROOF. We first consider the above distance function on the lifts of $\partial A_n = \partial\Sigma_0 \sqcup \partial\Sigma_n$.

First, by construction, the map $\widetilde{h_n}$ coincides with the lift of the model map $\widetilde{m}_{R,\alpha}$ on the lift of $\partial\Sigma_n$, so our attention turns to the distance between the two maps on the lift of $\partial\Sigma_0$.

To that end, choose a point, say $p_0$, on the boundary $\partial\Sigma_0$, and connect it by an arc $\gamma$ to a point on the boundary $\partial\Sigma_n$. Corollary 5.4 then asserts that the transverse measure of $\gamma$ with respect to the foliation $\mathcal{F}(q_n)$ is $L_n \sin\alpha$, up to an additive distortion $d_0 := T_0 + R\cos\alpha/2$ which is itself uniformly bounded (independent of $n$). Thus the image $\widetilde{h_n}(\tilde{\gamma})$ of the lift $\tilde{\gamma}$ of this arc under $\widetilde{h_n}$ has length $L_n \sin\alpha$, again up to an additive distortion of $d_0$. Since the length of the image of $\tilde{\gamma}$ by $\widetilde{m_{R,\alpha}}$ is exactly $L_n \sin\alpha$, so we obtain $d_T(\widetilde{h}_n(p_0), \widetilde{m_{R,\alpha}}(p_0) < d_0$. As the choice of $p_0$ on the lift of $\partial\Sigma_0$ was arbitrary,we can conclude that the distance function is uniformly bounded on the lift of that boundary component of $A_n$ as well.

However the distance function $d_T$ is subharmonic, as well as invariant under deck transformations. Hence the Maximum Principle applies to the function $d_T$ on the quotient $A_n$, and we find that $d_T$ is bounded in the interior of the strip by $d_0$, which is independent of $n$. □

We have already proved the uniform convergence $\widetilde{h}_n \to \widetilde{h}$, hence the images of the lifts of $\partial\Sigma_0$ under $\widetilde{h}_n$ are uniformly bounded. Hence the limiting harmonic map $h$ is asymptotic to (*i.e.* bounded distance from) the model map $m_{R,\alpha}$ on the punctured disk $U \setminus P = \bigcup\limits_n A_n$, which is a topological end of the surface $\Sigma \setminus P$.

By Lemma 4.4, the argument $\alpha$ of the residue $a$ at the pole agrees with that of the model map $m_{R,\alpha}$, and so does its real part which is determined by the transverse measure of the foliation $F$ (see equation (2.3) in §1).

When this transverse measure is positive, the above real part is $R\cos\alpha$. In this case, knowing $R\cos\alpha$ and $\alpha$ determines $R$, the modulus $|a|$ of the residue as well. This shows that the map $h$ has the required residue at the pole, in this case.

When the transverse measure is zero, we are in the case of a closed center at the pole, and the only parameter to determine is $R$. The diameter in $T$ of the image of a lift of $A_n$ under the model map $\tilde{m}_{R,\pm\frac{\pi}{2}}$ is $L_n$, which is related to the modulus of the annulus by $L_n = R\text{mod}(A_n)$ (see (5.2). Since $\text{mod}(A_n) \to \infty$ as $n \to \infty$, by Lemma 5.8 we have

$$\frac{\text{diam}\left(\widetilde{h}_n(\widetilde{A}_n)\right)}{\text{mod}(A_n)} = \frac{L_n + O(1)}{\text{mod}(A_n)} \to R \text{ as } n \to \infty$$

and this specifies $R$ uniquely.

Hence the map $h$ has the required residue at the pole, concluding the proof of (b).

*Proof of (c).* Recall that by construction, each map $\widetilde{h}_n$ collapses along a foliation that is equivalent to $\widetilde{F}|_{\widetilde{\Sigma}_n}$. Also, note that like the Dehn-Thurston parameters, the "intersection numbers" of a measured foliation with *all* simple closed curves (that is, the transverse measures of the curves) also serve to characterize the measured foliation - see, for example, [**FLP12**].

Let $\widetilde{F}'$ be the collapsing foliation of $\widetilde{h}$, the uniform limit of $\widetilde{h}_n$ as $n \to \infty$. By the equivariance of $\widetilde{h}$, this descends to a measured foliation $F'$ on the surface $\Sigma \setminus P$. Similarly, the collapsing foliation of each $\widetilde{h}_n$ descends to define a measured foliation on $\Sigma_n$, that we call $F'_n$.

To verify that the collapsing foliation $F'$ is equivalent to $F$, one needs to check that the intersection number of $F'$ with any simple closed curve, and the loop around the puncture, agrees with that of $F$.

First, the intersection numbers agree for the loop linking the puncture, since the parameters $R, \alpha$ of the model map were chosen so that the transverse measure of the loop is $R\cos\alpha$.

Second, any given simple closed curve $\gamma$ on $\Sigma$ lies in a compact set $K \subset \Sigma_m$ for $m$ sufficiently large, and hence its intersection number with $F'_n$ coincides with $i(\gamma, F)$ for all $n \geq m$. However, since the convergence $\widetilde{h}_n \to \widetilde{h}$ is uniform on a lift of the compact set $K$, the collapsing foliations of $\widetilde{h}_n$ converge to $\widetilde{F}'$ on $K$ (as is standard, the $C^0$-convergence of the sequence of holomorphic Hopf differentials can be bootstrapped to a $C^1$-convergence by the Cauchy integral formula), and hence the intersection numbers with $\gamma$ converge to $i(\gamma, F')$. Thus this limiting intersection number $i(\gamma, F')$ equals $i(\gamma, F)$, concluding the argument.

## References

[ALPS16] D. Alessandrini, L. Liu, A. Papadopoulos, and W. Su, *The horofunction compactification of Teichmüller spaces of surfaces with boundary*, Topology Appl. **208** (2016), 160–191, DOI 10.1016/j.topol.2016.05.011. MR3506976

[DDW00] G. Daskalopoulos, S. Dostoglou, and R. Wentworth, *On the Morgan-Shalen compactification of the* SL(2, **C**) *character varieties of surface groups*, Duke Math. J. **101** (2000), no. 2, 189–207, DOI 10.1215/S0012-7094-00-10121-4. MR1738182

[DW07] G. D. Daskalopoulos and R. A. Wentworth, *Harmonic maps and Teichmüller theory*, Handbook of Teichmüller theory. Vol. I, IRMA Lect. Math. Theor. Phys., vol. 11, Eur. Math. Soc., Zürich, 2007, pp. 33–109, DOI 10.4171/029-1/2. MR2349668

[FLP12] A. Fathi, F. Laudenbach, and V. Poénaru, *Thurston's work on surfaces*, Mathematical Notes, vol. 48, Princeton University Press, Princeton, NJ, 2012. Translated from the 1979 French original by Djun M. Kim and Dan Margalit. MR3053012

[GW] S. Gupta and M. Wolf, *Quadratic differentials, half-plane structures, and harmonic maps to trees*, Comment. Math. Helv. **91** (2016), no. 2, 317–356, DOI 10.4171/CMH/388. MR3493374

[GW16] S. Gupta and M. Wolf, *Quadratic differentials, half-plane structures, and harmonic maps to trees*, Comment. Math. Helv. **91** (2016), no. 2, 317–356, DOI 10.4171/CMH/388. MR3493374

[HM79] J. Hubbard and H. Masur, *Quadratic differentials and foliations*, Acta Math. **142** (1979), no. 3-4, 221–274, DOI 10.1007/BF02395062. MR523212

[Jos91] J. Jost, *Two-dimensional geometric variational problems*, Pure and Applied Mathematics (New York), John Wiley & Sons, Ltd., Chichester, 1991. A Wiley-Interscience Publication. MR1100926

[Kap09] M. Kapovich, *Hyperbolic manifolds and discrete groups*, Modern Birkhäuser Classics, Birkhäuser Boston, Inc., Boston, MA, 2009. Reprint of the 2001 edition. MR2553578

[KS93] N. J. Korevaar and R. M. Schoen, *Sobolev spaces and harmonic maps for metric space targets*, Comm. Anal. Geom. **1** (1993), no. 3-4, 561–659, DOI 10.4310/CAG.1993.v1.n4.a4. MR1266480

[Liu08] J. Liu, *Jenkins-Strebel differentials with poles*, Comment. Math. Helv. **83** (2008), no. 1, 211–240, DOI 10.4171/CMH/123. MR2365413

[Mes02] C. Mese, *Uniqueness theorems for harmonic maps into metric spaces*, Commun. Contemp. Math. **4** (2002), no. 4, 725–750, DOI 10.1142/S0219199702000828. MR1938491

[MP98] M. Mulase and M. Penkava, *Ribbon graphs, quadratic differentials on Riemann surfaces, and algebraic curves defined over* $\overline{\mathbf{Q}}$, Asian J. Math. **2** (1998), no. 4, 875–919, DOI 10.4310/AJM.1998.v2.n4.a11. Mikio Sato: a great Japanese mathematician of the twentieth century. MR1734132

[PH92] R. C. Penner and J. L. Harer, *Combinatorics of train tracks*, Annals of Mathematics Studies, vol. 125, Princeton University Press, Princeton, NJ, 1992. MR1144770

[Sch] R. M. Schoen, *Analytic aspects of the harmonic map problem*, Seminar on nonlinear partial differential equations (Berkeley, Calif., 1983), Math. Sci. Res. Inst. Publ., vol. 2, Springer, New York, 1984, pp. 321–358, DOI 10.1007/978-1-4612-1110-5_17. MR765241

[Sch84] ———, *Analytic aspects of the harmonic map problem*, Seminar on nonlinear partial differential equations (Berkeley, Calif., 1983), Math. Sci. Res. Inst. Publ., vol. 2, Springer, New York, 1984, pp. 321–358.

[Str84] K. Strebel, *Quadratic differentials*, Ergebnisse der Mathematik und ihrer Grenzgebiete (3) [Results in Mathematics and Related Areas (3)], vol. 5, Springer-Verlag, Berlin, 1984. MR743423

[Sun03] X. Sun, *Regularity of harmonic maps to trees*, Amer. J. Math. **125** (2003), no. 4, 737–771. MR1993740

[Thu] Dylan Thurston, *On geometric intersection of curves in surfaces (draft)*, `http://pages.iu.edu/%7Edpthurst/DehnCoordinates.pdf`.

[Wol89] M. Wolf, *The Teichmüller theory of harmonic maps*, J. Differential Geom. **29** (1989), no. 2, 449–479. MR982185

[Wol91] M. Wolf, *Infinite energy harmonic maps and degeneration of hyperbolic surfaces in moduli space*, J. Differential Geom. **33** (1991), no. 2, 487–539. MR1094467

[Wol95] M. Wolf, *Harmonic maps from surfaces to* **R**-*trees*, Math. Z. **218** (1995), no. 4, 577–593, DOI 10.1007/BF02571924. MR1326987

[Wol96] M. Wolf, *On realizing measured foliations via quadratic differentials of harmonic maps to* **R**-*trees*, J. Anal. Math. **68** (1996), 107–120, DOI 10.1007/BF02790206. MR1403253

Department of Mathematics, Indian Institute of Science, Bangalore 560012, India
*E-mail address*: subhojoy@math.iisc.ernet.in

Department of Mathematics, Rice University, Houston, Texas 77005-1892
*E-mail address*: mwolf@rice.edu

Contemporary Mathematics
Volume **696**, 2017
http://dx.doi.org/10.1090/conm/696/14022

# Quasi-isometric rigidity of the class of convex-cocompact Kleinian groups

Peter Haïssinsky

ABSTRACT. We provide a streamlined proof that the class of convex-cocompact Kleinian groups are quasi-isometrically rigid, relying on the quasi-isometric invariance of strong accessibility for word hyperbolic groups.

## 1. Introduction

A *Kleinian group* $G$ is a discrete subgroup of $\mathbb{P}SL_2(\mathbb{C})$. It acts properly discontinuously on the hyperbolic 3-space $\mathbb{H}^3$ via orientation-preserving isometries and it acts on the Riemann sphere $\widehat{\mathbb{C}}$ via Möbius transformations. The latter action is usually not properly discontinuous: there is a canonical and invariant partition

$$\widehat{\mathbb{C}} = \Omega_G \sqcup \Lambda_G$$

where $\Omega_G$ denotes the *ordinary set*, which is the largest open set of $\widehat{\mathbb{C}}$ on which $G$ acts properly discontinuously, and where $\Lambda_G$ denotes the *limit set*, which is the minimal $G$-invariant compact subset of $\widehat{\mathbb{C}}$.

As Poincaré observed, we may identify the Riemann sphere with the boundary at infinity of the three-dimensional hyperbolic space [**Poi**]. Explicitly, let us consider the open unit ball in $\mathbb{R}^3$ as a model of $\mathbb{H}^3$ and the unit sphere $\mathbb{S}^2$ for the Riemann sphere. One obtains in this way an action of a Kleinian group $G$ on the closed unit ball. With this identification in mind, the group $G$ preserves the convex hull $\mathrm{Hull}(\Lambda_G)$ of its limit set in $\mathbb{H}^3$. The group $G$ is *convex-cocompact* if its action is cocompact on $\mathrm{Hull}(\Lambda_G)$.

This class of groups plays an essential role in the classifcation of compact 3-manifolds. When $G$ is torsion-free, we may associate a 3-manifold $M_G = (\mathbb{H}^3 \cup \Omega_G)/G$, canonically endowed with a complete hyperbolic structure in its interior, which is called the *Kleinian manifold* associated to $G$. If $G$ is a torsion-free Kleinian group, then $G$ is convex-cocompact if and only if $M_G$ is compact. Note that $M_G$ is also orientable, irreducible and its fundamental group is infinite (as soon as $M_G$ is not a closed ball) and contains no subgroup isomorphic to $\mathbb{Z} \oplus \mathbb{Z}$. Conversely,

2010 *Mathematics Subject Classification.* Primary 20F67; Secondary 20E08, 30L, 37F30, 57M60.

*Key words and phrases.* Hyperbolic groups, cube complexes, hyperbolic 3-manifolds, convex-cocompact Kleinian groups, JSJ-decomposition.

This work was partially supported by the ANR project "GDSous/GSG" no. 12-BS01-0003-01.

Perel'man proved that an orientable, irreducible compact 3-manifold with infinite fundamental group which contains no subgroup isomorphic to $\mathbb{Z} \oplus \mathbb{Z}$ is isomorphic to some $M_G$, where $G$ is a convex-cocompact Kleinian group (when $M$ is Haken, this was previously established by Thurston).

Characterizing this class of convex-cocompact Kleinian groups is a natural and interesting problem. From the point of view of geometric group theory, one tries to understand the properties of a group by studying the different actions it admits on metric spaces. To start with, a finitely generated group acts by left-translation on the Cayley graph $X$ associated to any of its finite generating sets. If such a graph is equipped with the length metric which makes each edge isometric to the segment $[0, 1]$, then the action of $G$ becomes *geometric*: the group $G$ acts by isometries (the action is distance-preserving), properly discontinuously (for any compact subsets $K$ and $L$ of $X$, at most finitely many elements $g$ of $G$ will satisfy $g(K) \cap L \neq \emptyset$) and cocompactly (the orbit space $X/G$ is compact).

The classification of finitely generated groups up to *quasi-isometry* is a central issue: a quasi-isometry between metric spaces $X$ and $Y$ is a map $\varphi : X \to Y$ such that there are constants $\lambda > 1$ and $c > 0$ such that:

- (quasi-isometric embedding) for all $x, x' \in X$, the two inequalities
$$\frac{1}{\lambda} d_X(x, x') - c \leq d_Y(\varphi(x), \varphi(x')) \leq \lambda d_X(x, y) + c$$
hold and
- the $c$-neighborhood of the image $f(X)$ covers $Y$.

This defines in fact an equivalence relation on metric spaces. Note that any two locally finite Cayley graphs of the same group are quasi-isometric; this fact enables us to consider the quasi-isometry class of a finitely generated group (through the class of its locally finite Cayley graphs). More generally, Švarc-Milnor's lemma asserts that there is only one geometric action of a group on a proper geodesic metric space up to quasi-isometry [**GdlH**, Prop. 3.19].

On the other hand, quasi-isometric groups may be very different one from another. For instance, it is not clear that a group quasi-isometric to a linear group is itself linear.

Our main result says that a group $G$ is virtually a convex-cocompact Kleinian group if it looks like one:

THEOREM 1.1 (quasi-isometric rigidity). *The class of convex-cocompact Kleinian groups is quasi-isometrically rigid. More precisely, a finitely generated group quasi-isometric to a convex-cocompact Kleinian group contains a finite index subgroup isomorphic to a (possibly different) convex-cocompact Kleinian group.*

This theorem was derived in [**Haï**] from an indirect approach, using rather heavy material. In the present paper, we would like to offer a more direct proof, avoiding many technical parts of the aformentioned paper, relying instead on three-dimensional topological methods when possible. We assume nonetheless some familiarity with Bass-Serre theory.

There are several cases which were already known prior to this work.

Let us assume that $G$ is a finitely generated group quasi-isometric to a convex-cocompact Kleinian group $K$. The following cases are already known.

(1) If $\Lambda_K = \emptyset$ then $K$ is finite, so $G$ is finite as well, and we may consider the trivial group.

(2) If $\Lambda_K$ consists of two points, then $K$ contains an infinite cyclic subgroup of finite index, so $G$ as well [**SW**], hence $G$ contains a finite index subgroup isomorphic to the cyclic and Kleinian group $\langle z \mapsto 2z\rangle$.
(3) If $\Lambda_K$ is a Cantor set, then $K$ is virtually free, so $G$ as well [**Sta, SW**] and there is a finite index free subgroup $H$ which is isomorphic to the fundamental group of a handlebody. This implies that $H$ is isomorphic to a convex-cocompact Kleinian group —a so-called Schottky group.
(4) If $\Lambda_K$ is homeomorphic to the unit circle, then $G$ contains a finite index subgroup isomorphic to a cocompact Fuchsian group according to [**CJ**, **Gab**].
(5) If $\Lambda_K = \widehat{\mathbb{C}}$ then $G$ contains a finite index subgroup isomorphic to a cocompact Kleinian group [**CC**]; see also Corollary 3.6.
(6) If $\Lambda_K$ is homeomorphic to the Sierpiński carpet, then $G$ is commensurable to $K$ [**Fri**]; this also follows straightfowardly from [**BKM**], see Corollary 3.10.

Our method of proof provides us with the following result.

THEOREM 1.2. *Let $G$ be a word hyperbolic group and let us assume that its boundary is homeomorphic to the limit set of a convex-cocompact Kleinian group which contains no subset homeomorphic to the Sierpiński carpet. Then $G$ contains a finite index subgroup isomorphic to a convex-cocompact Kleinian group.*

In our setting, an *elementary group* is a group which is either finite or virtually cyclic. We will assume that $G$ is non-elementary with limit set different from a circle and a sphere, unless specifically stated.

**Outline of the paper.**— In the next section, we quickly review properties of 3-manifolds and their relationships with convex-cocompact Kleinian groups. Section 3 is devoted to word hyperbolic groups. After recalling their definition and main properties, we explain how the previously known cases of quasi-isometric rigidity may be established, using analytic properties of their boundaries. We conclude the section with a word on quasiconvex subgroups. Sections 4 and 5 are devoted to canonical splittings of word hyperbolic groups inspired by 3-manifolds. It is proved that strong accessibility is invariant under quasi-isometries (Theorem 5.2). The proof of the main theorem is given in Section 6. We introduce the notion of a regular JSJ-decomposition for groups which enables one to build manifolds and we adapt the notions of acylindrical pared manifolds to groups. Starting from a group $G$ quasi-isometric to a convex-cocompact Kleinian group $K$, the proof consists in building a 3-manifold with fundamental group isomorphic to (a finite-index subgroup of) $G$ and in applying Thurston's uniformization theorem. We proceed by induction on the hierarchy of $G$ over elementary groups provided by $K$. The initial cases are essentially already known. For the inductive step, we split the group over elementary groups and apply the induction hypothesis to obtain a finite collection of pared manifolds that are to be glued together. The gluing is made possible thanks to the separability of quasiconvex subgroups provided by the theory of Wise, Agol et al. on word hyperbolic groups acting on CAT(0) cube complexes (Theorem 6.2); it enables us to replace $G$ by a subgroup of finite index $G'$ such that the splitting is sufficiently regular so that the pared submanifolds can be pieced together to yield a Haken compact manifold with fundamental group isomorphic to $G'$. We conclude the paper with an example given by Kapovich and Kleiner which shows that taking a finite index subgroup can be necessary. We also provide some examples showing that the residual limit set of Kleinian groups does not depend only on its isomorphism class: this justifies the efforts done in the inductive step.

**Acknowledgements.**— I would like to thank the organizers of the Ahlfors-Bers Colloquium 2014 for their invitation. It is also my pleasure to thank Cyril Lecuire and Luisa Paoluzzi for the many conversations we have had on this work and for their comments on preliminary versions of this text. I also thank Matías Carrasco for discussions on JSJ-splittings for word hyperbolic groups. Last but not least, I am grateful to the referee for her or his comments.

## 2. Hyperbolizable 3-manifolds

We quickly review definitions and properties of 3-manifolds and of convex-cocompact Kleinian groups. Basic references include [**Thu2**, **Mor**, **Mad**]. The following exposition is inspired by [**And**].

Recall that a Kleinian group $K$ is convex-cocompact if $M_K = (\mathbb{H}^3 \cup \Omega_K)/K$ is compact. Ahlfors showed that if the limit set is not the whole Riemann sphere then it has measure 0. Furthermore, each connected component of its limit set is locally connected [**AnM**].

An orientable 3-manifold $M$ is *hyperbolizable* if there exists a convex-cocompact Kleinian group such that $M$ is homeomorphic to $M_K$ (this whole presentation rules out tori in $\partial M$ since they are not relevant to the present work). We say that $M$ is *uniformized* by $K$. Note that $K$ is isomorphic to the fundamental group of $M$, and that it is necessarily word hyperbolic if it contains no subgroup isomorphic to $\mathbb{Z}\oplus\mathbb{Z}$, see the next section. Moreover, the boundary $\partial M$ is a union of finitely many hyperbolic compact surfaces.

Let $M$ be a compact hyperbolizable 3-manifold with boundary. A surface $S$ is *properly embedded* in $M$ if $S$ is compact and orientable and if either $S \cap \partial M = \partial S$ or $S$ is contained in $\partial M$. A *compression disk* $D$ is an embedded disk such that $\partial D \subset \partial M$ and $\partial D$ is homotopically non trivial in $\partial M$. A properly embedded surface $S$ is *incompressible* if $S$ is not homeomorphic to the 2-sphere and either $S$ is a compression disk or if the inclusion $i : S \to M$ gives rise to an injective morphism $i_* : \pi_1(S, x) \to \pi_1(M, x)$. The *double* of a manifold $M$ with boundary is the union of $M$ and of a copy of itself glued along its boundary. A surface $S$ in $M$ is *boundary incompressible* if its double is incompressible in the double of $M$. A *Haken manifold* is a manifold which contains an incompressible surface. In our situation, as $\partial M \neq \emptyset$, $M$ is always Haken as soon as it fundamental group is infinite.

We say that $M$ has *incompressible boundary* if each component of $\partial M$ is incompressible. This is equivalent to the connectedness of the limit set of the group uniformizing $M$.

A surface $S$ in $M$ is *non-peripheral* if it is properly embedded and if the inclusion $i : S \to M$ is not homotopic to a map $f : S \to M$ such that $f(S) \subset \partial M$. A surface $S$ is *essential* if it is properly embedded, incompressible, boundary incompressible and non-peripheral. An *acylindrical compact manifold* has incompressible boundary and no essential annuli: the limit set of any uniformizing group is homeomorphic to the Sierpiński carpet.

A compact *pared manifold* $(M, P)$ is given by a 3-manifold $M$ as above together with a finite collection of pairwise disjoint incompressible annuli $P \subset \partial M$ such that any cylinder in $C \subset M$ with boundaries in $P$ can be homotoped relatively to its boundary into $P$. We say the paring is *acylindrical* if $\partial M \setminus P$ is incompressible and every incompressible cylinder disjoint from $P$ and with boundary curves in $\partial M$ can be homotoped into $\partial M \setminus P$ relatively to $\partial M$.

If $M$ has compressible boundary, it can be cut along *compression disks* into finitely many pieces each of which has incompressible boundary. Given a compact hyperbolizable manifold $M$ with incompressible boundary, we may cut it into finitely many pieces along essential annuli so that the remaining pieces are acylindrical pared manifolds. Forgetting the paring, we may then iterate the above procedure by looking for compression disks and essential annuli.

It is a consequence of Haken's finiteness theorem that this procedure has to stop in finite time, yielding a finite number of compact manifolds each of which is either a ball or an acylindrical manifold. The collection of all the manifolds that are obtained from $M$ defines a *hierarchy* of the manifold and of its fundamental group.

We will use the following form of Thurston's uniformization theorem for Haken manifolds:

THEOREM 2.1. *Let $M$ be a compact irreducible orientable Haken 3-manifold with word hyperbolic fundamental group. Then $M$ is hyperbolizable.*

## 3. Hyperbolicity

Background on word hyperbolic groups include [**Gro**, **GdlH**, **KB**].

Let $X$ be a metric space. It is *geodesic* if any pair of points $\{x, y\}$ can be joined by a (geodesic) segment i.e, a map $\gamma : [0, d(x, y)] \to X$ such that $\gamma(0) = x$, $\gamma(d(x, y)) = y$ and $d(\gamma(s), \gamma(t)) = |t - s|$ for all $s, t \in [0, d(x, y)]$. The metric space $X$ is *proper* if closed balls of finite radius are compact.

A *triangle* $\Delta$ in a metric space $X$ is given by three points $\{x, y, z\}$ and three segments joining them two by two. Given a constant $\delta \geq 0$, the triangle $\Delta$ is *$\delta$-thin* if any side of the triangle is contained in the $\delta$-neighborhood of the two others.

DEFINITION 3.1 (Hyperbolic spaces and groups). *A geodesic metric space is hyperbolic if there exists $\delta \geq 0$ such that every triangle is $\delta$-thin. A group $G$ is word hyperbolic if it acts geometrically on a proper, geodesic hyperbolic metric space.*

Basic examples of hyperbolic spaces are the simply connected Riemannian manifolds $\mathbb{H}^n$ of sectional curvature $(-1)$ and $\mathbb{R}$-trees. In particular since the action of a convex-compact Kleinian group $G$ is cocompact on $\mathrm{Hull}\,\Lambda_G$, the group $G$ is word hyperbolic.

**Shadowing lemma.**— A *quasigeodesic* is the image of an interval by a quasi-isometric embedding. The shadowing lemma asserts that, given $\delta$, $\lambda$ and $c$, there is a constant $H = H(\delta, \lambda, c)$ such that, for any $(\lambda, c)$-quasigeodesic $q$ in a proper geodesic $\delta$-hyperbolic metric space $X$, there is a geodesic $\gamma$ at Hausdorff distance at most $H$.

It follows from the shadowing lemma that, among geodesic metric spaces, hyperbolicity is invariant under quasi-isometry : if $X$, $Y$ are two quasi-isometric geodesic metric spaces, then $X$ is hyperbolic if and only if $Y$ is hyperbolic.

**Compactification.**— A proper geodesic hyperbolic space $X$ admits a canonical compactification $X \sqcup \partial X$ at infinity. This compactification can be defined by looking at the set of rays i.e., isometric embeddings $r : \mathbb{R}_+ \to X$, up to bounded Hausdorff distance. The topology is induced by the uniform convergence on compact subsets. The boundary can be endowed with a family of *visual distances* $d_v$ compatible with its topology i.e., which satisfy

$$d_v(a,b) \asymp e^{-\varepsilon d(w,(a,b))}$$

where $w \in X$ is any choice of a base point, $\varepsilon > 0$ is a *visual parameter* chosen small enough, and $(a,b)$ is any geodesic asymptotic to rays defining $a$ and $b$.

If $\Phi : X \to Y$ is a quasi-isometry between two proper geodesic hyperbolic spaces, then the shadowing lemma implies that $\Phi$ induces a homeomorphism $\phi : \partial X \to \partial Y$. This means that a word hyperbolic group $G$ admits a topological boundary $\partial G$ defined by considering the boundary of any proper geodesic metric space on which $G$ acts geometrically.

In the case of a convex-compact Kleinian group $K$, a model for the boundary $\partial K$ is given by its limit set $\Lambda_K$.

**3.1. Analytic aspects.** A general principle asserts that a word hyperbolic group is determined by its boundary. More precisely, Paulin proved that the quasi-isometry class of a word hyperbolic group is determined by its boundary equipped with its quasiconformal structure [**Pau**]. Let us recall some definitions. A homeomorphism $h : X \to Y$ between metric spaces is called *quasi-Möbius* [**Väi**] if there exists a homeomorphism $\theta : \mathbb{R}_+ \to \mathbb{R}_+$ such that, for any distinct points $x_1, x_2, x_3, x_4 \in X$,

$$[h(x_1) : h(x_2) : h(x_3) : h(x_4)] \le \theta([x_1 : x_2 : x_3 : x_4])$$

where

$$[x_1 : x_2 : x_3 : x_4] = \frac{|x_1 - x_2| \cdot |x_3 - x_4|}{|x_1 - x_3| \cdot |x_2 - x_4|}.$$

Quasi-Möbius maps are stable under composition. The boundary $\partial G$ of a word hyperbolic group $G$ is endowed with a *conformal gauge* i.e., a family of metrics which are pairwise quasi-Möbius equivalent. These metrics are exactly those metrics compatible with the topology of $\partial G$ for which the action of $G$ is *uniformly quasi-Möbius*, meaning that the distortion control $\theta$ is independent of $g \in G$. Visual distances are examples of metrics of the gauge.

Quasi-isometries provide natural examples of quasi-Möbius maps as well:

THEOREM 3.2. *A $(\lambda, c)$-quasi-isometry between proper, geodesic, hyperbolic metric spaces extends as a $\theta$-quasi-Möbius map between their boundaries, where $\theta$ only depends on $\lambda, c$, the hyperbolicity constants and the visual parameters.*

This result takes its roots in the work of Efremovich and Tihomirova [**ET**]; see also [**Mag**] where quasi-isometries are explicitly defined and where Theorem 3.2 is proved for real hyperbolic spaces. The key fact of its proof is that the crossratio $[x_1 : x_2 : x_3 : x_4]$ is measured by the distance between geodesics $(x_1, x_2)$ and $(x_3, x_4)$, a quantity quasi-preserved by quasi-isometries, cf. [**Pau**, Prop. 4.5] for a proof in the present setting.

Paulin's result reads

THEOREM 3.3 (Paulin [**Pau**]). *Two non-elementary word hyperbolic groups are quasi-isometric if and only if there is quasi-Möbius homeomorphism between their boundaries.*

**3.2. Convergence actions.** If $G$ is a word hyperbolic group acting geometrically on a proper hyperbolic metric space $X$, then its action extends as a uniformly quasi-Möbius action on $\partial X$. The action of $G$ is *a convergence action* i.e., its diagonal action on the set of distinct triples is properly discontinuous, and is moreover *uniform* i.e., its action is also cocompact on the set of distinct triples. As for Kleinian groups, the *limit set* $\Lambda_G$ is by definition the unique minimal closed invariant subset of $X$. It is empty if $G$ is finite. These properties characterize word hyperbolic groups and their boundaries:

THEOREM 3.4 (Bowditch [**Bow2**]). *Let $G$ be a convergence group acting on a perfect metrizable space $X$. The action of $G$ is uniform on $\Lambda_G$ if and only if $G$ is word hyperbolic and there exists an equivariant homeomorphism between $\Lambda_G$ and the boundary at infinity $\partial G$ of $G$.*

In our setting, we start with a finitely generated group $G$ quasi-isometric to a convex-cocompact Kleinian group $K$. Since $K$ is convex-cocompact, $K$ is word hyperbolic, hence $G$ as well. Thus, Theorem 3.4 implies that $\partial G$ is homeomorphic to $\Lambda_K$. Moreover, we obtain in this way a quasi-Möbius action of the group $G$ on $\Lambda_K \subset \widehat{\mathbb{C}}$. Note that the action of $G$ need not be related to the action of $K$. Let $F < G$ be the kernel of the action of $G$ on $\partial G$ and $G' = G/F$. Since the action of $G$ is properly discontinuous, $F$ is finite and $G'$ is quasi-isometric to $G$, hence $\partial G'$ is canonically identified with $\partial G$.

**3.3. Quasi-Möbius groups versus Möbius groups.** We recall some results which justify the quasi-isometric rigidity of some of the cases listed in the introduction. Let us begin with a general result which served as a motivation for the whole work.

THEOREM 3.5 (Sullivan, [**Sul**]). *A countable group of uniformly quasi-Möbius transformations on $\widehat{\mathbb{C}}$ is conjugate to a group of Möbius transformations.*

COROLLARY 3.6 (Cannon and Cooper, [**CC**]). *If $G$ is quasi-isometric to $\mathbb{H}^3$, then $G$ contains a finite index subgroup isomorphic to a cocompact Kleinian group.*

PROOF. According to the previous discussion, $G$ acts on $\widehat{\mathbb{C}}$ as a uniform quasi-Möbius group. Hence Theorem 3.5 implies that the action of $G$ is conjugate to a group of Möbius transformations. This action extends as an action by isometries on $\mathbb{H}^3$. Since the action is a uniform convergence action, we may check as follows that $G$ is a cocompact Kleinian group: to any triple on $\widehat{\mathbb{C}}$, we associate the center in $\mathbb{H}^3$ of the ideal triangle with those points as vertices. We may check that since the action of $G$ is properly discontinuous and cocompact on the set of triples, it is also properly discontinuous and cocompact on $\mathbb{H}^3$.

If the action of $G$ on its boundary is faithful, then the corollary is proved. Otherwise, we have just proved that it admits a quotient $G' = G/F$ by a finite subgroup $F$ isomorphic to a cocompact Kleinian group. To prove that $G$ itself is virtually isomorphic to a convex-cocompact Kleinian group, we need to invoke the theory of word hyperbolic groups acting on CAT(0) cube complexes that we state as Lemma 3.7 below. Therefore, we may find $G'' < G$ of finite index such that $G'' \cap F = \{e\}$ implying that its action is faithful. □

LEMMA 3.7. *Let $G$ be a finitely generated group which contains a normal finite subgroup $F$ such that $G/F$ is isomorphic to a convex-cocompact Kleinian group. Then $G$ is residually finite.*

A proof goes as follows: according to [**BW**], the group $G/F$ acts geometrically on a CAT(0) cube complex, so $G$ as well. Hence $G$ contains a finite index subgroup which acts specially in the sense of Haglund and Wise [**HW**] according to Agol [**Ago**]. Therefore, $G$ is residually finite [**HW**].

We now recall the rigidity of groups acting on the circle.

THEOREM 3.8 (Hinkkanen, Markovic, [**Hin1, Hin2, Mak**]). *Any uniformly quasi-Möbius group of homeomorphisms on the unit circle is quasisymmetrically conjugate to a group of Möbius transformations.*

Let us now pass to groups acting on Sierpiński carpets.

THEOREM 3.9 (Bonk, Merenkov, Kleiner, [**BKM**]). *Any quasi-Möbius selfmap of a round Sierpiński carpet of measure zero is the restriction of a Möbius transformation.*

COROLLARY 3.10 (Frigerio, [**Fri**]). *If $G$ is quasi-isometric to the fundamental group $K$ of a compact hyperbolic 3-manifold with totally geodesic boundary, then $G$ is commensurable to $K$.*

PROOF. Since $K$ is the fundamental group of a compact hyperbolic 3-manifold with totally geodesic boundary, its boundary is a measure zero round Sierpiński carpet. Let $G_M$ denote the set of quasi-Möbius selfhomeomorphisms of $\Lambda_K$. Note that $G_M$ contains $K \cup G'$, where $G'$ consists of the largest quotient which has a faithful action on its boundary as above.

According to Theorem 3.9, the action of $G_M$ extends to an action of Möbius transformations on $\widehat{\mathbb{C}}$. It is clearly discrete since any sequence which tends uniformly to the identity will have to eventually stabilize at least three circles, implying that such a sequence is eventually the identity. Therefore, $G_M$ is a convergence group, and since it contains $K$, its action is uniform on $\Lambda_K$, hence it is word hyperbolic. Since the boundaries of $K$ and $G_M$ coincide, $K$ has finite index in $G_M$ [**KS**]. The same holds for $G'$.

We conclude as above: if $G = G'$, then the corollary is proved. Otherwise, we apply Lemma 3.7 to conclude that $G$ is residually finite. Therefore, we may find $G'' < G$ of finite index such that $G'' \cap F = \{e\}$ implying that its action is faithful. □

Thus, we assume throughout this paper that $G$ is none of the above classes.

**3.4. Quasiconvex subgroups.** Basic references include [**KS**]. Let $X$ be a proper, geodesic, hyperbolic metric space. A $K$-*quasiconvex* subset $Y \subset X$ has the property that any geodesic segment joining two points of $Y$ remains in the $K$-neighborhood of $Y$. Note that quasiconvexity is a property invariant under quasi-isometries. A subgroup $H$ of a hyperbolic group $G$ is *quasiconvex* if $H$ is quasiconvex in any locally finite Cayley graph of $G$.

If a group $G$ acts properly discontinuously by isometries on a proper geodesic hyperbolic metric space, its *limit set* $\Lambda_G$ is defined as the set of accumulation points at infinity of any orbit $G(x)$, $x \in X$. This set is $G$-invariant.

PROPOSITION 3.11. *Let $\Phi : G_1 \to G_2$ be a quasi-isometry, $\phi : \partial G_1 \to \partial G_2$ its boundary map, and let $H_j \subset G_j$ be quasiconvex subgroups. If $\phi(\Lambda_{H_1}) = \Lambda_{H_2}$ then $\Phi(H_1)$ is at bounded distance from $H_2$ so $H_1$ and $H_2$ are quasi-isometric.*

PROOF. If $H_1$ is finite, then $\Lambda_{H_1}$ is empty, so $\Lambda_{H_2}$ is also empty and $H_2$ is finite as well. Otherwise, it follows from [**KS**, Prop. 3.4] that $H_j$ is at bounded distance from the convex hull $\mathcal{C}(\Lambda_{H_j})$, $j = 1, 2$, consisting of the union of geodesics with endpoints in the respective limit sets. By the shadowing lemma, $\Phi(\mathcal{C}(\Lambda_{H_1}))$ is quasiconvex, at bounded distance from $\mathcal{C}(\Lambda_{H_2})$, implying the proposition. □

Another proof could have been obtained from Theorem 3.3.

## 4. Canonical splittings

We describe well-known splittings of word hyperbolic groups which correspond to splitting a manifold along compression disks and essential annuli. In each case, the Bass-Serre theory will provide us with a simplicial action of $G$ on a simplicial tree $T$ with no edge inversions such that $T/G$ is finite and all vertex (stabilizers) groups and edge (stabilizers) groups are quasiconvex. Note that it is enough to check that the edge groups are quasiconvex to ensure the vertex groups are as well [**Bow1**, Prop. 1.2]. We refer to [**Ser**, **SW**] for a general introduction to the Bass-Serre theory.

In both cases, the splitting will be maximal in a precise sense.

**4.1. Splittings over finite groups.** Let $G$ be a non-elementary word hyperbolic group. If it is not one-ended then it splits over a finite group [**Sta**]. By [**Dun**], there is a maximal splitting over finite groups such that each vertex group is finite or one-ended; when $G$ is torsion-free, it leads to a free product of a free group with finitely many one-ended groups.

More precisely, the group $G$ acts simplicially on a simplicial tree $T$ with no edge inversions such that $T/G$ is finite and

- all edge groups are finite;
- all vertex groups are finite or one-ended.

Moreover, the one-ended vertex groups stabilise a nontrivial connected component of the boundary. Conversely, any nontrivial component of $\partial G$ is stabilised by a one-ended vertex group.

We call this splitting a *DS-splitting.*

PROPOSITION 4.1. *Let $G$ and $H$ be two quasi-isometric word hyperbolic groups. Then each infinite vertex group of the DS-splitting of $G$ is quasi-isometric to an infinite vertex group of the DS-splitting of $H$, and vice-versa.*

This proposition holds very generally for finitely presented groups [**PW**]. We provide an elementary proof in our setting.

PROOF. Let $\Phi : G \to H$ be a quasi-isometry. According to Theorem 3.2, it extends as a quasi-Möbius homeomorphism $\phi : \partial X \to \partial Y$. In particular, $\phi$ maps connected components of $\partial X$ to components of $\partial Y$. Therefore, if $G_v$ is a one-ended vertex of $G$ in its DS-splitting with limit set $\Lambda_v$, then $\phi(\Lambda_v)$ is a component of $\partial H$, hence is stabilised by a one-ended vertex group $H_v$. Proposition 3.11 implies that $G_v$ and $H_v$ are quasi-isometric. □

**4.2. Splittings over 2-ended groups.** We summarize briefly the JSJ decomposition of a non-Fuchsian one–ended word hyperbolic group $G$ following Bowditch [**Bow1**].

There exists a canonical simplicial minimal action of $G$ on a simplicial tree $T = (V, E)$ without edge inversions such that $T/G$ is a finite graph and which enjoys the following properties, cf. [**Bow1**, Thm 5.28, Prop. 5.29]. If $v$ is a vertex (resp. $e$ an edge), we will denote by $G_v$ (resp. $G_e$) its stabilizer, and by $\Lambda_v$ (resp. $\Lambda_e$) the limit set of $G_v$ (resp. $G_e$). Let $E_v$ denote the set of edges incident to $v \in T$. Every vertex and edge group is quasiconvex in $G$. Each edge group $G_e$ is two-ended and $\partial G \setminus \Lambda_e$ is not connected. A vertex $v$ of $T$ belongs to exactly one of the following three exclusive types.

> **Type I (elementary).**— The vertex has bounded valence in $T$. Its stabilizer $G_v$ is two-ended, and the connected components of $\partial G \setminus \Lambda_v$ are in bijection with the edges incident to $v$.
> **Type II (surface).**— The limit set $\Lambda_v$ is cyclically separating and the stabilizer $G_v$ of such a vertex $v$ is a non-elementary virtually free group canonically isomorphic to a convex-cocompact Fuchsian group. The incident edges are in bijection with the peripheral subgroups of that Fuchsian group.
> **Type III (rigid).**— Such a vertex $v$ does not belong to any class above. Its stabilizer $G_v$ is a non-elementary quasiconvex subgroup. Every local cut point of $\partial G$ in $\Lambda_v$ is in the limit set of an edge stabilizer incident to $v$ and $\Lambda_v$ is a maximal closed subset of $\partial G$ with the property that it cannot be separated by two points of $\partial G$.

By construction, no two vertices of the same type are adjacent, nor surface type and rigid can be adjacent either. This implies that every edge contains a vertex of elementary type. The action of $G$ preserves the types. Therefore, the edges incident to a vertex $v$ of surface type or rigid type are split into finitely many $G_v$-orbits.

PROPOSITION 4.2. *Let $G$ and $H$ be two quasi-isometric word hyperbolic groups. Then each vertex and edge group of the JSJ-splitting of $G$ is quasi-isometric to a vertex and edge group of the JSJ-splitting of $H$ respectively, and vice-versa.*

PROOF. We proceed as in Proposition 4.1. Let $\Phi : G \to H$ be a quasi-isometry. According to Theorem 3.2, it extends as a quasi-Möbius homeomorphism $\phi : \partial X \to \partial Y$. In particular, $\phi$ maps the local cut points of $\partial X$ to the local cut points of $\partial Y$. Therefore, since the JSJ decomposition is built from the local cut points of the boundaries of the groups, $\phi$ will map limit sets of edge groups to limit sets of edge groups, limit sets of vertex groups to limit sets of vertex groups of the same type. Thus, if $G_v$ is a vertex group of $G$ in its JSJ-splitting with limit set $\Lambda_v$, then $\phi(\Lambda_v)$ is the limit set of a vertex group $H_v$ in the JSJ-decomposition of $H$. Proposition 3.11 implies that $G_v$ and $H_v$ are quasi-isometric. □

## 5. Strong accessibility

What follows is inspired by similar properties of compact 3-manifolds, cf. § 2. We first show how to combine the DS- and JSJ- splittings together, following Drutu and Kapovich [**DK**].

PROPOSITION 5.1. *Let $G$ be a word hyperbolic group. If $G$ is not quasi-isometric to a Fuchsian group, then $G$ acts on a simplicial tree $T$ with no edge inversions such that*

- *the quotient $T/G$ is finite;*
- *the edge stabilizers are elementary groups;*
- *there is an equivariant projection $p : T \to T_{DS}$ such that $p^{-1}(v)$ provides the JSJ-decomposition of $G_v$, for each non-Fuchsian one-ended vertex.*

*Moreover, this splitting is maximal with respect to elementary groups.*

PROOF. Let us consider the DS-splitting of $G$: it is given by a graph of groups $\mathcal{G}_{DS} = (\Gamma_{DS}, \{G_v\}, \{G_e\}, G_e \hookrightarrow G_{t(e)})$, whose fundamental group is isomorphic to $G$ and where the edge groups are all finite and the vertex groups are either finite or one-ended. Let $G_v$ be a non-Fuchsian one-ended vertex group, and let us consider its JSJ-decomposition given by a graph of groups $\mathcal{G}_v = (\Gamma_v, \{H_s\}, \{H_a\}, H_a \hookrightarrow H_{t(a)})$. We wish to refine $\mathcal{G}_{DS}$ and insert $\mathcal{G}_v$.

Let $e$ be an incident edge to $v$ in $\Gamma_{DS}$. Note that $G_e$ is a subgroup of $G_v$ hence acts on the Bass-Serre tree $T_v$ associated to $\mathcal{G}_v$. Since $G_e$ is finite, it admits a fixed vertex point $s' \in T_v$. Its stabilizer $H_{s'}$ is conjugate to a vertex group $H_s$ of $\mathcal{G}_v$ by an element $h_{s'} \in G_v$.

We may now define a new graph of groups $\mathcal{G} = (\Gamma, \{G_v\}, \{G_e\}, G_e \hookrightarrow G_{t(e)})$ as follows. We let $\Gamma$ be the union $(\Gamma \setminus \{v\}) \cup \Gamma_v$ that we tie together by attaching each incident $e$ in $\Gamma_{DS}$ to the vertex $s \in \Gamma_v$ defined above. One may check that the natural embeddings yield a graph of groups whose fundamental group is isomorphic to $G$; see [**DK**, § 18.6] for details.

Proceeding similarly for each non-Fuchsian one-ended vertex group in $\Gamma_{DS}$ provides us with the desired splitting. □

Let $G$ be a word hyperbolic group. Applying Proposition 5.1, we let $\mathcal{H}_1$ denote all the vertex groups in the quotient $T/G$ and $\mathcal{H}'_1 \subset \mathcal{H}_1$ be those which are non-Fuchsian.

We proceed inductively so that $\mathcal{H}_{n+1}$ consists of the set of isomorphic classes of vertex groups appearing in the decomposition of vertex groups in $\mathcal{H}'_n$ given by Proposition 5.1 and let $\mathcal{H}'_{n+1} \subset \mathcal{H}_{n+1}$ be those classes of groups that are non-Fuchsian.

A group $G$ is *strongly accessible* if the process ends i.e., $\mathcal{H}'_n$ consists of one-ended groups which are either Fuchsian or such that their boundary have no local cut points. The collection of conjugacy classes $\cup \mathcal{H}_n$ defines a *hierarchy* of the group $G$ and the smallest such $\ell$ with the property that $\mathcal{H}_\ell$ consists of groups which cannot be split further is the *depth* of the hierarchy.

If $G$ has no element of order two, then this process stops in finite time [**DP**, **LT**], exactly as for Haken manifolds which are cut into finitely many balls. We will not need these results here, but rely instead on the following quasi-isometric invariance property, which may be of independent interest.

THEOREM 5.2 (QI-invariance of strong accessibility). *Let $G_1$ and $G_2$ be two quasi-isometric word hyperbolic groups. Then $G_1$ and $G_2$ are either both strongly accessible or not. In the former case, the depth of their hierarchy coincide and, for all $n \geq 1$, each group in $\mathcal{H}_n(G_1)$ is quasi-isometric to a group in $\mathcal{H}_n(G_2)$, and vice-versa.*

PROOF. Let us proceed by induction on the depth. Let us assume that $G_1$ is either virtually free, virtually Fuchsian, or with no local cut points. Then the same holds for $G_2$ since the boundaries are homeomorphic. Let us now assume that $G_1$ has a hierarchy of depth $\ell \geq 1$. Let us apply Proposition 5.1 to $G_1$ and $G_2$. Note that Proposition 4.1 implies that $G_1$ and $G_2$ have the same vertex groups up to quasi-isometry in their respective DS-splittings. Moreover, Proposition 4.2 implies that the vertex groups appearing in the JSJ splittings from the above vertex groups are also the same up to quasi-isometry. Therefore, groups in $\mathcal{H}_1(G_1)$ are quasi-isometric to groups in $\mathcal{H}_1(G_2)$ and vice-versa. Furthermore, elements of $\mathcal{H}_1(G_1)$ have depth at most $\ell - 1$, so we may apply the induction hypothesis. □

## 6. Quasi-isometric rigidity

Let $G$ be a finitely generated group quasi-isometric to a convex-cocompact Kleinian group. The basic idea is to build a compact 3-manifold with fundamental group isomorphic to (a finite-index subgroup of) $G$ and then to apply Thurston's uniformization theorem. This construction will proceed by induction on the hierarchy of $G$, and will require to have a *regular* JSJ decomposition as defined below. In general, a given group does not admit such a decomposition. We will show that we may obtain this property for a suitable subgroup of finite-index. Its existence will be granted by the separability of its quasiconvex subgroups (Theorem 6.2).

**6.1. Regular JSJ decomposition.** Let $M = M_K$ be a hyperbolizable manifold with incompressible boundary so that $\Lambda_K$ is connected. Let us assume that $M$ is built from gluing several manifolds $M_1, \ldots, M_k$ to a solid torus $T$ along annuli whose cores are homologous to multiples of the core of $T$. In this case, these curves have to be all parallel in $\partial T$. By taking a suitable finite covering $M' = M_{K'}$ of $M$, we may then assume that each one of them generate the fundamental group of the torus. Thus, the fundamental group of such a lifted annulus $A$ is generated by a primitive element $g$ of $K'$. Its limit set $\Lambda_g$ splits $\Lambda_K$ into finitely many components, which are fixed under the action of $g$. This leads us to the following notion:

DEFINITION 6.1 (Regular JSJ decomposition). *Let $G$ be a one-ended word hyperbolic group and let us consider its JSJ decomposition. We say it is regular if the following properties hold:*

- *$G$ is torsion free;*
- *every two-ended group $H$ which appears as a vertex group is isomorphic to $\mathbb{Z}$ and stabilizes the components of $\partial G \setminus \Lambda_H$.*

Let us assume that the decomposition of a group $G$ is regular and let $G_v$ be a vertex group of elementary type $G_v$; since its action fixes the components of $\partial G \setminus \Lambda_v$, it is generated by a primitive element.

**6.2. Separability properties.** A subgroup $H$ is *separable* in its ambient group $G$ if, for any $g \in G \setminus H$, there is a finite index subgroup $G'$ in $G$ which contains $H$ but not $g$.

THEOREM 6.2. *Let $G$ be a finitely generated group quasi-isometric to a convex-cocompact Kleinian group. Then its quasiconvex subgroups are separable.*

This theorem uses the recent breakthroughs of Wise, Agol, and their collaborators on word hyperbolic groups acting on CAT(0) cube complexes [**Wis**, **Ago**]. We only sketch the proof.

PROOF. We use the QI-invariance of the strong accessibility together with Corollary 3.10 to prove that $G$ admits a so-called virtual quasiconvex hierarchy. This implies according to Wise that the group is a so-called virtual special group, hence its quasiconvex subgroups are separable according to [**HW**]. See [**Haï**, § 8] for details. □

Separability is used to "clean" our group as explained by the next proposition and its corollaries. See [**Haï**, Prop. 7.2] for a proof.

PROPOSITION 6.3. *Let $A' < A < G$ be groups with $[A : A'] < \infty$ and $A'$ separable in $G$. Then there exist subgroups $A''$ and $H$ with the following properties:*

(1) *$H$ is a normal subgroup of finite index in $G$;*
(2) *$A'' = H \cap A'$ is a normal subgroup of finite index in $A$ ;*
(3) *for all $g \in G$, $(gAg^{-1}) \cap H = gA''g^{-1}$.*

As a first application, we obtain from the finiteness of the conjugacy classes of torsion elements:

COROLLARY 6.4. *A residually finite word hyperbolic group is virtually torsion-free.*

Another simple application of Proposition 6.3 is the following reduction consequence.

COROLLARY 6.5. *Let $G$ be the fundamental group of a finite graph of groups $\mathcal{G} = (\Gamma, \{G_v\}, \{G_e\})$, where each edge group is quasiconvex in $G$. Let $A_v < G_v$ be subgroups of finite index. If the quasiconvex subgroups of $G$ are separable, then $G$ contains a normal finite index subgroup $G'$ such that, for any $g \in G$, $gG_vg^{-1} \cap G'$ is a finite index subgroup of $gA_vg^{-1}$.*

PROOF. It follows from [**Bow1**, Prop. 1.2] that the vertex groups are quasiconvex as well. We apply Proposition 6.3 to each triple $(A_v, G_v, G)$ to obtain finitely many finite index subgroups $G'_v$ of $G$ and we let $G'$ denote their intersection. The conclusion follows. □

We now derive the virtual regularity of the JSJ decomposition.

COROLLARY 6.6. *Let $G$ be a finitely generated group quasi-isometric to a convex-cocompact Kleinian group. Then $G$ contains a finite index subgroup such that its one-ended vertex groups in the DS-splitting have a regular JSJ-decomposition.*

PROOF. It follows from Theorem 6.2 that every elementary group is separable. In particular, $G$ contains a finite index subgroup $H'$ which is torsion-free by Corollary 6.4. It follows that its action on $\partial G$ is faithful. Let $K'$ be one of the factors of the DS-splitting of $H'$ which is one-ended but not Fuchsian.

Let us consider a set of representatives of elementary vertex groups $\{K'_v\}$ arising from its JSJ decomposition. Each one of them contains a cyclic subgroup $A_v$ of finite index which stabilizes all the components of $\partial K \setminus \Lambda_v$. Corollary 6.5 applied to $(A_v, G_v)$ produces a normal subgroup $H$ of finite index in $G$. By construction, the vertices appearing in the DS-splitting of $H$ admit a regular JSJ-decomposition; see [**Haï**, Thm 7.1] for details. □

**6.3. Acylindrical pared groups.** We adapt the notion of acylindrical pared 3-manifolds to groups. This point of view was first developed by Otal for free groups in [**Ota**]. Let $G$ be a nonelementary word hyperbolic group. A *paring* will be given by a finite malnormal collection $P_G = \{C_1, \ldots, C_k\}$ of maximal 2-ended groups i.e., if $C_i \cap (gC_jg^{-1})$ is infinite for some $i, j \in \{1, \ldots, k\}$ and $g \in G$, then $i = j$ and $g \in C_i$ and such that $G$ does not split over them. We will say that the paring is *acylindrical* if, whenever $G$ splits over an elementary group $H$, the paring $P_G$ does not split to define a paring of vertex groups — up to conjugacy.

FACT 6.7. *Let $G$ be a non-Fuchsian one-ended word hyperbolic group with a regular JSJ decomposition. Let $v$ be a vertex from its JSJ decomposition of rigid type and $E_v$ denote the edges incident to $v$. Then $(G_v, \{G_e, e \in E_v/G_v\})$ is an acylindrical pared group.*

This fact follows from the maximality of the decomposition.

Note that an acylindrical pared compact hyperbolizable 3-manifold $(M, P_M)$ gives rise to a canonical acylindrical pared group $(K, P_K)$ by letting $K = \pi_1(M)$ and $P_K$ denote the fundamental groups generated by the core of the annuli composing $P_M$. A quasi-isometry between two acylindrical pared groups $(G, P_G)$ and $(H, P_H)$ will be given by a quasi-isometry $\Phi : G \to H$ such that, for any $C_j \in P_G$, $\Phi(C_j)$ is at bounded distance from a conjugate of an element from $P_H$, and if any element of $P_H$ is at bounded distance from the image of a conjugate of $P_G$. If $G'$ is a finite index subgroup of an acylindrical pared group $(G, P_G)$, then we associate an acylindrical paring to $G'$ by considering a finite set of representative classes of $\{G' \cap gCg^{-1},\ g \in G, C \in P_G\}$. We will say that an acylindrical pared manifold $(M, P_M)$ is quasi-isometric to an acylindrical pared group $(G, P_G)$ if there is a quasi-isometry between its canonical acylindrical pared group $(K, P_K)$ and $(G, P_G)$.

Let us say that an acylindrical pared group $(G, P_G)$ is *geometric* if there is a compact hyperbolic acylindrical pared 3-manifold $(M, P)$ such that its canonical acylindrical pared group is isomorphic to $(G, P_G)$ with $G = \pi_1(M)$.

Theorem 1.1 will follow from the following variant:

THEOREM 6.8 (pared quasi-isometric rigidity). *An acylindrical pared group $(G, P_G)$ is virtually geometric if it is quasi-isometric to a compact hyperbolic acylindrical pared 3-manifold $(M, P)$.*

The proof will be established by induction on the depth of the hierarchy of the group $G$.

The case of an acylindrical pared free group has been established by Otal [**Ota**]. Let us recall his theorem. Let $(G, P)$ be an acylindrical pared word hyperbolic group. Define on $\partial G$ an equivalence relation $\sim_P$ as follows: let $x \sim_P y$ if, either $x = y$ or if there is a subgroup $C \in P$, an element $g \in G$ such that $\{x, y\} = g(\Lambda_C)$. Set $\partial_P G = \partial G / \sim_P$ to be the quotient of $\partial G$ by this relation $\sim_P$. This is always Hausdorff compact, connected and locally connected set.

THEOREM 6.9 (Otal). *An acylindrical pared free group $(\mathbb{F}, P)$ is geometric if $\partial_P \mathbb{F}$ is planar.*

COROLLARY 6.10. *An acylindrical pared free group $(\mathbb{F}, P)$ quasi-isometric to an acylindrical pared handlebody $(M, P_M)$ is geometric.*

PROOF. Let $(K, P_K)$ be the canonical acylindrical pared group associated to $(M, P)$. According to [**Ota**, Proposition 0], $\partial_{P_K} K$ is planar, hence $\partial_P \mathbb{F}$ as well since the quasi-isometry $\Phi : (\mathbb{F}, P) \to (K, P_K)$ defines a homeomorphism $\phi : \partial\mathbb{F} \to \Lambda_K$ which maps the $\sim_P$-classes to $\sim_{P_K}$-classes. Therefore, Theorem 6.9 applies. □

We may now prove the main theorem.

PROOF. (Theorems 1.1 and 6.8) We first reduce the proof of Theorem 1.1 to the proof of Theorem 6.8. Let $G$ be a group quasi–isometric to a convex-cocompact Kleinian group $K$. We might as well assume that $G$ does not fall into one of the known cases mentioned in the introduction. Let us consider the splitting provided by Proposition 5.1 and its corresponding action $G$ on $T$. Propositions 4.1 and 4.2 imply that each vertex group is quasi-isometric to a convex-cocompact Kleinian group. By Fact 6.7, the rigid vertices are canonically pared and quasi-isometric to an acylindrical pared manifold. Assuming for the time being Theorem 6.8, we may even conclude that the rigid vertices are virtually geometric.

Theorem 6.2, Corollaries 6.6 and 6.5 enable us to assume that, up to a finite index subgroup, $G$ admits a regular JSJ decomposition with rigid vertex pared groups $(G_v, \{G_e, e \in E_v/G_v\})$ isomorphic to acylindrical pared manifolds $(M_v, P_v)$. If $v$ is of elementary type, then we associate a solid torus with parallel pairwise disjoint annuli on its boundary parametrized by the edges incident to $v$. If $v$ is of surface type, then Theorem 3.8 implies that $G_v$ is canonically isomorphic to a Fuchsian group $F_v$ such that $S_v = \mathbb{D}/F_v$ is a surface with holes parametrized by $E_v/G_v$; we then consider $M_v = S_v \times [0, 1]$ with annuli $A_e$ corresponding to the $I$-fibers of the holes. The finite graph $T/G$ provides us with a manual to build a 3-manifold $M$ with fundamental group isomorphic to $G$. We first glue the $M_v$'s along annuli in order to obtain the vertices of the DS-decomposition, and then glue the resulting manifolds along disks [**SW**]. Thurston's hyperbolization theorem for Haken manifolds shows that $M$ is hyperbolizable. This ends the proof of Theorem 1.1.

We now pass to the proof of Theorem 6.8. Let $(G, P_G)$ be a group quasi–isometric to a pared manifold $(M, P)$ and let $K$ be a convex-cocompact Kleinian group representing its fundamental group. Since $K$ is pared, its limit set is a proper subset of the sphere. According to Corollary 6.4 we may assume that $G$ is torsion-free. Theorem 5.2 implies that $G$ is strongly accessible so we may proceed by induction on the depth of its hierarchy.

If the depth is zero, then $G$ is either

- free, to which we associate a handlebody; or
- Fuchsian, to which we associate an I-bundle by Theorem 3.8; or
- a carpet group, to which we associate an acylindrical manifold by Corollary 3.10.

We easily obtain an acylindrical pared manifold when $G$ is free (Corollary 6.10), Fuchsian or a carpet group.

Let $G$ be of depth $n + 1$, with $n \geq 0$, and let us assume that the theorem is established up to rank $n$. As for Theorem 1.1, the induction hypothesis enables us to assume that $G$ admits a regular JSJ-decomposition and that each vertex group is the fundamental group of a pared manifold $(N_v, Q_v)$.

The graph of groups coming from Proposition 5.1 provides us with a manual to build a 3-manifold $N$ with fundamental group isomorphic to $G$ by gluing the

$N_v$'s along disks or annuli [**SW**]. In general, there are several choices for this last step. We need to make a suitable choice to make sure that $N$ will admit a paring associated to $P_G$, cf § 7.1 for justifying the necessity of this step. The choice of the gluings will be made according to the embedding of $\partial G$ in $\widehat{\mathbb{C}}$ provided by its quasi-isometry with $K$.

Let us explain how we use $K$ to construct $N$. It follows from [**Ohs**] that there exists a Kleinian group $K'$ isomorphic to $K$ such that each element of $P_K$ has become parabolic and its limit set is homeomorphic to $\partial_{P_K} K$. The canonical projection $\Lambda_K \to \partial_{P_K} K$ gives rise to a monotone map $p : (\widehat{\mathbb{C}}, \Lambda_K) \to (\widehat{\mathbb{C}}, \Lambda_{K'})$.

The quasi-isometry $\Phi : (G, P_G) \to (K, P_K)$ yields a homeomorphism $\phi : \partial G \to \Lambda_K$ which realizes a correspondence between the limit sets of $gCg^{-1}$, where $g \in G$ and $C \in P_G$, with the limit sets of $kDk^{-1}$, where $k \in K$ and $D \in P_K$. Thus, we obtain a homeomorphism $\varphi : \partial_{P_G} G \to \Lambda_{K'}$ which enables us to make $G$ act by homeomorphisms on $\Lambda_{K'}$.

Since $(K, P_K)$ is acylindrical, it follows that $M_{K'}$ has no essential annulus, so the closure of no component of $\Omega_{K'}$ separates $\Lambda_{K'}$, cf. [**Thu1**, Theorem 2]. Therefore, the action of $G$ on $\Lambda_{K'}$ maps the boundaries of the components of $\Omega_{K'}$ to the boundaries of (possibly other) components of $\Omega_{K'}$. It follows that the action of $G$ on $\Lambda_K$ extends to a convergence action on $\widehat{\mathbb{C}}$ of $G$, cf. [**Haï**, Prop. 6.2].

This action enables us to realize geometrically the splitting of $G$ given by Proposition 5.1: by [**MS**, Theorem 3.2], one may first find finitely many simple closed curves on $\Omega_K/G$ whose lifts form a collection $\Gamma_1$ of pairwise disjoint Jordan curves on $\widehat{\mathbb{C}}$; see also [**AbM**]. These curves realize the DS-splitting of $G$. Furthermore, the JSJ splitting of the $G$-stabilizers of non-trivial components of $\Lambda_K$ different from circles can also be realised on $\widehat{\mathbb{C}}$ by considering all the hyperbolic geodesics in $\Omega_K$ joining the fixed points of the edge groups; this defines a second family $\Gamma_2$ of curves. Let $\Gamma = \Gamma_1 \cup \Gamma_2$: we define a tree $T$ by considering

(a) the vertices to be the components of $\widehat{\mathbb{C}} \setminus \cup_{\gamma\in\Gamma}\overline{\gamma}$; to which we need to add the elementary vertices of the JSJ splittings; the latter are the fixed points of loxodromic elements which split a component into at least three components or which are at the intersection of two components corresponding to vertices of surface and rigid types;

(b) edges correspond to curves $\gamma \in \Gamma$ which are in the common boundary of the vertices, or on the boundary of one component and joining the fixed points of an elementary vertex.

We leave the reader check that $G$ acts simplicially on $T$ and that its action is isomorphic to the one given by Proposition 5.1.

We may now build a manifold $N$ so that $\partial N$ is homeomorphic to $\Omega_K/G$. Let $v \in T$ be a vertex and $e \in E_v$ be an edge incident to $v$. This edge corresponds to a curve $\gamma \in \Gamma$ which defines a simple closed curve $\overline{\gamma} \subset \partial N_v$. Either it bounds a disk $D_e \subset \partial N_v$, or it is homotopically non-trivial and hence can be thickened to become an annulus $A_e \subset \partial N_v$. Now the graph of groups tells us how to glue the different vertex manifolds $N_v$'s, $v \in T/G$, along the disks and annuli associated to the class of edges $E_v/G_v$. Let $N$ be the manifold obtained by this construction. By construction, $\partial N$ is homeomorphic to $\Omega_K/G$.

It remains to check that $P_G$ provides a paring of $N$. Let us first observe that the elements of $P_G$ are maximal 2-ended subgroups of $G$, hence they are cyclic and generated by primitive elements. Note that $P_K$ gives rise to a collection of pairwise disjoint curves in $\Omega_K$ joining pairs of fixed points of loxodromic elements. These points are also fixed points by the conjugates of $P_G$ coming from the action of $G$ on $(\widehat{\mathbb{C}}, \Lambda_K)$. Since $\partial N$ is homeomorphic to $\Omega_K/G$, this implies that $P_G$ define curves on $\partial N$ which are pairwise disjoint. So $P_G$ defines a paring $Q$ on $N$. The paring is acylindrical since this is the case for $K$.

Thurston's hyperbolization theorem for Haken manifolds shows that $N$ is hyperbolizable. This proves that $(G, P_G)$ is virtually geometric. □

**6.4. Groups with homeomorphic boundaries.** Since the proof of Theorem 1.2 is essentially the same as Theorem 1.1, we only sketch the argument. The main observation is that the hierarchy provided by Proposition 5.1 only depends on the topology of the limit set. Therefore, Theorem 5.2 still holds in this setting: if $G$ is a word hyperbolic group with boundary homeomorphic to the one of a convex-cocompact Kleinian group, then $G$ is strongly accessible since $K$ is, the depth of their hierarchy coincide and, for all $n \geq 1$, the boundary of each group in $\mathcal{H}_n(G)$ is homeomorphic to the boundary of a group in $\mathcal{H}_n(K)$, and vice-versa. Assuming that $K$ is furthermore torsion free and that $\Lambda_K$ contains no subset homeomorphic to the Sierpiński case implies that the hierarchy's leaves contain only free groups, Fuchsian groups and elementary groups. Therefore, $G$ will admit a quasiconvex hierarchy in the sense of Wise, so the conclusion of Theorem 6.2 and its corollaries will also hold in our context. Theorem 1.2 may be established by induction on the length of the hierarchy. The initial step consists in saying that a word hyperbolic whose boundary is either empty, consists of two points, a Cantor set or a circle is virtually isomorphic to a Fuchsian group: this was already known, cf. the introduction. The induction step proceeds as above: first construct a finite index subgroup $G'$ in order to get regular JSJ decompositions of its maximal one-ended subgroups with geometric pared rigid subgroups, then apply the induction hypothesis to construct a compact 3-manifold with fundamental group isomorphic to $G'$ and apply Thurston's uniformization theorem.

## 7. Examples

We provide examples explaining some of the difficulties that have to be dealt with.

**7.1. Unstability of the residual limit set.** In the proof of Theorem 6.8, we were very careful in the construction of the manifold $M$ to ensure that $P_G$ would be associated to a paring. This question is related to the residual limit set $R_G$ of a Kleinian group, defined by Abikoff as those points of $\Lambda_G$ which are not on the boundary of any component of $\Omega_G$.

Let $G$ be a finitely generated Kleinian group. According to Abikoff [**Abi**], the residual limit set has two different kinds of points: those which correspond to singleton connected components $z$ of $\Lambda_G$ for which there are components $\Lambda_n$, $n \geq 1$, of $\Lambda_G$, and components $\Omega_n$ of $\widehat{\mathbb{C}} \setminus \Lambda_n$ such that $(\Omega_n)_n$ defines a nested family of disks and $\cap \Omega_n = \{z\}$, and those points which belong to a non trivial connected component of $\Lambda_G$. In both cases, they are characterized by the fact that we may find a nested sequence of Jordan curves $\gamma_n \subset \Lambda_G$, components $\Omega_n$ of $\widehat{\mathbb{C}} \setminus \gamma_n$ such that $\cap \Omega_n$ is a residual point.

We give two examples that show that a homeomorphism between the limit sets of two isomorphic convex-cocompact Kleinian groups need not preserve the residual limit sets.

Let us denote by $S_g$ a compact, closed, orientable surface of genus $g \geq 0$, and $S_g^*$ the complement of an embedded closed disk in $S_g$.

7.1.1. *The residual set of an infinitely ended Kleinian group.* Let us first consider a hyperbolic compact 3-manifold $M_1$ with at least two non-homeomorphic boundary components $S_1$ and $S_1'$. Let $M_2 = S_g \times [0,1]$. Let us pick embedded closed disks $D_1 \subset S_1 (\subset \partial M_1)$, $D_1' \subset S_1' (\subset \partial M_1)$ and $D_2 \subset S_g \times \{0\} (\subset \partial M_2)$. Let $M$ be the connected sum of $M_1$ and $M_2$ obtained by identifying $D_1$ and $D_2$ and $M'$ by identifying $D_1'$ with $D_2$. By van Kampen's theorem, both fundamental groups are isomorphic to the free product of the fundamental groups of both manifolds. Let $K$ and $K'$ be two convex-cocompact Kleinian groups uniformizing $M$ and $M'$ respectively.

Now, let us consider homotopically non-trivial curves $\gamma_1 \subset S_1 (\subset \partial M_1)$ and $\gamma_2 \subset S_g \times \{0\} (\subset \partial M_2)$. In $M$, their homotopy classes can be concatenated into a curve $\gamma$ in $\partial M$. Its class defines a loxodromic element of $K$ whose fixed points are non-residual, since it represents a boundary curve. On the other hand, in $M'$, their concatenation defines a loop which has to go through the interior of $M_1$ to connect both pieces. Thus, any lift in its universal cover $\widetilde{M}'$ has to cross infinitely many lifts of $M_1$. This means that, in $K'$, this curve represents a loxodromic element whose fixed points are in different nested intersections of lifts of $S_1$, hence in $R_{K'}$.

7.1.2. *The residual set of a one-ended Kleinian group.* We consider $N_g = S_g^* \times [0,1]$ for $g = 1, \ldots, 4$ and let us denote by $A_g$ the boundary of $N_g$ which is homeomorphic to an annulus. Let $N_0$ be a solid torus to which we consider 4 parallel and disjoint annuli $B_1, \ldots, B_4$ whose core curves represent generators of $\pi_1(N_0)$. We let $N$ be the manifold obtained by identifying $A_g$ with $B_g$ for $g = 1, \ldots, 4$, and $N'$ by identifying $A_1$ with $B_1$, $A_4$ with $B_4$, $A_2$ with $B_3$ and $A_3$ with $B_2$. Both fundamental groups are isomorphic by van Kampen's theorem. These manifolds are hyperbolizable by Thurston's theorem and one-ended by construction.

We now let $\gamma_j$ be a non trivial curve in $S_j^* \times \{0\}$ for $j = 1, 2$. Their concatenation defines in $N$ a loxodromic element represented by a boundary curve, so its fixed points are not in the residual set. In $N'$, since the cyclic order is different, this curve is now represented by a loxodromic element whose fixed points cannot lie on the boundary of a component, hence are residual.

**7.2. A non-Kleinian group quasi-isometric to a Kleinian group.** We recall a construction of Kapovich and Kleiner which provides us with a torsion-free word hyperbolic group which is quasi-isometric to a convex-cocompact Kleinian group but is not isomorphic to a convex-cocompact Kleinian group: this implies that it is necessary, in some cases, to pass to a subgroup of finite index, and that it is not an artifact of the proof.

Following [**KK**], let us consider a 2-torus with two holes $S = \mathbb{T}^2 \setminus (D_1 \cup D_2)$, and let $\alpha$ represent $\partial D_1$ and $\beta = \partial D_2$. Define $X$ as the complex obtained from $S$ by gluing $\beta$ to $\alpha$ by a degree 2 covering map $\beta \to \alpha$.

The fundamental group of $S$ is isomorphic to $\mathbb{F}_3 = < a, b, c >$ where we can choose $c$ to be represented by $\alpha$. The other boundary component can be expressed as $\beta = caba^{-1}b^{-1} = c[a,b]$.

The fundamental group of $X$ is obtained by an HNN-extension of $\pi_1(S)$:

$$\pi_1(X) = \mathbb{F}_3 *_{\mathbb{Z}} =< a, b, c, t | tc^2t^{-1} = c[a, b] > .$$

As such, $G = \pi_1(X)$ is a torsion-free group. It is also word-hyperbolic according to Bestvina-Feighn's combination theorem.

The JSJ decomposition is given by a graph with two vertices and two edges connecting the two vertices. The first vertex corresponds to the fundamental group of $S$, labeled by $\mathbb{F}_3$, the second to a cylinder with cyclic fundamental group $\mathbb{Z} =< \gamma >$. Note that $\mathbb{F}_3$ corresponds to a surface group in this decomposition.

The first edge identifies one boundary component to $\alpha$, yielding the morphism $c \mapsto \gamma$, and the second edge identifies the other boundary component to $\beta^2$, yielding $c[a, b] \mapsto \gamma^2$.

The limit set of $c$ splits $\partial G$ into three components, and its action exchanges the two components attached at $\beta$ and keep fixed the third one. Therefore, this decomposition is not regular. It turns out that $G$ cannot be a 3-manifold group because of the identification $\alpha^2 = \beta$ which should take its origin in an essential annulus or Möbius band.

Nonetheless $G$ contains a finite-index subgroup $G'$ with a regular JSJ-decomposition. This can be seen topologically by considering a finite covering $S' \to S$ which has degree 1 over $\alpha$, degree 2 over $\beta$ and such that there are twice the number of preimages of $\alpha$ than of $\beta$.

For the group $G'$, we may build an irreducible, orientable, Haken 3-manifold with fundamental group $G'$. See [**KK**, § 8] for more details.

## References


[Abi] W. Abikoff, *The residual limit sets of Kleinian groups*, Acta Math. **130** (1973), 127–144, DOI 10.1007/BF02392264. MR0404613

[AbM] W. Abikoff and B. Maskit, *Geometric decompositions of Kleinian groups*, Amer. J. Math. **99** (1977), no. 4, 687–697, DOI 10.2307/2373860. MR0480992

[Ago] I. Agol, *The virtual Haken conjecture*, Doc. Math. **18** (2013), 1045–1087. With an appendix by Agol, Daniel Groves, and Jason Manning. MR3104553

[And] J. W. Anderson, *Closed essential surfaces in hyperbolizable acylindrical 3-manifolds*, Pacific J. Math. **183** (1998), no. 1, 1–19, DOI 10.2140/pjm.1998.183.1. MR1616614

[AnM] J. W. Anderson and B. Maskit, *On the local connectivity of limit set of Kleinian groups*, Complex Variables Theory Appl. **31** (1996), no. 2, 177–183. MR1423249

[BW] N. Bergeron and D. T. Wise, *A boundary criterion for cubulation*, Amer. J. Math. **134** (2012), no. 3, 843–859, DOI 10.1353/ajm.2012.0020. MR2931226

[BKM] M. Bonk, B. Kleiner, and S. Merenkov, *Rigidity of Schottky sets*, Amer. J. Math. **131** (2009), no. 2, 409–443, DOI 10.1353/ajm.0.0045. MR2503988

[Bow1] B. H. Bowditch, *Cut points and canonical splittings of hyperbolic groups*, Acta Math. **180** (1998), no. 2, 145–186, DOI 10.1007/BF02392898. MR1638764

[Bow2] B. H. Bowditch, *A topological characterisation of hyperbolic groups*, J. Amer. Math. Soc. **11** (1998), no. 3, 643–667, DOI 10.1090/S0894-0347-98-00264-1. MR1602069

[CC] J. W. Cannon and D. Cooper, *A characterization of cocompact hyperbolic and finite-volume hyperbolic groups in dimension three*, Trans. Amer. Math. Soc. **330** (1992), no. 1, 419–431, DOI 10.2307/2154172. MR1036000

[CJ] A. Casson and D. Jungreis, *Convergence groups and Seifert fibered 3-manifolds*, Invent. Math. **118** (1994), no. 3, 441–456, DOI 10.1007/BF01231540. MR1296353

[DP] T. Delzant and L. Potyagailo, *Accessibilité hiérarchique des groupes de présentation finie* (French, with English summary), Topology **40** (2001), no. 3, 617–629, DOI 10.1016/S0040-9383(99)00078-6. MR1838998

[DK] C. Druţu and M. Kapovich. Lectures on geometric group theory. Lecture notes, 2013.

[Dun] M. J. Dunwoody, *The accessibility of finitely presented groups*, Invent. Math. **81** (1985), no. 3, 449–457, DOI 10.1007/BF01388581. MR807066

[ET] V. A. Efremovič and E. S. Tihomirova, *Equimorphisms of hyperbolic spaces* (Russian), Izv. Akad. Nauk SSSR Ser. Mat. **28** (1964), 1139–1144. MR0169121

[Fri] R. Frigerio, *Commensurability of hyperbolic manifolds with geodesic boundary*, Geom. Dedicata **118** (2006), 105–131, DOI 10.1007/s10711-005-9028-x. MR2239451

[Gab] D. Gabai, *Convergence groups are Fuchsian groups*, Ann. of Math. (2) **136** (1992), no. 3, 447–510, DOI 10.2307/2946597. MR1189862

[GdlH] E. Ghys and P. de la Harpe, editors. *Sur les groupes hyperboliques d'après Mikhael Gromov*, volume 83 of *Progress in Mathematics*. Birkhäuser Boston Inc., Boston, MA, 1990. Papers from the Swiss Seminar on Hyperbolic Groups held in Bern, 1988.

[Gro] M. Gromov, *Hyperbolic groups*, Essays in group theory, Math. Sci. Res. Inst. Publ., vol. 8, Springer, New York, 1987, pp. 75–263, DOI 10.1007/978-1-4613-9586-7_3. MR919829

[HW] F. Haglund and D. T. Wise, *Special cube complexes*, Geom. Funct. Anal. **17** (2008), no. 5, 1551–1620, DOI 10.1007/s00039-007-0629-4. MR2377497

[Haï] P. Haïssinsky, *Hyperbolic groups with planar boundaries*, Invent. Math. **201** (2015), no. 1, 239–307, DOI 10.1007/s00222-014-0552-x. MR3359053

[Hin1] A. Hinkkanen, *Uniformly quasisymmetric groups*, Proc. London Math. Soc. (3) **51** (1985), no. 2, 318–338, DOI 10.1112/plms/s3-51.2.318. MR794115

[Hin2] A. Hinkkanen, *The structure of certain quasisymmetric groups*, Mem. Amer. Math. Soc. **83** (1990), no. 422, iv+87, DOI 10.1090/memo/0422. MR948926

[KB] I. Kapovich and N. Benakli, *Boundaries of hyperbolic groups*, Combinatorial and geometric group theory (New York, 2000/Hoboken, NJ, 2001), Contemp. Math., vol. 296, Amer. Math. Soc., Providence, RI, 2002, pp. 39–93, DOI 10.1090/conm/296/05068. MR1921706

[KS] I. Kapovich and H. Short, *Greenberg's theorem for quasiconvex subgroups of word hyperbolic groups*, Canad. J. Math. **48** (1996), no. 6, 1224–1244, DOI 10.4153/CJM-1996-065-6. MR1426902

[KK] M. Kapovich and B. Kleiner, *Hyperbolic groups with low-dimensional boundary* (English, with English and French summaries), Ann. Sci. École Norm. Sup. (4) **33** (2000), no. 5, 647–669, DOI 10.1016/S0012-9593(00)01049-1. MR1834498

[LT] L. Louder and N. Touikan, *Strong accessibility for finitely presented groups*, Geom. Topol. **21** (2017), no. 3, 1805–1835, DOI 10.2140/gt.2017.21.1805. MR3650082

[Mad] A. Marden, *Outer circles. An introduction to hyperbolic 3-manifolds*, Cambridge University Press, Cambridge, 2007. MR2355387

[Mag] G. A. Margulis, *The isometry of closed manifolds of constant negative curvature with the same fundamental group* (Russian), Dokl. Akad. Nauk SSSR **192** (1970), 736–737. MR0266103

[Mak] V. Markovic, *Quasisymmetric groups*, J. Amer. Math. Soc. **19** (2006), no. 3, 673–715, DOI 10.1090/S0894-0347-06-00518-2. MR2220103

[MS] G. J. Martin and R. K. Skora, *Group actions of the* 2*-sphere*, Amer. J. Math. **111** (1989), no. 3, 387–402, DOI 10.2307/2374665. MR1002005

[Mor] J. W. Morgan, *On Thurston's uniformization theorem for three-dimensional manifolds*, The Smith conjecture (New York, 1979), Pure Appl. Math., vol. 112, Academic Press, Orlando, FL, 1984, pp. 37–125, DOI 10.1016/S0079-8169(08)61637-2. MR758464

[Ohs] K. Ohshika, *Geometrically finite Kleinian groups and parabolic elements*, Proc. Edinburgh Math. Soc. (2) **41** (1998), no. 1, 141–159, DOI 10.1017/S0013091500019477. MR1604337

[Ota] J.-P. Otal, *Certaines relations d'équivalence sur l'ensemble des bouts d'un groupe libre* (French), J. London Math. Soc. (2) **46** (1992), no. 1, 123–139, DOI 10.1112/jlms/s2-46.1.123. MR1180888

[PW] P. Papasoglu and K. Whyte, *Quasi-isometries between groups with infinitely many ends*, Comment. Math. Helv. **77** (2002), no. 1, 133–144, DOI 10.1007/s00014-002-8334-2. MR1898396

[Pau] F. Paulin, *Un groupe hyperbolique est déterminé par son bord* (French, with French summary), J. London Math. Soc. (2) **54** (1996), no. 1, 50–74, DOI 10.1112/jlms/54.1.50. MR1395067

[Poi] H. Poincaré, *Mémoire sur les groupes kleinéens* (French), Acta Math. **3** (1883), no. 1, 49–92, DOI 10.1007/BF02422441. MR1554613

[SW] P. Scott and T. Wall, *Topological methods in group theory*, Homological group theory (Proc. Sympos., Durham, 1977), London Math. Soc. Lecture Note Ser., vol. 36, Cambridge Univ. Press, Cambridge-New York, 1979, pp. 137–203. MR564422

[Ser] J.-P. Serre, *Arbres, amalgames,* $SL_2$ (French), Société Mathématique de France, Paris, 1977. Avec un sommaire anglais; Rédigé avec la collaboration de Hyman Bass; Astérisque, No. 46. MR0476875

[Sta] J. Stallings, *Group theory and three-dimensional manifolds*, Yale University Press, New Haven, Conn.-London, 1971. A James K. Whittemore Lecture in Mathematics given at Yale University, 1969; Yale Mathematical Monographs, 4. MR0415622

[Sul] D. Sullivan, *On the ergodic theory at infinity of an arbitrary discrete group of hyperbolic motions*, Riemann surfaces and related topics: Proceedings of the 1978 Stony Brook Conference (State Univ. New York, Stony Brook, N.Y., 1978), Ann. of Math. Stud., vol. 97, Princeton Univ. Press, Princeton, N.J., 1981, pp. 465–496. MR624833

[Thu1] W. Thurston, *Hyperbolic geometry and 3-manifolds*, Low-dimensional topology (Bangor, 1979), London Math. Soc. Lecture Note Ser., vol. 48, Cambridge Univ. Press, Cambridge-New York, 1982, pp. 9–25. MR662424

[Thu2] W. P. Thurston, *Three-dimensional manifolds, Kleinian groups and hyperbolic geometry*, Bull. Amer. Math. Soc. (N.S.) **6** (1982), no. 3, 357–381, DOI 10.1090/S0273-0979-1982-15003-0. MR648524

[Väi] J. Väisälä, *Quasi-Möbius maps*, J. Analyse Math. **44** (1984/85), 218–234, DOI 10.1007/BF02790198. MR801295

[Wis] D. T. Wise, *Research announcement: the structure of groups with a quasiconvex hierarchy*, Electron. Res. Announc. Math. Sci. **16** (2009), 44–55, DOI 10.3934/era.2009.16.44. MR2558631



Université Paul Sabatier, Institut de Mathématiques de Toulouse, 118 route de Narbonne, 31062 Toulouse Cedex 9, France

*Current address*: Université d'Aix-Marseille, Institut de Mathématiques de Marseille/CMI, 39 rue Frédéric Joliot-Curie, 13453 Marseille Cedex 13, France

*E-mail address*: phaissin@math.cnrs.fr

Contemporary Mathematics
Volume **696**, 2017
http://dx.doi.org/10.1090/conm/696/14023

# Variation of moduli under continuous motions

Yunping Jiang and Sudeb Mitra

*In memory of Professor Clifford J. Earle*

Abstract. We show that a continuous motion of a finite set $E$ in the Riemann sphere $\widehat{\mathbb{C}}$ over a simply connected parameter $V$ can be extended to a quasiconformal motion of $\widehat{\mathbb{C}}$ over $V$, and that such extension is unique up to isotopy rel $E$. This fact leads to results on continuity of the extremal length of a general family of curves, and in particular of the modulus of annuli and lengths of closed geodesics.

## 1. Statements of the results

The concept of a holomorphic motion was introduced in the paper [**14**], where the authors proved the $\lambda$-lemma, which connected holomorphic motions with quasiconformality. Since its inception, the question of extending holomorphic motions has been an important subject of research. The papers [**3**] and [**20**] studied partial results on extensions of holomorphic motions. The concept of quasiconformal motions was also introduced in the paper [**20**]. Later, Slodkowski showed that any holomorphic motion of a set in $\widehat{\mathbb{C}}$ over the open unit disk $\Delta$ can be extended to a holomorphic motion of $\widehat{\mathbb{C}}$ over $\Delta$. The paper [**3**] used theorems in Teichmüller theory to prove results on extensions of holomorphic motions. Subsequently, the paper [**7**] used Slodkowski's theorem to prove several important results in the theory of Teichmüller spaces.

Slodkowski's theorem does not hold for holomorphic motions over more general parameter spaces. The paper [**9**] gave a simple example of a holomorphic motion of a finite set $E$ in the sphere over the Teichmüller space of that set, that cannot be extended to a holomorphic motion of the sphere over the same parameter space. This is a fundamental difference between holomorphic motions and continuous motions. The main theorem in our present paper shows that a continuous motion of

---

2010 *Mathematics Subject Classification.* Primary 32G15; Secondary 30Cxx, 30C62, 30F60, 37F30.

*Key words and phrases.* Continuous motions, quasiconformal motions, Teichmüller spaces, extremal length.

The first author was partially supported by the collaboration grant from the Simons Foundation (No.199837), the CUNY collaborative incentive research grant (No.1861), an award from PSC-CUNY, a grant from the NSF of China (No.11171121), and a collaboration grant from Academy of Mathematics and Systems Science and the Morningside Center of Mathematics at the Chinese Academy of Sciences. The research of the second author was partially supported by a PSC-CUNY grant.

a finite set $E$ over a simply connected parameter space $V$ extends to a quasiconformal motion of $\widehat{\mathbb{C}}$ over $V$ (which is also a continuous motion of $\widehat{\mathbb{C}}$ over $V$). The technique is to use a result of Nag ([**18**]) and the idea of *Douady-Earle sections* for Teichmüller spaces (Lemma 5 in [**6**]). Moreover, we show that the extended quasiconformal motion is unique up to isotopy rel $E$.

In [**8**] Earle and Mitra showed that if a quadrilateral or a ring domain $D$ in the Riemann sphere $\widehat{\mathbb{C}}$ varies by letting its boundary move holomorphically over a connected complex Banach manifold as the parameter space, then the modulus of $D$ is a real analytic function on the parameter space. They also showed that if $D$ is any region in $\widehat{\mathbb{C}}$ with at least three boundary points and its boundary undergoes a holomorphic motion over a simply connected complex manifold as the parameter space, then the length of the closed Poincaré geodesic in a given homotopy class is a real analytic function on the parameter space. See also the papers [**11**], [**12**], [**13**], and [**21**].

In our paper, we consider a slightly weaker case. We let a finite set $E$ (containing at least four points) of the boundary of a region in $\widehat{\mathbb{C}}$ move continuously with respect to a simply connected parameter space $V$, and we show that the extremal lengths of a family of curves in that region will vary continuously with respect to $V$. We also show that if the finite set $E$ undergoes a continuous motion over $V$, then the length of the closed Poincaré geodesic in a given homotopy class is a continuous function on $V$. Note that the initial region undergoes a continuous motion and these invariants are measured relative to the corresponding regions. These results follow as consequences of the main theorem of our paper.

The basic definitions are given in §2. In what follows, $V$ is a simply connected Hausdorff space with a basepoint $x_0$.

**The Main Theorem.**

*Suppose $E$ consists of $n+3$ distinct points $\{0, 1, \infty, \zeta_1, \ldots, \zeta_n\}$. Let $\phi : V \times E \to \widehat{\mathbb{C}}$ be a continuous motion. Then, there exists a quasiconformal motion $\widetilde{\phi} : V \times \widehat{\mathbb{C}} \to \widehat{\mathbb{C}}$ that extends $\phi$. Moreover, $\widetilde{\phi}$ is unique up to isotopy rel $E$.*

The following are immediate applications of our main theorem.

THEOREM 1. *Let $D$ be a region in $\widehat{\mathbb{C}}$ and $E$ be a finite subset of the boundary $\partial D$, such that $E$ contains at least four points. Suppose $\Gamma$ is a family of curves in $D$. Let $\phi : V \times E \to \widehat{\mathbb{C}}$ be a continuous motion. Then, for each $x$ in $V$, there exists a region $D_x$ and a family $\Gamma_x$ of curves in $D_x$ with $D_{x_0} = D$, $\Gamma_{x_0} = \Gamma$, such that $E_x := \phi_x(E) \subset \partial D_x$, and the extremal length $\Lambda(\Gamma_x)$ varies continuously with respect to $x$.*

THEOREM 2. *Under the assumptions of Theorem 1, let $\gamma$ be a closed Poincaré geodesic in $D$. For each $x$ in $V$, there is a region $D_x$ and a closed Poincaré geodesic $\gamma_x$ in $D_x$ with $D_{x_0} = D$, $\gamma_{x_0} = \gamma$, such that $E_x := \phi_x(E) \subset \partial D_x$, and the hyperbolic length of $\gamma_x$ varies continuously with respect to $x$.*

The following results are consequences of Theorems 1 and 2.

COROLLARY 1. *Suppose $D$ is a quadrilateral with four vertices $p_i$ for $i = 1, 2, 3, 4$. Suppose we have four continuously moving points $p_i(x)$ parametrized by $V$*

*such that* $p_i(x_0) = p_i$ *and* $p_i(x) \neq p_j(x)$ *for* $1 \leq i < j \leq 4$. *Then there is a family of quadrilaterals* $D_x$ *with vertices* $p_i(x)$ *for* $i = 1, 2, 3, 4$, *with* $D_{x_0} = D$ *such that the modulus* mod $(D_x)$ *varies continuously over* $V$.

COROLLARY 2. *Suppose* $A$ *is a topological annulus. Let* $E = \{p_i\}_{i=1}^n$ *be a subset of the boundary* $\partial A$ *consisting of* $n \geq 4$ *distinct points. Suppose we have* $n$ *continuously moving points* $p_i(x)$ *parametrized by* $V$ *such that* $p_i(x_0) = p_i$ *and* $p_i(x) \neq p_j(x)$ *for* $1 \leq i < j \leq n$. *Then there is a family of topological annuli* $A_x$ *with* $E_x = \{p_i(x)\}_{i=1}^n \subset \partial A_x$, *with* $A_{x_0} = A$ *such that the modulus* mod $(A_x)$ *varies continuously over* $V$. *Furthermore, the hyperbolic length of the unique closed Poincaré geodesic of* $A_x$ *varies continuously with respect to* $x$ *in* $V$.

The proofs of the Main Theorem, as well as Theorems 1, 2 and Corollaries 1, 2 will be given in §3.

**Acknowledgement.** We are extremely grateful to the referee for his/her careful and valuable comments and suggestions that helped us to significantly improve our previous version. In particular, we are very thankful for his/her suggestion that we should emphasize the Main Theorem as the centerpiece of our paper.

## 2. Background material

Let $D$ be a region in $\widehat{\mathbb{C}}$. Suppose $\Gamma$ is a family of curves in $D$. Each $\gamma \in \Gamma$ shall be a countable union of open arcs, closed arcs or closed curves, and every closed sub arc shall be rectifiable.

The extremal length $\Lambda(\Gamma)$ of a family $\Gamma$ of curves in a region $D$ in $\mathbb{C}$ is a conformal invariant defined as follows (see [**1**]). Let $\rho \geq 0$ be a measurable function on $D$ such that

$$\iint_D \rho(z)dxdy = 1, \quad z = x + iy.$$

Define

$$L(\rho) = \inf_{\gamma\in\Gamma} \int_\gamma \rho(z)|dz|.$$

The extremal length $\Lambda(\Gamma)$ is defined as

$$\Lambda(\Gamma) = \sup_\rho L(\rho)^2.$$

This quantity is quasi-invariant under quasiconformal mappings. In fact, if $f : D \to D'$ is a $K$-quasiconformal homeomorphism and $\Gamma' = f(\Gamma)$, then

$$K^{-1}\Lambda(\Gamma) \leq \Lambda(\Gamma') \leq K\Lambda(\Gamma).$$

In particular, if $f$ is conformal, then $\Lambda(\Gamma') = \Lambda(\Gamma)$.

A *quadrilateral* is a closed Jordan domain $D$ together with four distinct points $p_1$, $p_2$, $p_3$, and $p_4$ in positive cyclic order on its boundary. Here the $p_i$ are called the *vertices* of $D$. We say that an orientation-preserving homeomorphism $f$ maps a quadrilateral $D$ with vertices $p_1, p_2, p_3, p_4$ to another quadrilateral $D'$ with vertices $q_1, q_2, q_3, q_4$ if it is a homeomorphism from the region $D$ to the region $D'$ and maps each $p_i$ to $q_i$.

In a quadrilateral $D$, we call the two subarcs of $\partial D$ connecting $p_1$ to $p_4$ and $p_2$ to $p_3$ as *a-sides* and the two subarcs of $\partial D$ connecting $p_1$ to $p_2$ and $p_3$ to $p_4$ as *b-sides*.

An example of a quadrilateral is a rectangle $R = [0,a] \times [0,b]$ with four vertices labeled so it has $a$-sides $[0,a] \times \{0\}$ and $[0,a] \times \{b\}$ and $b$-sides $\{0\} \times [0,b]$ and $\{1\} \times [0,b]$.

By the Riemann mapping theorem, there is a conformal map $f$ from any quadrilateral $D$ to a rectangle $R$ mapping $a$-sides to $a$-sides and $b$-sides to $b$-sides. The ratio $\mathrm{mod}(D) = a/b$ is called the *modulus* of $D$ and is a conformal invariant.

A topological annulus $A$ is a doubly connected region such that each of the two components of $\widehat{\mathbb{C}} \setminus A$ contains at least two points. An example of a topological annulus is a round annulus $RA(r) = \{z \in \mathbb{C} \mid 1 \leq |z| \leq r\}$.

For every topological annulus $A$, there is a conformal map $f$ from $A$ to a round annulus $RA(r)$. The number $r$ is a conformal invariant. The modulus of $A$ is defined as $\mathrm{mod}(A) = (1/2\pi)\log r$.

DEFINITION 1. Let $E$ be a subset of $\widehat{\mathbb{C}}$, and let $X$ be a connected Hausdorff space with basepoint $x_0$. A *motion of $E$ over $X$* is a map $\phi\colon X \times E \to \widehat{\mathbb{C}}$ satisfying

(i) $\phi(x_0, z) = z$ for all $z$ in $E$, and

(ii) for all $x$ in $X$ the map $z \mapsto \phi(x,z)$ from $E$ to $\widehat{\mathbb{C}}$ is injective.

We say that $X$ is the *parameter space* of the motion $\phi$, and we often write the restriction $\phi(x,\cdot)$ as $\phi_x(\cdot)$ for each $x$ in $X$.

We will always assume that 0, 1, and $\infty$ belong to $E$ and that the motion $\phi$ is *normalized*; i.e. 0, 1, and $\infty$ are fixed points of the map $\phi(x,\cdot)$ for every $x$ in $X$. This can always be achieved after a suitable change of coordinates: Choose three distinct points $a$, $b$, $c$ in $E$. For each $x \in X$, let $S_x$ be the Möbius transformation that maps $\phi_x(a)$, $\phi_x(b)$, $\phi_x(c)$ to 0, 1, $\infty$ respectively. Let $\widehat{E} = S_{x_0}(E)$. Then the map $\widehat{\phi} : X \times \widehat{E} \to \widehat{\mathbb{C}}$ defined by

$$\widehat{\phi}(x, S_{x_0}(z)) = S_x(\phi(x,z))$$

is a normalized motion.

If $E$ is a proper subset of $\widetilde{E}$ and $\phi : X \times E \to \widehat{\mathbb{C}}$ and $\widetilde{\phi} : X \times \widetilde{E} \to \widehat{\mathbb{C}}$ are two motions, we say that $\widetilde{\phi}$ *extends* $\phi$ if $\widetilde{\phi}(x,z) = \phi(x,z)$ for all $(x,z) \in X \times E$.

DEFINITION 2. Let $X$ be a connected Hausdorff space with basepoint $x_0$, and let $E$ be a subset of $\widehat{\mathbb{C}}$ that contains the points 0, 1, and $\infty$. A motion $\phi$ of $E$ over $X$ is called a *continuous motion* if:

(i) the map $\phi : X \times E \to \widehat{\mathbb{C}}$ is continuous, and

(ii) for each $x$ in $X$, the map $\phi(x,\cdot)$ is a homeomorphism onto its image.

For any map $\phi : X \times E \to \widehat{\mathbb{C}}$, any $x$ in $X$, and any quadruple $a,b,c,d$ of points in $E$, let $\phi_x(a,b,c,d)$ denote the cross-ratio of the four points $\phi(x,a)$, $\phi(x,b)$, $\phi(x,c)$, $\phi(x,d)$, defined by

$$\phi_x(a,b,c,d) = \frac{(\phi_x(a) - \phi_x(c))(\phi_x(b) - \phi_x(d))}{(\phi_x(a) - \phi_x(d))(\phi_x(b) - \phi_x(c))} \tag{2.1}$$

It is obvious that condition (ii) in Definition 1 holds if and only if $\phi_x(a,b,c,d)$ belongs to the thrice-punctured sphere $\widehat{\mathbb{C}} \setminus \{0,1,\infty\}$ for all $x$ in $X$ and all quadruples $a,b,c,d$ of distinct points in $E$.

Let $\rho$ be the Poincaré metric on $\widehat{\mathbb{C}} \setminus \{0,1,\infty\}$. In their paper [**20**], Sullivan and Thurston introduced the following notion.

DEFINITION 3. A *quasiconformal motion* is a motion $\phi : X \times E \to \widehat{\mathbb{C}}$ of $E$ over $X$ with the following additional property:

(iii) given any $x$ in $X$ and any $\epsilon > 0$, there exists a neighborhood $U_x$ of $x$ such that for any quadruple of distinct points $a, b, c, d$ in $E$, we have

$$\rho\big(\phi_y(a,b,c,d), \phi_{y'}(a,b,c,d)\big) < \epsilon \quad \text{for all } y \text{ and } y' \text{ in } U_x.$$

Quasiconformal motions of $\widehat{\mathbb{C}}$ in the above sense can be characterized alternatively as follows.

Let $M(\mathbb{C})$ be the open unit ball of the complex Banach space $L^\infty(\mathbb{C})$. Each $\mu$ in $M(\mathbb{C})$ is the Beltrami coefficient of a unique normalized quasiconformal homeomorphism $w^\mu$ of $\widehat{\mathbb{C}}$ (which means, as usual, that $w^\mu$ satisfies the equation $\bar{\partial} w^\mu = \mu \partial w^\mu$ and fixes the three points 0, 1, and $\infty$). The basepoint of $M(\mathbb{C})$ is the zero function.

We define a map $\Psi_{\widehat{\mathbb{C}}} : M(\mathbb{C}) \times \widehat{\mathbb{C}} \to \widehat{\mathbb{C}}$ by:

$$\Psi_{\widehat{\mathbb{C}}}(\mu, z) = w^\mu(z).$$

It is easy to see that this is a continuous motion of $\widehat{\mathbb{C}}$ over $M(\mathbb{C})$. In fact, the continuity of $\Psi_{\widehat{\mathbb{C}}}$ follows from the fact that $w^{\mu_n} \to w^\mu$ uniformly in the spherical metric if $\mu_n \to \mu$ in $M(\mathbb{C})$.

It was proved in [**16**] that

PROPOSITION 1. *A motion $\phi$: $X \times \widehat{\mathbb{C}} \to \widehat{\mathbb{C}}$ is quasiconformal if and only if*
*(a) the map $\phi_x$: $\widehat{\mathbb{C}} \to \widehat{\mathbb{C}}$ is quasiconformal for each $x$ in $X$, and*
*(b) the map from $X$ to $M(\mathbb{C})$ that sends $x$ to the Beltrami coefficient $\mu_x$ of $\phi_x$ is continuous.*

Recall that a homeomorphism of $\widehat{\mathbb{C}}$ is called normalized if it fixes the points 0, 1 and $\infty$.

DEFINITION 4. Let $E \subset \widehat{\mathbb{C}}$ where $E$ is a closed set that has the points 0, 1, and $\infty$. Two normalized quasiconformal self-mappings $f$ and $g$ of $\widehat{\mathbb{C}}$ are said to be $E$-equivalent if $f^{-1} \circ g$ is isotopic to the identity rel $E$. The *Teichmüller space* $T(E)$ is the set of all $E$-equivalence classes of normalized quasiconformal self-mappings of $\widehat{\mathbb{C}}$. The basepoint of $T(E)$ is the $E$-equivalence class of the identity map.

We define the projection

$$P_E : M(\mathbb{C}) \to T(E)$$

by setting $P_E(\mu)$ equal to the $E$-equivalence class of $w^\mu$, written as $[w^\mu]_E$. Clearly, $P_E$ maps the basepoint of $M(\mathbb{C})$ to the basepoint of $T(E)$.

In his doctoral dissertation ([**10**]), G. Lieb proved that $T(E)$ is a complex Banach manifold such that the projection map $P_E$ is a holomorphic split submersion. For more details, see [**8**]. Using Lemma 5 and Proposition 7 in [**6**], one can show the following:

PROPOSITION 2. *There is a continuous basepoint preserving map $s$ from $T(E)$ to $M(\mathbb{C})$ such that $P_E \circ s$ is the identity map on $T(E)$.*

See Proposition 7.22 in [**8**]. We call this the *Douady-Earle section* (or the *barycentric section*) for $T(E)$.

There exists a continuous motion $\Psi_E$ of $E$ over $T(E)$, defined as follows:

$$\Psi_E(P_E(\mu), z) = w^{\mu}(z) \text{ for } \mu \in M(\mathbb{C}) \text{ and } z \in E.$$

By Definition 4, the map $\Psi_E$ is well-defined. Continuity of $\Psi_E$ follows from Proposition 1 and Lemma 17 of [**2**].

The case when $E$ is finite will be especially important in our paper. In this case, its complement $\Omega = \widehat{\mathbb{C}} \setminus E$ is the Riemann sphere with punctures at the points of $E$. Since $T(E)$ and the classical Teichmüller space $Teich(\Omega)$ are quotients of $M(\mathbb{C})$ by the same equivalence relation, $T(E)$ can be naturally identified with $Teich(\Omega)$ (see Example 3.1 in [**15**] for the details).

The following lemma was proved in [**15**]. We include its easy proof for the convenience of the reader. In what follows, $B$ is a path-connected Hausdorff space with a basepoint and $\mathcal{H}(\widehat{\mathbb{C}})$ is the group of homemorphisms of $\widehat{\mathbb{C}}$ onto itself, with the topology of uniform convergence in the spherical metric. This topology makes $\mathcal{H}(\widehat{\mathbb{C}})$ a topological group. Let $E$ be a closed set in $\widehat{\mathbb{C}}$ that contains the points 0, 1, and $\infty$.

LEMMA 1. *Let $h : B \to \mathcal{H}(\widehat{\mathbb{C}})$ be a continuous map such that $h(t)(e) = e$ for all $t$ in $B$ and for all $e$ in $E$. If $h(t_0)$ is isotopic to the identity rel $E$ for some $t_0$ in $B$, then $h(t)$ is isotopic to the identity rel $E$ for all $t$ in $B$.*

PROOF. Let $t$ be any point in $B$. Choose a path $\gamma : [0,1] \to B$ such that $\gamma(0) = t_0$ and $\gamma(1) = t$. The map $(s, z) \mapsto h(\gamma(s))(z)$ from $[0,1] \times \widehat{\mathbb{C}}$ to $\widehat{\mathbb{C}}$ is an isotopy rel $E$ between $h(t_0)$ and $h(t)$. □

PROPOSITION 3. *Let $\phi : B \times E \to \widehat{\mathbb{C}}$ be a continuous motion. If $\widetilde{\phi}$ and $\widetilde{\psi}$ are continuous motions of $\widehat{\mathbb{C}}$ over $B$ that extend $\phi$, then, for each $x$ in $B$, the homeomorphisms $\widetilde{\phi}_x$ and $\widetilde{\psi}_x$ are isotopic rel $E$.*

PROOF. Define maps $f$ and $g$ from $B$ to $\mathcal{H}(\widehat{\mathbb{C}})$ by $f(x)(z) = \widetilde{\phi}(x, z)$ and $g(x)(z) = \widetilde{\psi}(x, z)$ for $x$ in $B$ and $z$ in $\widehat{\mathbb{C}}$. The continuity of $\widetilde{\phi}$ and $\widetilde{\psi}$ implies that $f$ and $g$ are continuous by Theorem 5 in [**4**]. Therefore, the map $h : B \to \mathcal{H}(\widehat{\mathbb{C}})$ defined by

$$h(x) = (g(x))^{-1} \circ f(x) \text{ for } x \in B$$

is continuous. Then $h(x)$ fixes $E$ pointwise for every $x$ in $B$, and $h(x_0)$ is the identity map on $\widehat{\mathbb{C}}$. By Lemma 1, $h(x)$ is isotopic to the identity rel $E$ for each $x$ in $B$. □

The next proposition is similar to Corollary and Definition 2.4 in [**8**]. There, it was proved for holomorphic motions, and quasiconformal trivializations. We include it here, mainly to emphasize that the result holds under weaker conditions, i.e. for continuous motions. The proof is, verbatim, the same.

PROPOSITION 4. *Let $\phi : B \times E \to \widehat{\mathbb{C}}$ be a continuous motion that extends to a quasiconformal motion $\widetilde{\phi} : B \times \widehat{\mathbb{C}} \to \widehat{\mathbb{C}}$. If $\gamma$ is a closed Poincaré geodesic in a component $D$ of the complement $\widehat{\mathbb{C}} \setminus E$, the homotopy class of $\widetilde{\phi}_x(\gamma)$ in $\widetilde{\phi}_x(D)$ contains a unique closed Poincaré geodesic.*

PROOF. The region $\widetilde{\phi}_x(D)$ has a Poincaré metric because $E$ contains at least three points. A homotopy class of closed curves in $\widetilde{\phi}_x(D)$ contains a Poincaré geodesic if and only if no curve in the class lies inside a disk or in a punctured disk in $\widetilde{\phi}_x(D)$. This property is preserved by quasiconformal mappings and so the homotopy class of $\widetilde{\phi}_x(\gamma)$ in $\widetilde{\phi}_x(D)$ contains a Poincaré geodesic whenever $\gamma$ is a Poincaré geodesic in $D$. Finally, there is at most one Poincaré geodesic in a given homotopy class of closed curves. □

## 3. The proofs

We begin with a few preliminary results. Let $X$ be a connected Hausdorff space with a basepoint.

PROPOSITION 5. *A continuous motion $\phi : X \times E \to \widehat{\mathbb{C}}$ extends to a quasiconformal motion $\widetilde{\phi} : X \times \widehat{\mathbb{C}} \to \widehat{\mathbb{C}}$ if and only if there exists a basepoint-preserving continuous map $f : X \to T(E)$ such that $\phi(x, z) = \Psi_E(f(x), z)$ for all $(x, z) \in X \times E$.*

PROOF. Suppose there exists a basepoint-preserving continuous map $f : X \to T(E)$ such that $\phi(x, z) = \Psi_E(f(x), z)$ for all $(x, z) \in X \times E$. Define $\widetilde{f} = s \circ f$, where $s$ is the Douady-Earle section given in Proposition 2; then, $\widetilde{f} : X \to M(\mathbb{C})$ is a basepoint preserving continuous map. Let

$$\widetilde{\phi}(x, z) = w^{\widetilde{f}(x)}(z) \qquad (x, z) \in X \times \widehat{\mathbb{C}}.$$

By Proposition 1, $\widetilde{\phi}$ is a quasiconformal motion of $\widehat{\mathbb{C}}$ over $X$.

Also, for all $(x, z) \in X \times E$, we have

$$\phi(x, z) = \Psi_E(f(x), z) = \Psi_E(P_E(s(f(x))), z) = \Psi_E(P_E(\widetilde{f}(x)), z) = w^{\widetilde{f}(x)}(z),$$

which shows that $\widetilde{\phi}$ extends $\phi$.

Next, suppose there exists a quasiconformal motion $\widetilde{\phi} : X \times \widehat{\mathbb{C}} \to \widehat{\mathbb{C}}$ extending $\phi$. Then, by Proposition 1, the map $\widetilde{\phi}_x : \widehat{\mathbb{C}} \to \widehat{\mathbb{C}}$ is quasiconformal, for each $x$ in $X$. Let $\widetilde{f}(x)$ be the Beltrami coefficient of $\widetilde{\phi}_x$; by Proposition 1, $\widetilde{f} : X \to M(\mathbb{C})$ is a basepoint-preserving continuous map. Now, set $f = P_E \circ \widetilde{f}$. Clearly, $f$ is a basepoint-preserving continuous map. Note that

$$\Psi_E(f(x), z) = \Psi_E(P_E(\widetilde{f}(x)), z) = w^{\widetilde{f}(x)}(z) = \widetilde{\phi}(x, z) = \phi(x, z)$$

for all $(x, z) \in X \times E$. □

We next recall the following result that is a key ingredient in the proof of the main theorem of our paper.

THEOREM 3 (Nag [**18**]). *Given $n > 0$, choose a point $(\zeta_1, \ldots, \zeta_n)$ in the domain*

$$Y_n = \{(z_1, \ldots, z_n) \in \mathbb{C}^n : z_i \neq z_j \text{ for } i \neq j \text{ and } z_i \neq 0, 1 \text{ for all } i = 1, \ldots, n\}$$

*and let $E = \{0, 1, \infty, \zeta_1, \ldots, \zeta_n\}$. Then, the map $p : T(E) \to Y_n$ defined by*

$$p([w^\mu]_E) = (w^\mu(\zeta_1), \ldots, w^\mu(\zeta_n)) \text{ for all } \mu \in M(\mathbb{C})$$

*is a holomorphic universal covering.*

**Proof of the main theorem.** We define a map $F : V \to Y_n$ by

$$F(x) = (\phi(x, \zeta_1), \ldots \phi(x, \zeta_n)).$$

Clearly, the map $F$ is a basepoint preserving continuous map. Since $V$ is simply connected, by Theorem 3, this map $F$ has a unique basepoint preserving lift

$$f : V \to T(E)$$

under $p$ that is also a continuous map. The definition of the universal covering map $p$ then shows that $f^*(\Psi_E) = \phi$.

It now follows by Proposition 5, that there exists a quasiconformal motion $\widetilde{\phi} : V \times \widehat{\mathbb{C}} \to \widehat{\mathbb{C}}$ that extends $\phi$.

Finally, by Proposition 3, it follows that $\widetilde{\phi}$ is unique up to isotopy rel $E$. □

The following is an example of a continuous motion of a four-point set over the punctured unit disk, that cannot be extended to a quasiconformal motion of the sphere.

**Example.** Let $\Delta^* := \{z \in \mathbb{C} : 0 < |z| < 1\}$ and choose some basepoint $a$ in $\Delta^*$. Let $E := \{0, 1, \infty, a\}$. Set $\phi(t, z) = z$ for all $(t, z)$ in $\Delta^* \times \{0, 1, \infty\}$ and $\phi(t, a) = t$ for all $t$ in $\Delta^*$. Then, clearly, $\phi$ is a continuous motion of $E$ over $\Delta^*$. Suppose there exists a quasiconformal motion $\widetilde{\phi} : V \times \widehat{\mathbb{C}} \to \widehat{\mathbb{C}}$ that extends $\phi$. By [**16**], $\widetilde{\phi}$ is a continuous motion. For each $\zeta$ in $\mathbb{C} \setminus \{0\}$, let $\gamma_\zeta : [0, 2\pi] \to \mathbb{C} \setminus \{0\}$ be the closed curve

$$\gamma_\zeta(\theta) = \widetilde{\phi}(ae^{i\theta}, \zeta)$$

for $\theta$ in $[0, 2\pi]$. Since $\widetilde{\phi}$ is a continuous motion, the winding number of $\gamma_\zeta$ about zero is a continuous function of $\zeta$. But that winding number is zero when $\zeta = 1$ and one when $\zeta = a$.

Remark 1. This appeared in [**17**] for holomorphic motions. We thank Clifford Earle for showing us this example. For other such examples, the reader is referred to the paper [**5**], where we introduce the concepts of *trace monodromy* and *monodromy* of a holomorphic motion.

Proof of Theorem 1. By the main theorem, there exists a quasiconformal motion $\widetilde{\phi} : V \times \widehat{\mathbb{C}} \to \widehat{\mathbb{C}}$, that extends $\phi$. By Proposition 1, for each $x$ in $V$, the map $\widetilde{\phi}_x : \widehat{\mathbb{C}} \to \widehat{\mathbb{C}}$ is quasiconformal and its Beltrami coefficient $\mu_x$ varies continuously with respect to $x$. Set $D_x = \widetilde{\phi}_x(D)$ and $\Gamma_x = \widetilde{\phi}_x(\Gamma)$. Evidently, $E_x = \phi_x(E) = \widetilde{\phi}_x(E) \subset \partial D_x$.

It remains to show that $\Lambda(\Gamma_y) \to \Lambda(\Gamma_x)$ whenever $y \to x$. Note that the composition $\Phi = \widetilde{\phi}_x \circ (\widetilde{\phi}_y)^{-1}$ sends $\Gamma_y$ to $\Gamma_x$. Let $K(\Phi)$ denote its dilatation. By the quasi-invariance of the extremal length (see [**1**])

$$\frac{1}{K(\Phi)}\Lambda(\Gamma_y) \leq \Lambda(\Gamma_x) \leq K(\Phi)\Lambda(\Gamma_y).$$

As $y \to x$, $K(\Phi) \to 1$ (by Proposition 1), and hence $\Lambda(\Gamma_y) \to \Lambda(\Gamma_x)$. □

For the proof of Theorem 2, we need the following lemma (see Lemma 5.1 in [**8**] and also [**19**]):

Lemma 2. *Suppose $D$ is a region in the Riemann sphere and $\gamma$ is a closed Poincaré geodesic in $D$. For each $\mu$ in $M(\mathbb{C})$, let $L^\mu(\gamma)$ be the length of the unique closed Poincaré geodesic that is homotopic to $w^\mu(\gamma)$ in $w^\mu(D)$. Then, the function $\mu \mapsto L^\mu(\gamma)$ is real analytic.*

Proof of Theorem 2. Since $\phi$ is a continuous motion of the finite set $E$ over $V$, it follows from our main theorem that $\phi$ extends to a quasiconformal motion $\widetilde{\phi}$ of $\widehat{\mathbb{C}}$ over $V$. By Proposition 1, for each $x$ in $V$, the map $\widetilde{\phi}_x : \widehat{\mathbb{C}} \to \widehat{\mathbb{C}}$ is quasiconformal, and its Beltrami coefficient $\mu_x$ depends continuously on $x$. The map $x \mapsto \mu_x$ is a basepoint preserving continuous map $V \to M(\mathbb{C})$. By Proposition 4, the isotopy class of $\widetilde{\phi}_x(\gamma) = w^{\mu_x}(\gamma)$ contains a unique Poincaré geodesic $\gamma_x$, whose length we denote by $L^{\mu_x}(\gamma)$. By Lemma 2, the function $\mu_x \mapsto L^{\mu_x}(\gamma)$ is real analytic. It follows easily that the function $x \mapsto L^{\mu_x}(\gamma)$ is continuous on $V$. □

Proof of Corollary 1. Consider the family $\Gamma$ of all curves connecting $b$-sides of a quadrilateral $D$. It is well known (see [**1**]) that

$$\text{mod }(D) = \Lambda(\Gamma).$$

The result now follows from Theorem 1. □

Proof of Corollary 2. Let $\Gamma$ be the family of all closed curves in $A$ separating the two components of $\widehat{\mathbb{C}} \setminus A$. Then we have

$$\text{mod }(A) = \Lambda(\Gamma)^{-1}.$$

Moreover, the hyperbolic length of the unique closed Poincaré geodesic in $A$ is $\pi\Lambda(\Gamma)$. The result now follows from Theorem 2. □

## References


[1] L. V. Ahlfors, *Lectures on quasiconformal mappings*, 2nd ed., University Lecture Series, vol. 38, American Mathematical Society, Providence, RI, 2006. With supplemental chapters by C. J. Earle, I. Kra, M. Shishikura and J. H. Hubbard. MR2241787

[2] L. Ahlfors and L. Bers, *Riemann's mapping theorem for variable metrics*, Ann. of Math. (2) **72** (1960), 385–404, DOI 10.2307/1970141. MR0115006

[3] L. Bers and H. L. Royden, *Holomorphic families of injections*, Acta Math. **157** (1986), no. 3-4, 259–286, DOI 10.1007/BF02392595. MR857675

[4] R. Arens, *Topologies for homeomorphism groups*, Amer. J. Math. **68** (1946), 593–610, DOI 10.2307/2371787. MR0019916

[5] M. Beck, Y. Jiang, S. Mitra, and H. Shiga, *Extending holomorphic motions and monodromy*, Ann. Acad. Sci. Fenn. Math. **37** (2012), no. 1, 53–67, DOI 10.5186/aasfm.2012.3713. MR2920423

[6] A. Douady and C. J. Earle, *Conformally natural extension of homeomorphisms of the circle*, Acta Math. **157** (1986), no. 1-2, 23–48, DOI 10.1007/BF02392590. MR857678

[7] C. J. Earle, I. Kra, and S. L. Krushkal′, *Holomorphic motions and Teichmüller spaces*, Trans. Amer. Math. Soc. **343** (1994), no. 2, 927–948, DOI 10.2307/2154750. MR1214783

[8] C. J. Earle and S. Mitra, *Variation of moduli under holomorphic motions*, In the tradition of Ahlfors and Bers (Stony Brook, NY, 1998), Contemp. Math., vol. 256, Amer. Math. Soc., Providence, RI, 2000, pp. 39–67, DOI 10.1090/conm/256/03996. MR1759669

[9] Y. Jiang and S. Mitra, *Douady-Earle section, holomorphic motions, and some applications*, Quasiconformal mappings, Riemann surfaces, and Teichmüller spaces, Contemp. Math., vol. 575, Amer. Math. Soc., Providence, RI, 2012, pp. 219–251, DOI 10.1090/conm/575/11400. MR2933902

[10] G. S. Lieb, *Holomorphic motions and Teichmüller space*, Ph.D. dissertation, Cornell University (1990).

[11] R. Kühnau, *Der konforme Modul schmaler Vierecke* (German, with German summary), Math. Nachr. **175** (1995), 193–198, DOI 10.1002/mana.19951750110. MR1355017

[12] R. Kühnau, *Der konforme Modul von Vierecken* (German, with English and German summaries), Analysis and topology, World Sci. Publ., River Edge, NJ, 1998, pp. 483–495. MR1667827

[13] R. Kühnau, *The conformal module of quadrilaterals and of rings*, Handbook of complex analysis: geometric function theory. Vol. 2, Elsevier Sci. B. V., Amsterdam, 2005, pp. 99–129, DOI 10.1016/S1874-5709(05)80007-1. MR2121858

[14] R. Mañé, P. Sad, and D. Sullivan, *On the dynamics of rational maps*, Ann. Sci. École Norm. Sup. (4) **16** (1983), no. 2, 193–217. MR732343

[15] S. Mitra, *Teichmüller spaces and holomorphic motions*, J. Anal. Math. **81** (2000), 1–33, DOI 10.1007/BF02788984. MR1785276

[16] S. Mitra, *Extensions of holomorphic motions to quasiconformal motions*, In the tradition of Ahlfors-Bers. IV, Contemp. Math., vol. 432, Amer. Math. Soc., Providence, RI, 2007, pp. 199–208, DOI 10.1090/conm/432/08309. MR2342817

[17] S. Mitra, *Extensions of holomorphic motions*, Israel J. Math. **159** (2007), 277–288, DOI 10.1007/s11856-007-0047-1. MR2342482

[18] S. Nag, *The Torelli spaces of punctured tori and spheres*, Duke Math. J. **48** (1981), no. 2, 359–388. MR620255

[19] M. Seppälä and T. Sorvali, *Geometry of Riemann surfaces and Teichmüller spaces*, North-Holland Mathematics Studies, vol. 169, North-Holland Publishing Co., Amsterdam, 1992. MR1202043

[20] D. P. Sullivan and W. P. Thurston, *Extending holomorphic motions*, Acta Math. **157** (1986), no. 3-4, 243–257, DOI 10.1007/BF02392594. MR857674

[21] O. Teichmüller, *Beweis der analytischen Abhängigkeit des konformen Moduls einer analytischen Ringflächenschar von den Parametern* (German), Deutsche Math. **7** (1944), 309–336. MR0018760

Department of Mathematics, Queens College of the City University of New York, Flushing, New York 11367-1597 – and – Department of Mathematics, Graduate Center of the City University of New York, New York, New York 10016

*E-mail address*: yunping.jiang@qc.cuny.edu

Department of Mathematics, Queens College of the City University of New York, Flushing, New York 11367-1597 – and – Department of Mathematics, Graduate Center of the City University of New York, New York, New York 10016

*E-mail address*: sudeb.mitra@qc.cuny.edu

Contemporary Mathematics
Volume **696**, 2017
http://dx.doi.org/10.1090/conm/696/14025

# Systolic surfaces of arithmetic hyperbolic 3-manifolds

Benjamin Linowitz and Jeffrey S. Meyer

Abstract. In this paper we examine the geometry of minimal surfaces of arithmetic hyperbolic 3-manifolds. In particular, we give bounds on the totally geodesic 2-systole, construct infinitely many incommensurable manifolds with the same initial geometric genus spectrum in which volume and 1-systole are controlled, and analyze the growth of the genera of minimal surfaces across commensurability classes. These results have applications to the study of how Heegard genus grows across commensurability classes.

## 1. Introduction

Recent years have seen significant research in the geometry of $\pi_1$-injective surfaces of hyperbolic 3-manifolds. The existence of $\pi_1$-injective surfaces was initially established for cusped finite volume hyperbolic 3-manifolds in [**7**], for closed arithmetic hyperbolic 3-manifolds in [**18**], and then finally for all closed hyperbolic 3-manifolds in [**15**]. These results have had profound implications, in particular the resolution of Thurston's virtual Haken conjecture [**1**, **2**]. A natural continuation of this line of inquiry is to give an explicit description of the types of $\pi_1$-injective surfaces that can appear within a given hyperbolic 3-manifold $M$.

The goal of this paper is to understand the minimal $\pi_1$-injective surfaces of $M$ when volume and 1-systole are controlled. We consider three distinct types of minimality. The **systolic genus** of $M$, denoted $\mathrm{sysg}(M)$, is the minimal genus of a $\pi_1$-injective surface. The **2-systole** of $M$, denoted $\mathrm{sys}_2(M)$, is the minimal area of a $\pi_1$-injective surface. The **totally geodesic 2-systole** of $M$, denoted $\mathrm{sys}_2^{TG}(M)$, is the minimal area of an immersed totally geodesic surface. Since hyperbolic 3-manifolds are atoroidal, $\mathrm{sysg}(M) \geq 2$. Furthermore, as totally geodesic surfaces are $\pi_1$-injective, $\mathrm{sys}_2(M) \leq \mathrm{sys}_2^{TG}(M)$. Finally, we denote by $\mathrm{sys}_1(M)$ the **1-systole** of $M$; that is, the least length of a closed geodesic on $M$.

Before stating our first result we introduce a piece of notation. Given two commensurable Kleinian groups $\Gamma_1$ and $\Gamma_2$, we define the **generalized index** of $\Gamma_2$ in $\Gamma_1$ to be $[\Gamma_1 : \Gamma_2] := \frac{[\Gamma_1:\Gamma_1\cap\Gamma_2]}{[\Gamma_2:\Gamma_1\cap\Gamma_2]}$. For a commensurability class of arithmetic Kleinian groups, we denote by $\Gamma_{\mathcal{O}}$ a group with minimal covolume in the class and by $\Gamma_{\mathcal{O}}^1$ a group contained in the class which is derived from the elements of reduced norm one in a maximal order contained in the class' invariant quaternion algebra. These groups and their construction will be described in detail in Section 3.

2010 *Mathematics Subject Classification.* Primary 57M50; Secondary 53C42.
*Key words and phrases.* hyperbolic manifolds, totally geodesic surfaces, systole.

THEOREM A. *Let $M$ be an arithmetic hyperbolic 3-manifold which contains a finite area, properly immersed, totally geodesic surface. Then*

$$\mathrm{sys}_2^{TG}(M) > c\left(\frac{\mathrm{Vol}(M)}{[\Gamma_{\mathcal{O}} : \pi_1(M)]}\right)^{1/2}$$

*where $c > 0$ is a constant depending only on the invariant trace field of $M$. If $\pi_1(M)$ contains a subgroup of the form $\Gamma^1_{\mathcal{O}}$ then we additionally have the inequality*

$$\mathrm{sys}_2^{TG}(M) < c'\mathrm{Vol}(M)^{1/2},$$

*where $c' > 0$ is a constant depending only on the invariant trace field of $M$.*

An immediate consequence of Theorem A is that if $\pi_1(M) = \Gamma_{\mathcal{O}}$ then $\mathrm{sys}_2^{TG}(M)$ has order of magnitude $\mathrm{Vol}(M)^{1/2}$. In fact, because $[\Gamma_{\mathcal{O}} : \Gamma^1_{\mathcal{O}}]$ depends only on the groups' commensurability class, $\mathrm{sys}_2^{TG}(M)$ has order of magnitude $\mathrm{Vol}(M)^{1/2}$ whenever

$$\Gamma_{\mathcal{O}} \supseteq \pi_1(M) \supseteq \Gamma^1_{\mathcal{O}}.$$

REMARK. If $\pi_1(M)$ is a maximal arithmetic group or a congruence subgroup of a maximal arithmetic group then the generalized index appearing in Theorem A can easily be made explicit using Borel's classification of maximal arithmetic subgroups of $\mathrm{PSL}_2(\mathbf{C})$ and the computation of their covolumes [**4**].

The proof of Theorem A makes extensive use of Borel's computation of the covolumes of maximal arithmetic subgroups of $\mathrm{PSL}_2(\mathbf{R})$ and $\mathrm{PSL}_2(\mathbf{C})$ and the close relationship between the invariant quaternion algebra of an arithmetic hyperbolic 3-manifold and those of its totally geodesic surfaces. An easy consequence of the proof of Theorem A is the following.

COROLLARY 1.1. *For every $X > 0$ there exist infinitely many arithmetic hyperbolic 3-manifolds with $\mathrm{sys}_2^{TG}(M) > X$.*

Let $M$ be a closed hyperbolic 3-manifold. To each free homotopy class $[S]$ in $M$ of a $\pi_1$-injective, properly immersed, closed surface, there is a well-defined minimal area, which we denote $\mathrm{area}(S)$ [**25, 26**]. We define the **area set** of $M$ to be the set $\mathrm{A}(M)$ of all such areas. For such a $\pi_1$-injective surface $S$ of genus $g$, the inequalities of Uhlenbeck and Thurston (cf., [**13**, Lemma 6]) give

$$2\pi(g-1) \leq \mathrm{area}(S) \leq 4\pi(g-1).$$

By the work of Thurston, [**27**, 8.8.6] there are only finitely many $\pi_1$-injective, properly immersed, closed surfaces up to free homotopy of a fixed genus. It follows that the area set is a discrete subset of the real numbers. Furthermore, each area can only be realized by finitely many free homotopy classes of surfaces.

Every totally geodesic surface is the area minimizing representative in its homotopy class, and we denote the set of all areas contributed by totally geodesic surfaces $\mathrm{A}^{TG}(M)$. An interesting, though likely difficult, problem would be to study the distribution of $\mathrm{A}^{TG}(M)$ within $\mathrm{A}(M)$. For $N \geq 1$, we define $\mathrm{A}^{TG}(M)[N]$ to be the set of the first $N$ areas in $\mathrm{A}^{TG}(M)$. The geometric genus set of $M$, denoted $\mathrm{GS}(M)$, is the set of isometry classes of finite area, properly immersed, totally geodesic surfaces of $M$ up to free homotopy. We define

$$\mathrm{GS}(M)[N] := \{Y \in \mathrm{GS}(M) : \mathrm{area}(Y) \in \mathrm{A}^{TG}(M)[N]\}.$$

Our next theorem revisits a construction of pairs of incommensurable arithmetic hyperbolic 3-manifolds with the same set of commensurability classes of surfaces up to an arbitrary threshold that appeared in [**20**, Section 7]. In particular we construct an infinite family of incommensurable arithmetic hyperbolic 3-manifolds whose geometric genus spectra start off with the same $N$ terms and in which both volume and systole length are controlled.

THEOREM B. *Let $N \geq 1$. There exists an infinite sequence of incommensurable arithmetic hyperbolic 3-manifolds $M_1, M_2, \ldots$ such that $\mathrm{GS}(M_i)[N] = \mathrm{GS}(M_j)[N] \neq \emptyset$ for all $i, j$. In particular $\mathrm{sys}_2^{TG}(M_i) = \mathrm{sys}_2^{TG}(M_j)$ for all $i, j$. Furthermore, we have the estimates*

$$\mathrm{Vol}(M_n) < c_1 \left(n \log(2n)\right)^{3/2}, \qquad \textit{and} \qquad \mathrm{sys}_1(M_n) < c_2 \log(n),$$

*where $c_1, c_2$ are positive constants depending on $N$.*

Theorem B may be viewed as a higher dimensional analogue of a recent result of Millichap [**24**], who exhibited an infinite family of incommensurable hyperbolic 3-manifolds, all of whose complex length spectra begin with the same $2n+1$ lengths.

It is natural to ask about the minimum value assumed by $\mathrm{sysg}(M)$, $\mathrm{sys}_2(M)$ or $\mathrm{sys}_2^{TG}(M)$ as $M$ varies over the set of arithmetic hyperbolic 3-manifolds lying in a fixed commensurability class. Our next result will show that given any real number $x > 0$, almost every commensurability class of arithmetic hyperbolic 3-manifolds has the property that every representative $M$ satisfies $\mathrm{sysg}(M), \mathrm{sys}_2(M), \mathrm{sys}_2^{TG}(M) > x$. In order to make this precise we introduce some additional notation.

Let $k$ be a number field with a unique complex place and $\mathscr{C}$ be a commensurability class of arithmetic hyperbolic 3-manifolds having invariant trace field $k$. In Section 3 we will define a notion of volume on $\mathscr{C}$. This volume is in terms of the volume of a certain distinguished representative of the class which arises in a natural way in the construction of arithmetic lattices in $\mathrm{PSL}_2(\mathbf{C})$. Let $N_k(V)$ denote the number of commensurability classes of arithmetic hyperbolic 3-manifolds with volume less than $V$ and invariant trace field $k$. Additionally, let $N_k^g(V; x)$ (respectively $N_k^2(V; x)$ and $N_k^{TG}(V; x)$) denote the number of commensurability classes of arithmetic hyperbolic 3-manifolds with invariant trace field $k$, volume less than $V$ and which contain a representative $M$ satisfying $\mathrm{sysg}(M) < x$ (respectively $\mathrm{sys}_2(M) < x$ and $\mathrm{sys}_2^{TG}(M) < x$).

THEOREM C. *For all sufficiently large $x$ we have*

$$\lim_{V \to \infty} \frac{N_k^g(V; x)}{N_k(V)} = \lim_{V \to \infty} \frac{N_k^2(V; x)}{N_k(V)} = \lim_{V \to \infty} \frac{N_k^{TG}(V; x)}{N_k(V)} = 0.$$

The Heegard genus of a closed 3-manifold, denoted $\mathrm{Hg}(M)$, is the minimal genus of a Heegard surface. There has been considerable work towards understanding how Heegard genus increases along covers [**3, 8–10, 12, 17**]. In particular, in the case of congruence covers of a compact arithmetic 3-manifold, [**17**] and [**12**] independently have shown that the Heegard genus grows linearly with the degree of the covers. Since the Heegard genus of a 3-manifold is bounded below by its systolic genus, Theorem C implies that the Heegard genus grows across commensurability classes. More precisely, let $\mathrm{Hg}_k(V; x)$ denote the number of commensurability classes of arithmetic hyperbolic 3-manifolds with invariant trace field $k$, volume less than $V$ and which contain a representative $M$ satisfying $\mathrm{Hg}(M) < x$.

COROLLARY 1.2. *For all sufficiently large $x$ we have* $\lim_{V\to\infty} \frac{\mathrm{Hg}_k(V;x)}{N_k(V)} = 0.$

## 2. Notation

Let $k$ be a number field of degree $n_k$ and discriminant $\Delta_k$. We denote the ring of integers of $k$ by $\mathcal{O}_k$ and the Dedekind zeta function of $k$ by $\zeta_k(s)$. Given an ideal $I \subset \mathcal{O}_k$, we denote by $|I|$ the norm of $I$.

If $A$ is a quaternion algebra over $k$ then we denote by $\mathrm{Ram}(A)$ (respectively, $\mathrm{Ram}_f(A)$) the set of primes (respectively, finite primes) of $k$ which ramify in $A$. The discriminant of $A$, denoted $\mathrm{disc}(A)$, is defined to be the product of all primes in $\mathrm{Ram}_f(A)$.

We denote by $\mathbf{H}^2$ and $\mathbf{H}^3$ real hyperbolic spaces of dimensions 2 and 3, and will refer to lattices in $\mathrm{PSL}_2(\mathbf{R})$ as Fuchsian and lattices in $\mathrm{PSL}_2(\mathbf{C})$ as Kleinian.

## 3. Arithmetic Fuchsian and Kleinian groups and their covolumes

In this section we briefly describe the construction of arithmetic lattices in $\mathrm{PSL}_2(\mathbf{R})$ and $\mathrm{PSL}_2(\mathbf{C})$. For a more detailed treatment we refer the reader to [**23**].

Let $\ell$ be a totally real number field and $B$ a quaternion algebra over $\ell$ which is unramified at a unique real prime $v$. We therefore have an isomorphism $B \otimes_\ell \ell_v \cong \mathrm{M}_2(\mathbf{R})$. Let $\mathcal{O}$ be a maximal order of $B$ and $\mathcal{O}^1$ be the subgroup of $\mathcal{O}^*$ consisting of those elements with reduced norm one. We denote by $\Gamma^1_{\mathcal{O}}$ the image in $\mathrm{PSL}_2(\mathbf{R})$ of $\mathcal{O}^1$ and by $\Gamma_{\mathcal{O}}$ the image in $\mathrm{PSL}_2(\mathbf{R})$ of $N(\mathcal{O})$, where $N(\mathcal{O})$ is the normalizer in $B^*$ of $\mathcal{O}$. The groups $\Gamma_{\mathcal{O}}$ and $\Gamma^1_{\mathcal{O}}$ are discrete subgroups of $\mathrm{PSL}_2(\mathbf{R})$ which are cocompact whenever $B$ is a division algebra. The coareas of these groups are given by the following formulas (see Borel [4] and Chinburg and Friedman [**5**, Proposition 2.1]) :

$$\mathrm{coarea}(\Gamma^1_{\mathcal{O}}) = \frac{8\pi\,|\Delta_\ell|^{\frac{3}{2}}\,\zeta_\ell(2)}{(4\pi^2)^{n_\ell}} \prod_{\mathfrak{p}\in\mathrm{Ram}_f(B)} (|\mathfrak{p}|-1), \tag{1}$$

and

$$\mathrm{coarea}(\Gamma_{\mathcal{O}}) = \frac{4\pi\,|\Delta_\ell|^{\frac{3}{2}}\,\zeta_\ell(2)}{(4\pi^2)^{n_\ell}t(B)} \prod_{\mathfrak{p}\in\mathrm{Ram}_f(B)} \frac{|\mathfrak{p}|-1}{2}, \tag{2}$$

where $t(B)$ is the type number of $B$. It is known that the group $\Gamma_{\mathcal{O}}$ has minimal covolume within its commensurability class. Any group which is commensurable with a group $\Gamma^1_{\mathcal{O}}$ as constructed above is called an **arithmetic Fuchsian group**.

We now describe the construction of arithmetic lattices in $\mathrm{PSL}_2(\mathbf{C})$. Let $k$ be a number field with a unique complex place $v$ and $A$ a quaternion algebra over $k$ which is ramified at all real places of $k$. Let $\mathcal{O}$ be a maximal order of $A$ and define subgroups $\Gamma^1_{\mathcal{O}}$ and $\Gamma_{\mathcal{O}}$ of $\mathrm{PSL}_2(\mathbf{C})$ (constructed as above) via the identification $A \otimes_k k_v \cong \mathrm{M}_2(\mathbf{C})$. The groups $\Gamma^1_{\mathcal{O}}$ and $\Gamma_{\mathcal{O}}$ are discrete subgroups of $\mathrm{PSL}_2(\mathbf{C})$ of finite covolume which are cocompact whenever $A \not\cong \mathrm{M}_2(\mathbf{Q}(\sqrt{-d}))$ for some imaginary quadratic field $\mathbf{Q}(\sqrt{-d})$. Furthermore, $\Gamma_{\mathcal{O}}$ has minimal covolume within its commensurability class. Any discrete subgroup of $\mathrm{PSL}_2(\mathbf{C})$ which is commensurable with a group of the form $\Gamma^1_{\mathcal{O}}$ as constructed above is called an **arithmetic Kleinian group**. The work of Borel [4] (see also [**5**, Proposition 2.1]) shows that the covolumes of $\Gamma^1_{\mathcal{O}}$ and $\Gamma_{\mathcal{O}}$ are given by

$$\text{covol}(\Gamma^1_{\mathcal{O}}) = \frac{|\Delta_k|^{\frac{3}{2}}\,\zeta_k(2)}{(4\pi^2)^{n_k-1}} \prod_{\mathfrak{p}\in \text{Ram}_f(A)} (|\mathfrak{p}|-1)\,, \tag{3}$$

and

$$\text{covol}(\Gamma_{\mathcal{O}}) = \frac{2\pi^2\,|\Delta_k|^{\frac{3}{2}}\,\zeta_k(2)}{(4\pi^2)^{n_k} t(A)} \prod_{\mathfrak{p}\in \text{Ram}_f(A)} \frac{|\mathfrak{p}|-1}{2}. \tag{4}$$

A pair of arithmetic lattices $\Gamma_1, \Gamma_2$ derived from quaternion algebras $(k_1, A_1)$, $(k_2, A_2)$ will be commensurable (in the wide sense) precisely when $k_1 \cong k_2$ and $A_1 \cong A_2$ [**23**, Theorem 8.4.1]. We will make use of this fact throughout this paper.

Given a commensurability class $\mathscr{C}$ of arithmetic hyperbolic 3-manifolds arising from the arithmetic data $(k, A)$, we define the **volume** $V_{\mathscr{C}}$ of $\mathscr{C}$ to be $V_{\mathscr{C}} := \text{covol}(\Gamma^1_{\mathcal{O}})$, where $\mathcal{O}$ is a maximal order of $A$. It follows from the work of Borel [**4**] that there are only finite many commensurability classes of bounded volume.

## 4. Proof of Theorem A

We now prove Theorem A.

Let $\Gamma = \pi_1(M)$, and denote by $k = k\Gamma$ be the invariant trace field of $M$ and by $A = A\Gamma$ the invariant quaternion algebra of $M$. The results of [**23**, Chapter 9.5] show that all totally geodesic surfaces in $M$ arise from quaternion algebras $B$ over $\ell$, where $\ell$ is the maximal totally real subfield of $k$ (and satisfies $[k : \ell] = 2$) and $B$ is a quaternion algebra defined over $\ell$ which is unramified at a unique real place of $\ell$ (the one lying under the complex place of $k$) and has the property that $A \cong B \otimes_\ell k$. The requirement that $A \cong B \otimes_\ell k$ imposes severe restrictions on the ramification of $B$, as the following result [**23**, Theorem 9.5.5] makes clear.

THEOREM 4.1. *Let $k$ be a number field with a unique complex place and $A$ be a quaternion algebra over $k$ which is ramified at all real places. Let $k$ be such that $[k : \ell] = 2$, where $\ell = k \cap \mathbf{R}$. Let $B$ be a quaternion algebra over $\ell$ which is ramified at all real places except the identity. Then $A \cong B \otimes_\ell k$ if and only if* $\text{Ram}_f(A)$ *consists of $2r$ distinct ideals $\{\mathfrak{P}_1, \mathfrak{P}_1', \dots, \mathfrak{P}_r, \mathfrak{P}_r'\}$, where*

$$\mathfrak{P}_i \cap \mathcal{O}_\ell = \mathfrak{P}_i' \cap \mathcal{O}_\ell = \mathfrak{p}_i$$

*and* $\text{Ram}_f(B) \supseteq \{\mathfrak{p}_1, \dots, \mathfrak{p}_r\}$ *with* $\text{Ram}_f(B) \setminus \{\mathfrak{p}_1, \dots, \mathfrak{p}_r\}$ *consists of primes of $\mathcal{O}_\ell$ which are either inert or ramified in the extension $k/\ell$.*

As $\text{GS}(M) \neq \emptyset$, the set $\text{Ram}_f(A)$ is of the form $\{\mathfrak{P}_1, \mathfrak{P}_1', \dots, \mathfrak{P}_r, \mathfrak{P}_r'\}$ described in Theorem 4.1. Every non-elementary Fuchsian subgroup of $\Gamma$ is contained in an arithmetic Fuchsian group [**23**, Theorem 9.5.2] which arises from a quaternion algebra $B$ over $\ell$ which is ramified at $\{\mathfrak{p}_1, \dots, \mathfrak{p}_r\}$ by Theorem 4.1. Note that if $B$ is an indefinite quaternion algebra over $\ell$ then the type number $t(B)$ of $B$ is equal to $[\ell(B) : \ell]$ where $\ell(B)$ is the maximal abelian extension of $\ell$ of exponent 2 which is unramified outside the real places in $\text{Ram}(B)$ and in which all primes of $\text{Ram}_f(B)$ split completely (see [**6**, p. 39]). Observe that $\ell(B)$ is contained in the narrow class field of $\ell$, which has degree over $\ell$ at most $2^{n_\ell} h_\ell$. Section 3 and equation (2) now show that

$$\text{sys}_2^{TG}(M) \geq \frac{4\pi\,|\Delta_\ell|^{\frac{3}{2}}\,\zeta_\ell(2)}{(16\pi^2)^{n_\ell} h_\ell} \prod_{i=1}^{r} \frac{|\mathfrak{p}_i|-1}{2} = c_\ell \prod_{i=1}^{r} \frac{|\mathfrak{p}_i|-1}{2}, \tag{5}$$

where $c_\ell > 0$ is a constant depending only on $\ell$ (and hence on $k$ as $\ell = k \cap \mathbf{R}$ is the maximal totally real subfield of $k$).

Recall that $\mathrm{Ram}_f(A) = \{\mathfrak{P}_1, \mathfrak{P}_1', \ldots, \mathfrak{P}_r, \mathfrak{P}_r'\}$ and $|\mathfrak{P}_i| = |\mathfrak{P}_i'| = |\mathfrak{p}_i|$ for $i = 1, \ldots, r$. It follows that

$$\prod_{\substack{\mathfrak{p} \in \mathrm{Ram}_f(A) \\ |\mathfrak{p}|>2}} \frac{|\mathfrak{p}|-1}{2} \geq \left( \prod_{i=1}^{r} \frac{|\mathfrak{p}_i|-1}{2} \right)^2,$$

hence the first part of the theorem follows from equations (5) and (4).

We now suppose that $\pi_1(M)$ contains a subgroup of the form $\Gamma_{\mathcal{O}}^1$ and will derive an upper bound for $\mathrm{sys}_2^{TG}(M)$. That $\mathrm{GS}(M) \neq \emptyset$ implies that $\mathrm{Ram}_f(A) = \{\mathfrak{P}_1, \mathfrak{P}_1', \ldots, \mathfrak{P}_r, \mathfrak{P}_r'\}$ where the primes $\mathfrak{P}_i$ and $\mathfrak{P}_i'$ both lie above the prime $\mathfrak{p}_i$ of $\mathcal{O}_\ell$. Let $\mathfrak{p}$ be a prime of $\mathcal{O}_\ell$ which is inert in the extension $k/\ell$. The effective Chebotarev density theorem [**19**] shows that there exists an absolute constant $C > 0$ such that we may assume $|\mathfrak{p}| < |\Delta_\ell|^C$. We now define a quaternion algebra $B$ over $\ell$ by declaring it to be ramified at all real places except the identity and at the finite primes $\{\mathfrak{p}_1, \ldots, \mathfrak{p}_r\}$ if $r + n_\ell - 1$ is even and $\{\mathfrak{p}_1, \ldots, \mathfrak{p}_r\} \cup \{\mathfrak{p}\}$ if $r + n_\ell - 1$ is odd. Theorem 4.1 shows that $A \cong B \otimes_\ell k$. Let $\mathcal{O}_B$ be a maximal order of $B$, chosen so that $\mathcal{O}_B \otimes_{\mathcal{O}_\ell} \mathcal{O}_k \cong \mathcal{O}$. Therefore $\Gamma_{\mathcal{O}_B}^1 \subset \Gamma_{\mathcal{O}}^1 \subset \pi_1(M)$ and $\mathrm{sys}_2^{TG}(M) \leq \mathrm{coarea}(\Gamma_{\mathcal{O}_B}^1)$. Borel [**4**] (see also [**23**, Chapter 11.6]) has shown that $[\Gamma_{\mathcal{O}} : \Gamma_{\mathcal{O}}^1]$ depends only on $k$, hence it suffices to show that $\mathrm{coarea}(\Gamma_{\mathcal{O}_B}^1) \leq c_k \, \mathrm{covol}(\Gamma_{\mathcal{O}}^1)^{\frac{1}{2}}$ for some constant $c_k > 0$ depending only on $k$. Notice that $|\mathfrak{p}|$ has been bounded by a quantity which depends only on $\ell$, and that

$$\left( \prod_{i=1}^{r} (|\mathfrak{p}_i| - 1) \right)^2 = \prod_{i=1}^{r} (|\mathfrak{P}_i| - 1)(|\mathfrak{P}_i'| - 1) = \prod_{\mathfrak{q} \in \mathrm{Ram}_f(A)} (|\mathfrak{q}| - 1).$$

The upper bound for $\mathrm{sys}_2^{TG}(M)$ now follows from equations (1) and (3).

We conclude this section by proving Corollary 1.1.

Let $k$ be a number field with a unique complex place and whose maximal totally real subfield $\ell$ satisfies $[k : \ell] = 2$. Infinitely many primes of $\ell$ split in the quadratic extension $k/\ell$, hence there exist finitely many primes $\mathfrak{p}$ of $\mathcal{O}_\ell$ such that

$$|\mathfrak{p}| > X \left( \frac{(8\pi^2)^{n_\ell} h_\ell \, |\Delta_\ell|^{-\frac{3}{2}}}{2\pi \zeta_\ell(2)} \right) + 1.$$

Fix one such prime $\mathfrak{p}$. Let $A$ be the quaternion algebra over $k$ which is ramified at all real primes of $k$ and at the two finite primes lying above $\mathfrak{p}$. Let $\Gamma$ be any torsion-free arithmetic Kleinian group arising from $(A, k)$. The results of [**23**, Chapter 9.5] show that any non-elementary Fuchsian subgroup of $\Gamma$ must arise from an indefinite quaternion algebra $B$ over $\ell$ which is ramified at $\mathfrak{p}$ (and potentially other primes). Note that if $t(B)$ is the type number of an indefinite quaternion algebra over $\ell$ then we have already seen that $t(B) \leq 2^{n_\ell} h_\ell$. The corollary now follows from Section 3 and in particular equation (2).

## 5. Proof of Theorem B

Our proof will make use of the following refinement of Linnik's theorem on primes in arithmetic progressions, the proof of which follows from [**14**, Corollary 18.8, p. 442] (see also [**22**, Proposition 8.1]).

LEMMA 5.1. *There is an absolute constant $L > 0$ such that for every pair of coprime integers $a, q$ with $q \geq 2$ the $n$-th prime $p \equiv a \pmod{q}$ satisfies*

$$p \leq q^L \cdot n \log(2n).$$

Let $m$ be a natural number such that the $m$-th prime $p_m$ satisfies $p_m > 2^{10}3^2 p_{N+1}$.

Let $L_1, L_2, \ldots$ be a sequence of imaginary quadratic fields of discriminants $-q_1, -q_2, \ldots$ (where all of the $q_i$ are prime) in which the rational prime 13 splits and in which $p_1, p_2, p_3, p_4, p_5, p_7, \ldots, p_{m-1}$ are inert. We assume as well that none of the $L_i$ are contained in one of the cyclotomic fields $\mathbf{Q}(\sqrt{-1}), \mathbf{Q}(\sqrt{-3}), \mathbf{Q}(\zeta_5)$. Note that these fields may be constructed by selecting the primes $q_1, q_2, \ldots$ to lie within a certain residue class modulo $C = 4p_1 \cdots p_{m-1}$. It now follows from Lemma 5.1 that we may take

$$|\Delta_{L_i}| < C^L \cdot i \log(2i) \tag{6}$$

for some absolute constant $L > 0$.

For every $i \geq 1$ let $A_i$ be the quaternion algebra over $L_i$ such that $\operatorname{disc}(A_i) = 13\mathcal{O}_{L_i}$. Let $\mathcal{O}_i$ be a maximal order of $A_i$ and observe that $\Gamma_{\mathcal{O}_i}$ is torsion-free as no quadratic cyclotomic extension of $L_i$ embeds into $A_i$ (note that 13 splits in both $\mathbf{Q}(i)/\mathbf{Q}$ and $\mathbf{Q}(\sqrt{-3})/\mathbf{Q}$, hence neither $L_j(i)$ nor $L_j(\sqrt{-3})$ embed into $A_j$).

A consequence of Theorem 4.1 is that if a quaternion algebra $B$ over $\mathbf{Q}$ satisfies $B \otimes_{\mathbf{Q}} L_i \cong A_i$ but $B \otimes_{\mathbf{Q}} L_j \not\cong A_j$ is that $\operatorname{Ram}(B)$ must contain a prime $p$ which splits in $L_i$ but not $L_j$. By construction this prime must satisfy $p \geq p_m$. Let $B_1, B_2, \ldots, B_5, B_7, \ldots, B_{N+1}$ be indefinite rational quaternion algebras for which $\operatorname{Ram}(B_i) = \{13, p_i\}$ and $\mathcal{O}_{B_i}$ be a maximal order of $B_i$. We now deduce from equations (1) and (2) that every arithmetic Fuchsian derived from $B$ has coarea larger than all of the $\Gamma_{\mathcal{O}_{B_i}}$. Because the $B_i$ embed into all of the quaternion algebras $A_1, A_2, \ldots$, we see that $\operatorname{GS}(\mathbf{H}^3/\Gamma_{\mathcal{O}_i})[N] = \operatorname{GS}(\mathbf{H}^3/\Gamma_{\mathcal{O}_j})[N]$ for all $i, j$. The assertion about $\operatorname{Vol}(M_n)$ now follows from equations (6) and (3), along with the fact that $\zeta_{L_i}(2) < \zeta(2)^2 < 3$ for all $i$.

To conclude the proof of Theorem B, we must derive an upper bound for $\operatorname{sys}_1(\mathbf{H}^3/\Gamma_{\mathcal{O}_i})$. This upper bound is an immediate consequence of the well-known fact that if $M$ is a closed hyperbolic 3-manifold then $\operatorname{sys}_1(M) < \log(\operatorname{Vol}(M))$ (this follows from the injectivity radius of $M$ being $\frac{1}{2}\operatorname{sys}_1(M)$ and the fact that the volume of a ball of radius $r$ in $\mathbf{H}^3$ has order of magnitude $e^r$).

## 6. Proof of Theorem C

If $M$ is a closed hyperbolic 3-manifold then it is clear that $\operatorname{sys}_2^{TG}(M) \geq \operatorname{sys}_2(M)$, hence it suffices to prove Theorem C for $N_k^g(V; x)$ and $N_k^2(V; x)$.

We begin by proving the theorem for $N_k^g(V; x)$. In essence, our proof follows from two previously known results. The first is due to the first author and McReynolds, Pollack and Thompson [**21**] and concerns the number of commensurability classes of arithmetic hyperbolic 3-manifolds containing a representative with systole less than any fixed bound. The second result is an important systole inequality of Belolipetsky [**3**], whose proof relies on a high genus systole inequality of Gromov [**11**] (see also [**16**, p. 88]):

THEOREM 6.1 (Belolipetsky). *Let $M$ be a closed hyperbolic 3-manifold. For any $\epsilon > 0$, if the systole $\mathrm{sys}_1(M)$ of $M$ is sufficiently large then*

$$\mathrm{sysg}(M) \geq e^{(\frac{1}{2}-\epsilon)\,\mathrm{sys}_1(M)}. \tag{7}$$

Let $x_0 = x_0(\epsilon)$ be large enough so that (7) holds with $\epsilon = \frac{1}{4}$ whenever $\mathrm{sys}_1(M) > x_0$. If $x > x_0$ and $\mathrm{sysg}(M) < x$ then (7) shows that $\mathrm{sys}_1(M) < 4\log(x)$. Given a function $f(x)$, define $N_k(V; f(x))$ to be the number of commensurability classes of arithmetic hyperbolic 3-manifolds with invariant trace field $k$, volume less than $V$ and which contain a representative $M$ with systole $\mathrm{sys}_1(M) < f(x)$. We now see that

$$N_k^g(V;x) < N_k(V;x_0) + N_k(V; 4\log(x)),$$

as for a given commensurability class $\mathscr{C}$, either $\mathscr{C}$ contains a manifold $M$ with $\mathrm{sys}_1(M) \leq x_0$ in which case (7) need not apply, or else every manifold $M \in \mathscr{C}$ satisfies

$$\mathrm{sysg}(M) < x \Longrightarrow x_0 < \mathrm{sys}_1(M) \leq 4\log(x).$$

Our proof in the case of $N_k^g(V;x)$ now follows from [**21**, Corollary 1.2]. The remaining case of $N_k^2(V;x)$ follows from exactly the same argument, replacing the aforementioned systole inequality of Belolipetsky by [**3**, Corollary 2.3].

**Acknowledgments.** The first author was partially supported by an NSF RTG grant DMS-1045119 and an NSF Mathematical Sciences Postdoctoral Fellowship.

## References


[1] I. Agol, *The virtual Haken conjecture*, Doc. Math. **18** (2013), 1045–1087. With an appendix by Agol, Daniel Groves, and Jason Manning. MR3104553

[2] N. Bergeron and D. T. Wise, *A boundary criterion for cubulation*, Amer. J. Math. **134** (2012), no. 3, 843–859, DOI 10.1353/ajm.2012.0020. MR2931226

[3] M. Belolipetsky, *On 2-systoles of hyperbolic 3-manifolds*, Geom. Funct. Anal. **23** (2013), no. 3, 813–827, DOI 10.1007/s00039-013-0223-x. MR3061772

[4] A. Borel, *Commensurability classes and volumes of hyperbolic 3-manifolds*, Ann. Scuola Norm. Sup. Pisa Cl. Sci. (4) **8** (1981), no. 1, 1–33. MR616899

[5] T. Chinburg and E. Friedman, *The smallest arithmetic hyperbolic three-orbifold*, Invent. Math. **86** (1986), no. 3, 507–527, DOI 10.1007/BF01389265. MR860679

[6] T. Chinburg and E. Friedman, *An embedding theorem for quaternion algebras*, J. London Math. Soc. (2) **60** (1999), no. 1, 33–44, DOI 10.1112/S0024610799007607. MR1721813

[7] D. Cooper, D. D. Long, and A. W. Reid, *Essential closed surfaces in bounded 3-manifolds*, J. Amer. Math. Soc. **10** (1997), no. 3, 553–563, DOI 10.1090/S0894-0347-97-00236-1. MR1431827

[8] J. DeBlois, S. Friedl, and S. Vidussi, *Rank gradients of infinite cyclic covers of 3-manifolds*, Michigan Math. J. **63** (2014), no. 1, 65–81, DOI 10.1307/mmj/1395234359. MR3189468

[9] D. Girão, *Rank gradient of small covers*, Pacific J. Math. **266** (2013), no. 1, 23–29, DOI 10.2140/pjm.2013.266.23. MR3105775

[10] D. Girão, *Rank gradient in co-final towers of certain Kleinian groups*, Groups Geom. Dyn. **8** (2014), no. 1, 143–155, DOI 10.4171/GGD/220. MR3209706

[11] M. Gromov, *Filling Riemannian manifolds*, J. Differential Geom. **18** (1983), no. 1, 1–147. MR697984

[12] M. Gromov, *Singularities, expanders and topology of maps. I. Homology versus volume in the spaces of cycles*, Geom. Funct. Anal. **19** (2009), no. 3, 743–841, DOI 10.1007/s00039-009-0021-7. MR2563769

[13] J. Hass, *Acylindrical surfaces in 3-manifolds*, Michigan Math. J. **42** (1995), no. 2, 357–365, DOI 10.1307/mmj/1029005233. MR1342495

[14] H. Iwaniec and E. Kowalski, *Analytic number theory*, American Mathematical Society Colloquium Publications, vol. 53, American Mathematical Society, Providence, RI, 2004. MR2061214

[15] J. Kahn and V. Markovic, *Immersing almost geodesic surfaces in a closed hyperbolic three manifold*, Ann. of Math. (2) **175** (2012), no. 3, 1127–1190, DOI 10.4007/annals.2012.175.3.4. MR2912704

[16] M. G. Katz, *Systolic geometry and topology*, Mathematical Surveys and Monographs, vol. 137, American Mathematical Society, Providence, RI, 2007. With an appendix by Jake P. Solomon. MR2292367

[17] M. Lackenby, *Heegaard splittings, the virtually Haken conjecture and property* $(\tau)$, Invent. Math. **164** (2006), no. 2, 317–359, DOI 10.1007/s00222-005-0480-x. MR2218779

[18] M. Lackenby, *Surface subgroups of Kleinian groups with torsion*, Invent. Math. **179** (2010), no. 1, 175–190, DOI 10.1007/s00222-009-0215-5. MR2563763

[19] J. C. Lagarias, H. L. Montgomery, and A. M. Odlyzko, *A bound for the least prime ideal in the Chebotarev density theorem*, Invent. Math. **54** (1979), no. 3, 271–296, DOI 10.1007/BF01390234. MR553223

[20] B. Linowitz, D. B. McReynolds, P. Pollack, L. Thompson, *Counting and effective rigidity in algebra and geometry*, `http://arxiv.org/abs/1407.2294`.

[21] B. Linowitz, D. B. McReynolds, P. Pollack, L. Thompson, *Systoles of Arithmetic Hyperbolic Surfaces and* 3*-manifolds*, `http://arxiv.org/abs/1504.05257`. To appear in Math. Res. Lett.

[22] B. Linowitz, J. S. Meyer, and P. Pollack, *The length spectra of arithmetic hyperbolic 3-manifolds and their totally geodesic surfaces*, New York J. Math. **21** (2015), 955–972. MR3425629

[23] C. Maclachlan and A. W. Reid, *The arithmetic of hyperbolic 3-manifolds*, Graduate Texts in Mathematics, vol. 219, Springer-Verlag, New York, 2003. MR1937957

[24] C. Millichap, *Mutations and short geodesics in hyperbolic* 3*-manifolds*, `http://arxiv.org/abs/1406.6033`.

[25] J. Sacks and K. Uhlenbeck, *Minimal immersions of closed Riemann surfaces*, Trans. Amer. Math. Soc. **271** (1982), no. 2, 639–652, DOI 10.2307/1998902. MR654854

[26] R. Schoen and S. T. Yau, *Existence of incompressible minimal surfaces and the topology of three-dimensional manifolds with nonnegative scalar curvature*, Ann. of Math. (2) **110** (1979), no. 1, 127–142, DOI 10.2307/1971247. MR541332

[27] W. Thurston, *The Geometry and Topology of 3-Manifolds.* Lecture Notes, Princeton University Math. Dept. (1978).

Department of Mathematics, University of Michigan, Ann Arbor, Michigan 48109
*E-mail address*: `linowitz@umich.edu`

Department of Mathematics, University of Oklahoma, Norman, Oklahoma 73019
*E-mail address*: `jmeyer@math.ou.edu`

Contemporary Mathematics
Volume **696**, 2017
http://dx.doi.org/10.1090/conm/696/14024

# Extremal length functions are log-plurisubharmonic

Hideki Miyachi

ABSTRACT. In this paper, we show that the extremal length functions on Teichmüller space are log-plurisubharmonic. As a corollary, we obtain an alternative proof of Liu and Su's result on the plurisubharmonicity of extremal length functions. We also obtain alternative proofs of Krushkal's results that a function defined by the Teichmüller distance from a reference point is plurisubharmonic, and the Teichmüller space is hyperconvex. To show the log-plurisubharmonicity, we give an explicit formula for the Levi form of the extremal length functions in the generic case.

CONTENTS



## 1. Introduction

**1.1. Results.** A positive function $U$ on a complex manifold $N$ is said to be *log-plurisubharmonic* if $\log U$ is plurisubharmonic. Any log-plurisubharmonic function is plurisubharmonic (cf. §5.3).

Let $\mathcal{T}_{g,m}$ be the Teichmüller space of Riemann surfaces of analytically finite type $(g, m)$ with $2g - 2 + m > 0$. One of the aims of this paper is to give the following theorem, which improves a result by Liu and Su in [**23**].

THEOREM 1.1 (Log-plurisubharminicity). *Extremal length functions on $\mathcal{T}_{g,m}$ are log-plurisubharmonic.*

---

2010 *Mathematics Subject Classification.* Primary 32G15, 30F45, 31C10.
*Key words and phrases.* Teichmüller theory, Teichmüller distance, Extremal length, Plurisubharmonic functions.
This work was supported by Grant-in-Aid for Scientific Research (C) 16K05202.

Moreover, we will also observe that for generic measured foliations, the log-extremal length functions are real analytic strictly plurisubharmonic functions on $\mathcal{T}_{g,m}$ (cf. Theorem 5.2). In [**37**], Yeung proved that the reciprocals of the hyperbolic length functions are plurisuperharmonic. Thus, our work here is recognized as a comparable result with his result in a dictionary of analogies between notions and results associated to Extremal length geometry and Hyperbolic geometry (cf. [**29**, §1]. See also §5.3).

When $3g-3+m=1$, Theorem 1.1 follows from the direct calculation (cf. §7.2). Hence, in the discussion in all subsequent sections except for the appendix, we always assume that $3g-3+m\geq 2$. We will prove Theorem 1.1 in §5.3. In fact, Theorem 1.1 is derived from a stronger property of the extremal length functions than that given in Theorem 1.1 (cf. Theorem 5.3). Namely, we will show that the extremal length function $u(x)=\mathrm{Ext}_x(F)$ for any measured foliation $F$ satisfies the differential inequality

$$\sqrt{-1}\,u\,\partial\overline{\partial}u-2\sqrt{-1}\partial u\wedge\overline{\partial}u\geq 0 \tag{1.1}$$

in the sense of currents under the standard complex structure on $\mathcal{T}_{g,m}$ (cf. Theorem 5.2). However, the property in Theorem 1.1 is also important as we see below (cf. Corollaries 1.1 and 1.2).

To show Theorem 1.1, we will give an explicit formula of the Levi forms of extremal length functions in the generic case (cf. Theorem 5.1).

The following is an immediate consequence of Theorem 1.1, because the set of log-plurisubharmonic functions on a complex manifold forms a semiring (cf. [**17**, Colollary 2.6.9]).

COROLLARY 1.1 (Log-plurisubharmonicity of positive polynomials). *Let $F_1,\cdots,F_n$ be measured foliations. For any polynomial $P$ of $n$-variables with positive coefficients, the function*

$$\mathcal{T}_{g,m}\ni x\mapsto P(\mathrm{Ext}_x(F_1),\cdots,\mathrm{Ext}_x(F_n))\in\mathbb{R}$$

*is log-plurisubharmonic.*

**1.2. Teichmüller distance is plurisubharmonic.** It is known that the least upper semicontinuous majorant of a family of plurisubharmonic funtions is either the constant function $+\infty$ or plurisubharmonic (cf. Theorem 5 in [**21**, Chapter II, §2]). Kerckhoff's formula (2.3) implies that the Teichmüller distance $d_T(x_0,\cdot)$ from a reference point $x_0\in\mathcal{T}_{g,m}$ coincides with the half of the least upper semicontinuous majorant of the family of log-extremal length functions. Therefore, we obtain an alternative proof of Krushkal's result [**20**, Corollary 3] as follows.

COROLLARY 1.2 (Teichmüller distance is plurisubharmonic). *For $x_0\in\mathcal{T}_{g,m}$, the distance function*

$$\mathcal{T}_{g,m}\ni x\mapsto d_T(x_0,x)$$

*is plurisubharmonic.*

**1.3. Convexities for Teichmüller space.** A subset $K$ in a complex manifold $N$ is said to be *disk-convex* in $N$ if for any continuous mapping $g\colon\overline{\mathbb{D}}\to N$, holomorphic in $\mathbb{D}$, $f(\partial\mathbb{D})\subset K$ implies $f(\overline{\mathbb{D}})\subset K$ (cf. [**26**]). For $\epsilon>0$ and a measured foliation $F$, we call the set of the form $\{x\in\mathcal{T}_{g,m}\mid \mathrm{Ext}_x(F)<\epsilon\}$ the *$\epsilon$-horoball for $F$*. From the maximum modulus principle for (pluri)subharmonic functions, we obtain the following.

COROLLARY 1.3 (Disk convexity). *For $\epsilon > 0$ and a measured foliation $F$, the $\epsilon$-horoball for $F$ is disk-convex in $\mathcal{T}_{g,m}$. The metric ball with respect to the Teichmüller distance is also disk-convex in $\mathcal{T}_{g,m}$.*

About the convexity of metric balls, Lenzhen and Rafi [**22**] observed that the metric ball with respect to the Teichmüller distance is quasi-convex. Furthermore, they also showed that the extremal length function is not convex along a Teichmüller geodesic in general. This means that the intersection between a horoball of the extremal length (or a metric ball) and a Teichmüller disk is simply connected, but may not be hyperbolically convex in general.

A complex manifold $N$ is said to be *hyperconvex* (in the sense of Stehlé) if it admits a continuous plurisubharmonic exhaustion function $\rho\colon N \to [-\infty, 0)$ in the sense that $\{x \in N \mid \rho(x) < c\}$ is relatively compact in $M$ for every $c < 0$ (cf. [**32**]. See also [**28**]). Take two essentially complete measured foliations $F$ and $G$ which are *filling* in the sense that

$$i(F, H) + i(G, H) > 0 \tag{1.2}$$

for all $H \in \mathcal{MF} - \{0\}$. In Theorem 5.3 given later, we will show that

$$\rho\colon \mathcal{T}_{g,m} \ni x \mapsto -\frac{1}{\mathrm{Ext}_x(F) + \mathrm{Ext}_x(G) + 1} \tag{1.3}$$

is a real analytic, negative, strictly plurisubharmonic exhaustion function with lower bounds. Thus we obtain the following.

COROLLARY 1.4 (Hyperconvexitiy). *Teichmüller space is hyperconvex.*

Corollary 1.4 is already given by Krushkal in [**19**]. The hyperconvexity of Teichmüller space also follows from the completeness of the Carathéodory distance (cf. [**1**] and [**6**]. See also [**37**]). As an immediate corollary of the hyperconvexity (or Corollary 1.2), by virtue of Oka's theorem, we deduce the following, which is first proven by Bers and Ehrenpreis (see [**2**]).

COROLLARY 1.5 (Holomorphic convexity). *Teichmüller space is a domain of holomorphy.*

In fact, Corollary 1.5 is observed by many people (cf. [**6**], [**31**] and [**36**]). The author should remark that Liu and Su also gave Corollary 1.5 in [**23**] by constructing a plurisubharmonic exhaustion from the extremal length functions.

**Acknowledgements.** The author thanks the referee for his/her useful comments.

## 2. Notation

**2.1. Teichmüller space.** Let $\Sigma_{g,m}$ be a compact orientable surface of genus $g$ with $m$-disks removed with $2g-2+m > 0$. The *Teichmüller space* $\mathcal{T}_{g,m}$ of Riemann surfaces of analytically finite type $(g, m)$ is the set of equivalence classes of pairs $(M, f)$ consisting of a Riemann surface $M$ of analytically finite type $(g, m)$ and an orientation preserving homeomorphism $f\colon \mathrm{Int}(\Sigma_{g,m}) \to M$. Two marked Riemann surfaces $(M_1, f_1)$ and $(M_2, f_2)$ are *equivalent* if there is a conformal mapping $h\colon M_1 \to M_2$ which homotopic to $f_2 \circ f_1^{-1}$.

The Teichmüller space admits a canonical distance $d_T$, which we call the *Teichmüller distance*. The Teichmüller distance is originally defined as the logarithm of the infimum of the maximal dilatations of quasiconformal mappings respecting the markings (cf. [**14**, §5]). Kerckhoff [**16**] gave a geometric description of the Teichmüller distance via the extremal length, which we will recall in (2.3) below.

**2.2. Measured foliations.** Let $\mathcal{S}$ be the set of homotopy classes of non-trivial and non-peripheral simple closed curves on $\Sigma_{g,m}$. Let $\mathcal{WS}$ be the set of weighted simple closed curves, that is, the set of formal products $t\alpha$ of non-negative number $t$ and $\alpha \in \mathcal{S}$. The closure $\mathcal{MF}$ of the embedding

$$\mathcal{WS} \ni t\alpha \mapsto [\mathcal{S} \ni \beta \mapsto t\, i(\alpha, \beta)] \in \mathbb{R}_+^{\mathcal{S}}$$

is called the *space of measured foliations* on $\Sigma_{g,m}$, where $i(\cdot,\cdot)$ is the geometric intersection number on $\mathcal{S}$. We consider $\mathcal{WS}$ as a subset of $\mathcal{MF}$. We identify $1 \cdot \alpha \in \mathcal{WS}$ with $\alpha \in \mathcal{S}$. Any measured foliation is described as a pair consisting of a singular foliation and a transverse measure (cf. [**8**, Exposé 5]). For instance, $t\alpha \in \mathcal{MF}$ is described with a foliated annulus with core $\alpha$ which is identified with an annulus $[0, t] \times [0, 1]/(x, 0) \sim (x, 1)$ and whose complement is a graph on $\Sigma_{g,m}$, and the transverse measure is defined by $|dx|$. It is known that the geometric intersection number $i(\alpha, \beta)$ for $\alpha, \beta \in \mathcal{S}$ extends continuously to $\mathcal{MF} \times \mathcal{MF}$ with $i(t\alpha, s\beta) = ts\, i(\alpha, \beta)$ (cf. [**3**]).

If we fix a complete hyperbolic structure on the interior of $\Sigma_{g,m}$, any measured foliation canonically corresponds to a measured lamination. A *measured lamination* is a pair of a geodesic lamination, which called the *support*, and a transverse measure, where a *geodesic lamination* a is a compact set in the interior fo $\Sigma_{g,m}$ which consists of disjoint complete simple geodesics (cf. [**30**, §1.7]). A measured foliation is said to be *essentially complete* if the support of the corresponding measured lamination is maximal, that is, the complement consists of ideal triangles or ideal punctured monogons if $(g, m) \neq (1, 1)$, a punctured bigon otherwise (cf. [**33**, Definition 9.5.1, Propositions 9.5.2 and 9.5.4]). In the interior of $\Sigma_{g,m}$, each singularity of the associated foliation of an essentially complete measured foliation is a three prong singularity. At any puncture, the associated foliation has a one prong singularity (cf. [**30**, Epilogue]).

**2.3. Extremal length.** Let $M$ be a Riemann surface and let $A$ be a doubly connected domain on $M$. If $A$ is conformally equivalent to a round annulus $\{1 < |z| < R\}$, we define the *modulus* of $A$ by

$$\mathrm{Mod}(A) = \frac{1}{2\pi} \log R.$$

The *extremal length* of a simple closed curve $\alpha$ on $M$ is defined by

$$\mathrm{Ext}_M(\alpha) = \inf \left\{ \frac{1}{\mathrm{Mod}(A)} \mid \text{the core curve of } A \subset M \text{ is homotopic to } \alpha \right\}. \tag{2.1}$$

In [**16**], Kerckhoff showed that if we define the extremal length of $t\alpha \in \mathcal{WS}$ by

$$\mathrm{Ext}_M(t\alpha) = t^2 \mathrm{Ext}_M(\alpha),$$

then the extremal length function $\mathrm{Ext}_M$ on $\mathcal{WS}$ extends continuously to $\mathcal{MF}$. For $F \in \mathcal{MF}$, the extremal length function on $\mathcal{T}_{g,m}$ is defined by

$$\mathrm{Ext}_x(F) = \mathrm{Ext}_M(f(F)).$$

It is known that the function

$$\mathcal{T}_{g,m} \times \mathcal{MF} \ni (x, F) \mapsto \mathrm{Ext}_x(F)$$

is continuous. Furthermore, the extremal length function $\mathrm{Ext}_x\colon \mathcal{MF} \to \mathbb{R}$ is a proper function and satisfies the quasiconformal distortion property:

$$\mathrm{Ext}_y(F) \leq e^{2d_T(x,y)}\mathrm{Ext}_x(F)$$

for any $F \in \mathcal{MF}$ and $x, y \in \mathcal{T}_{g,m}$. This distortion property implies the continuity of the function

$$\mathcal{T}_{g,m} \times \mathcal{MF} \ni (x, F) \mapsto \mathrm{Ext}_x(F). \tag{2.2}$$

**2.4. Kerckhoff's formula and Minsky's inequality.** The Teichmüller distance $d_T$ is expressed with the extremal length, which we call *Kerckhoff's formula*:

$$\begin{aligned} d_T(x_1, x_2) &= \frac{1}{2}\log \sup_{\alpha\in\mathcal{S}} \frac{\mathrm{Ext}_{x_2}(\alpha)}{\mathrm{Ext}_{x_1}(\alpha)} \\ &= \frac{1}{2}\sup_{\alpha\in\mathcal{S}} (\log \mathrm{Ext}_{x_2}(\alpha) - \log \mathrm{Ext}_{x_1}(\alpha)) \end{aligned} \tag{2.3}$$

for $x_1, x_2 \in \mathcal{T}_{g,m}$ (see [**16**]). Minsky [**27**] observed the following inequality, which we recently call *Minsky's inequality*:

$$i(F, G)^2 \leq \mathrm{Ext}_x(F)\,\mathrm{Ext}_x(G) \tag{2.4}$$

for $x \in \mathcal{T}_{g,m}$ and $F, G \in \mathcal{MF}$. If two measured foliation $F$ and $G$ are filling in the sense of (1.2), we have

$$\mathrm{Ext}_x(F) + \mathrm{Ext}_x(G) \geq \frac{i(F,H)^2 + i(G,H)^2}{\mathrm{Ext}_x(H)}$$

for all $H$ and hence we deduce

$$\mathrm{Ext}_x(F) + \mathrm{Ext}_x(G) \geq e^{2d_T(x_0,x)} \min\{i(F,H)^2 + i(G,H)^2 \mid \mathrm{Ext}_{x_0}(H) = 1\}.$$

This implies that the function $x \mapsto \mathrm{Ext}_x(F) + \mathrm{Ext}_x(G)$ is proper on $\mathcal{T}_{g,m}$.

**2.5. Quadratic differentials.** Let $\mathcal{Q}_{g,m}$ be the space of holomorphic quadratic differentials with finite $L^1$-norm. Namely, the space $\mathcal{Q}_{g,m}$ is a holomorphic vector bundle over $\mathcal{T}_{g,m}$ and the fiber $\mathcal{Q}_x$ at $x = (M, f) \in \mathcal{T}_{g,m}$ is the space of holomorphic quadratic differentials on $M$ with finite $L^1$-norm (cf. §2.7). Any $q \in \mathcal{Q}_x$ extends meromorphically to the completion $\overline{M}$ by filling punctures and has (at most) simple poles at punctures. A quadratic differential $q \in \mathcal{Q}_{g,m}$ is said to be *generic* if it satisfies the following two conditions:

(1) $q$ has a simple pole at every puncture.
(2) all zeroes of $q$ are simple.

We can easily check that the set of all generic holomorphic quadratic differentials is an open and dense subset in $\mathcal{Q}_{g,m}$.

**2.6. Hubbard-Masur differentials.** For $q \in \mathcal{Q}_{g,m}$, there is a unique measured foliation $F$ such that

$$i(F, \alpha) = \inf_{\alpha'\sim f(\alpha)} \int_{\alpha'} |\mathrm{Re}\sqrt{q}|.$$

for all $\alpha \in \mathcal{S}$, where $(M, f) = \pi(q) \in \mathcal{T}_{g,m}$. We call $F$ the *vertical foliation* and denote it by $\mathbf{v}(q)$. We call $\mathbf{h}(q) = \mathbf{v}(-q)$ the *horizontal foliation* of $q$. It is known that for any $F \in \mathcal{MF}$ and $x = (M, f) \in \mathcal{T}_{g,m}$, there is a unique holomorphic quadratic differential $q_{F,x}$ such that $\mathbf{v}(q_{F,x}) = F$. We call $q_{F,x}$ the *Hubbard-Masur differential* for $F$ on $x$ (cf. [**13**]). If $F$ is essentially complete, the Hubbard-Masur differential $q_{F,x}$ is generic for all $x \in \mathcal{T}_{g,m}$, since any measured foliation which is realized by a holomorphic quadratic differential whose vertical foliation is equivalent to $F$ is isotopic to $F$ (cf. [**13**, §2] and [**5**, §2.5]).

The Hubbard-Masur differential $q_{F,x} = q_{F,x}(z)dz^2$ for $F$ on $x = (M, f)$ satisfies

$$\mathrm{Ext}_x(F) = \|q_{F,x}\| = \int_M |q_{F,x}| = \int_M |q_{F,x}(z)| \cdot \frac{i}{2} dz \wedge d\overline{z}. \tag{2.5}$$

**2.7. Infinitesimal theory.** Teichmüller space also has a canonical complex structure, which induced from the deformation by quasiconformal mappings. A *Beltrami differential* is, by definition, an $L^\infty$-measurable $(-1, 1)$-form. The holomorphic tangent space is described as the quotient of the complex Banach space of the Beltami differentials (cf. [**14**, Theorem 7.5]). The holomorphic cotangent space is identified with the space of holomorphic quadratic differentials. The natural pairing between holomorphic tangent space and cotangent space is given by

$$\langle v, q \rangle = \int_M \dot{\mu} q = \int_M \dot{\mu}(z) q(z) \cdot \frac{i}{2} dz \wedge d\overline{z}$$

for $v = [\dot{\mu}] \in T_x\mathcal{T}_{g,m}$ and $q \in T_x^*\mathcal{T}_{g,m} \cong \mathcal{Q}_x$ and $x = (M, f) \in \mathcal{T}_{g,m}$.

CONVENTION 1. *Henceforth, for* $(p_i, q_i)$ *forms* $\varphi_i = \varphi_i(z) dz^{p_i} d\overline{z}^{q_i}$ $(i = 1, 2)$ *on a Riemann surface* $M$ *with* $p_1 + p_2 = q_1 + q_2 = 1$, *we write*

$$\int_M \varphi_1 \varphi_2 = \int_M \varphi_1(z) \varphi_2(z) \cdot \frac{i}{2} dz \wedge d\overline{z}.$$

## 3. Coordinates via representations of the odd cohomology

**3.1. The tangent spaces to quadratic differentials.** Following [**13**], we describe the tangent space $T_{q_0}\mathcal{Q}_{g,m}$ as the first hypercohomolgy group $\mathbb{H}^1(L^\bullet)$ a complex of sheaves (cf. [**11**] or [**12**]). We will need the Kodaira-Spencer identification of the tangent space of Teichmüller space with the first cohomology group of the sheaf of holomorphic vector fields (for instance, see [**18**]. See also [**14**] and [**15**]).

Let $X$ and $q$ be a holomorphic vector field and a holomorphic quadratic differential on an open set of a Riemann surface $M$. Denote by $L_X q$ the Lie derivative of $q$ along $X$. Let $\Theta_M$ and $\Omega_M^{\otimes 2}$ be the sheaves of germs of holomorphic vector fields with zeroes at punctures and meromorphic quadratic differentials on $M$ with (at most) first order poles at punctures, respectively. Let $q_0 \in \mathcal{Q}_{g,m}$ ($q_0$ need not to be generic). The tangent space $T_{q_0}\mathcal{Q}_{g,m}$ is identified with the first hypercohomology group of the complex of sheaves

$$L^\bullet \colon \quad 0 \longrightarrow \Theta_M \xrightarrow{L_\cdot q_0} \Omega_M^{\otimes 2} \longrightarrow 0$$

(cf. [**13**, Proposition 4.5]). For the convenience, we recall the definition of the (first) hypercohomology group which we use here. The first cochain group is the direct sum $C^0(M, \Omega_M^{\otimes 2}) \oplus C^1(M, \Theta_M)$. Consider an appropriate covering $\mathcal{U} = \{U_i\}_i$ on $M$ such that $\mathbb{H}^1(L^\bullet) \cong \mathbb{H}^1(\mathcal{U}, L^\bullet)$ (see the proof of [**13**, Proposition 4.5]). A cochain $(\{\phi_i\}_i, \{X_{ij}\}_{i,j})$ in $C^0(\mathcal{U}, \Omega_M^{\otimes 2}) \oplus C^1(\mathcal{U}, \Theta_M)$ is said to be *cocycle* if it satisfies

$$\delta\{X_{ij}\}_{i,j} = X_{ij} + X_{jk} + X_{ki} = 0, \quad \delta\{\phi_i\}_i = \phi_i - \phi_j = L_{X_{ij}}(q_0).$$

A *coboundary* is a cochain $(\{\phi_i\}_i, \{X_{ij}\}_{i,j})$ of the form

$$X_{ij} = Z_i - Z_j = \delta\{Z_i\}_i, \quad \phi_i = L_{Z_i}(q_0)$$

$$
\begin{array}{ccccc}
0 & & & & \\
\uparrow & & & & \\
C^0(\mathcal{U}, \Omega_M^{\otimes 2}) & \xrightarrow{\delta} & C^1(\mathcal{U}, \Omega_M^{\otimes 2}) & & \\
{\scriptstyle L.q_0}\uparrow & & {\scriptstyle -L.q_0}\uparrow & & \\
C^0(\mathcal{U}, \Theta_M) & \xrightarrow{\delta} & C^1(\mathcal{U}, \Theta_M) & \xrightarrow{\delta} & C^2(\mathcal{U}, \Theta_M)
\end{array}
$$

FIGURE 1. Double complex for the tangent spaces to $\mathcal{Q}_{g,m}$

for some 0-cochain $\{Z_i\}_i \in C^0(\mathcal{U}, \Theta_M)$ (cf. Figure 1).

**3.2. Branched covering spaces associated with generic differentials.** Let $x = (M, f) \in \mathcal{T}_{g,m}$ and $q_0$ a generic holomorphic quadratic differential on $M$. We consider $M$ as a pair of topological surface $\Sigma_g$ and a complex strtucture on $\Sigma_g - \{\text{punctures of } M\}$.

Let $U$ be a contractible neighborhood of $q_0$ in $\mathcal{Q}^0_{g,m}$. For $q \in U$, let $A_q = \mathrm{Zero}(q) \cup \mathrm{Pole}(q) \subset \Sigma_g$ and $M_q$ the underlying surface of $q$. Then, we have a homomorphism

$$\rho_q \colon \pi_1(\Sigma_g - A_q) \to \mathbb{Z}/2\mathbb{Z} \subset \mathrm{Isom}(\mathbb{C}) \tag{3.1}$$

associated to the double covering space

$$\pi_q \colon \tilde{M}_q \to M_q$$

which is the Riemann surface of $\sqrt{q}$ over $\Sigma_g - A_q$, where $\mathbb{Z}/2\mathbb{Z}$ in (3.1) is recognized as a subgroup of the isometry group of $\mathbb{C}$ with the standard Euclidean metric generated by the $\pi$-rotation, where $\mathbb{C}$ is given the standard Euclidean metric. The square root $\sqrt{q}$ lifts to a well-defined holomorphic 1- form $\omega_q$ on $\tilde{M}_q$.

Let $W = \Sigma_g - A_{q_0}$ and $W'$ be a compact subsurface of $W$ with smooth boundary such that the inclusion $W' \hookrightarrow W_0$ is homotopy equivalence and the complement $\Sigma_g - W'$ is a union of tiny open $|q_0|$-disks of centers in $A_0$ By taking sufficiently small $U$, we assume that $A_q \subset \Sigma_g - W'$ for any $q \in U$. Hence, by passing the inclusions $W' \hookrightarrow W$ and $W' \hookrightarrow \Sigma_g - A_q$, we get a canonical identification between $\pi_1(W)$ and $\pi_1(\Sigma_g - A_q)$ so that $\rho_q = \rho_0$ as a homomorphism from $\pi_1(W)$ to $\mathbb{Z}/2\mathbb{Z}$ for all $q \in U$.

**3.3. Backgrounds.** Let $q_0 \in \mathcal{Q}_{g,m}$ be a generic differential on $M_{q_0}$. Take a neighborhood $U$ of $q_0$ as the previous section. Then, $\mathbb{M} = \cup_{q \in U} \tilde{M}_q$ is a holomorphic family of closed Riemann surfaces over $U$, and the arrangement of the covering transformation $r_q \colon \tilde{M}_q \to \tilde{M}_q$ defines the holomorphic automorphism $\mathbf{r}$ of $\mathbb{M}$ of order 2. Taking a sufficiently small $U$ if necessary, we may assume that there is a covering $\{\mathcal{V}_i\}_i$ of $\mathbb{M}$ such that

(1) each $\mathcal{V}_i$ admits a biholomorphic mapping $\mathcal{A}_i$ onto the product domain $\mathbb{D} \times U$ such that its restriction to $\mathcal{V}_{i,q} = \mathcal{V}_i \cap \tilde{M}_q$ is a biholomorphism onto $\mathbb{D} \times \{q\}$ for each $q \in U$,
(2) for any finite $i_1, \cdots, i_k$ and $q \in U$, the intersection $\mathcal{V}_{i_1,q} \cap \cdots \cap \mathcal{V}_{i_k,q}$ is connected and simply connected,

(3) eacn $V_{i,q}$ contains at most one branch point of $\tilde{M}_q \to M_q$ and any branch point of $\tilde{M}_q \to M_q$ is contained in at most one open set in $\{V_{i,q}\}_i$, and
(4) the covering $\{\mathcal{V}_i\}_i$ is equivariant under $\mathbf{r}$ in the sense that $\{\mathbf{r}(\mathcal{V}_i)\}_i = \{\mathcal{V}_i\}_i$ and $\mathcal{A}_j \circ \mathbf{r} = \mathcal{A}_i$ when $\mathbf{r}(\mathcal{V}_i) = \mathcal{V}_j$.

Set $D = \{\lambda = s + it \mid |s|, |t| < \delta_0\}$. Let $\{q_\lambda\}_{\lambda \in D}$ be a differentiable family of holomorphic quadratic differenitals with $q_0 = q_0$ and $q_\lambda \in U$ for $\lambda \in D$. For any $\lambda \in D$, we let $\alpha_i(\lambda)\colon \mathcal{V}_{i,q_\lambda} \to \mathbb{D}$ by $\alpha_i(\lambda) = \mathcal{A}_i \mid_{\tilde{M}_{q_\lambda}}$ and set $V_{\lambda;ij} = \alpha_j(\lambda)(\mathcal{V}_{i,q_\lambda} \cap \mathcal{V}_{j,q_\lambda})$, and $\alpha_{ij}(\lambda) = \alpha_j(\lambda) \circ \alpha_i(\lambda)^{-1}\colon V_{\lambda;ji} \to V_{\lambda;ij}$. For simplicity, set $\alpha_i = \alpha_{0;i}$, $\alpha_{ij} = \alpha_{0;ij}$ and $V_{ij} = V_{0;ij}$. Consider the image of $\alpha_j$ is in the $z_j$-plane.

Let $\tilde{q}_{\lambda;i} dz_i^2$ be the representation of $(\pi_{q_\lambda})^*(q_\lambda)$ under the coordinate $\alpha_i(\lambda)$. Let

$$\tilde{\phi}_{\lambda;i} = \frac{\partial \tilde{q}_{\lambda;i}}{\partial \lambda}, \quad \tilde{X}_{\lambda;i} = -\frac{\partial \alpha_i(\lambda)}{\partial \lambda} \circ \alpha_i(\lambda)^{-1}, \tag{3.2}$$

on $\mathbb{D}$ for all $i$ and set

$$\tilde{X}_{\lambda;ij}(z_j) = \frac{\partial \alpha_{ij}(\lambda)}{\partial \lambda}(z_i),$$

where $z_j = \alpha_{ij}(\lambda)(z_i)$ and $z_i \in V_{\lambda;ij}$ (tildes mean objects (differentials or vector fields etc.) on the covering space $\tilde{M}_{q_0}$ which obtained as lifts of objects on $M_{q_0}$). Then,

$$\tilde{X}_{\lambda;i}(z_i)\frac{\partial}{\partial z_i}, \quad \tilde{X}_{\lambda;ij}(z_j)\frac{\partial}{\partial z_j}$$

are vector fields on appropriate sets of $\tilde{M}_{q_\lambda}$. Set $\tilde{q}_i = \tilde{q}_{0;i}$ and $\tilde{\phi}_i = \tilde{\phi}_{0;i}$. Then

$$\alpha_{ij}(\lambda)_*(\tilde{X}_{\lambda;i}) - \tilde{X}_{\lambda;j} = \tilde{X}_{\lambda;ij}, \tag{3.3}$$

$$\alpha_{ij}(\lambda)^*(\tilde{\phi}_{\lambda;i}) - \tilde{\phi}_{\lambda;j} = L_{\tilde{X}_{\lambda;ij}}(\tilde{q}_{\lambda;j}) \tag{3.4}$$

on $V_{\lambda;ij}$. We abbreviate these equations to

$$\tilde{X}_{\lambda;i} - \tilde{X}_{\lambda;j} = \tilde{X}_{\lambda;ij}, \quad \tilde{\phi}_{\lambda;i} - \tilde{\phi}_{\lambda;j} = L_{\tilde{X}_{\lambda;ij}}(\tilde{q}_\lambda). \tag{3.5}$$

for instance. By definition, $\omega_{q_\lambda}^2 = (\pi_{q_\lambda})^*(q_\lambda)$ for all $\lambda \in D$.

For any cocycle $\{X_{ij}\}_{ij} \in C^1(M, \Theta_M)$, we can always find a 0-cochain $\{X_i\}_i$ of the sheaf of $C^\infty$-vector fields which satisfies $X_i - X_j = X_{ij}$ on $U_i \cap U_j$ applying a partition of unity. Then, from (3.2), the $\overline{z}$-derivative

$$\dot{\nu} = -(\tilde{X}_i)_{\overline{z}} \tag{3.6}$$

defines a Beltrami differential on $M$ which represents the infinitesimal deformation corresponding to the tangent vector associated to the cohomology class of $\{X_{ij}\}_{i,j}$ in $H^1(M, \Theta_M)$. Compare with Equation (7.27) in [**14**, §7.2.4].

**3.4. Local coordinates via homomorphisms.** Let $H_1(\tilde{M}_q)^-$ be the odd part of the homology of $\tilde{M}_q$ with coefficient in $\mathbb{Z}$. Namely, $H_1(\tilde{M}_q)^-$ is the the eigenspace of the eigenvalue $-1$ of the linear automorphism $(r_q)_*$ on the first cohomology group of $\tilde{M}_q$ defined by the covering transformation $r_q$. In [**4**], Douady and Hubbard consider the following mapping

$$\mathrm{DH}\colon U \ni q \mapsto \chi_q \in \mathrm{Hom}(H_1(\tilde{M}_{q_0})^-, \mathbb{C}) \tag{3.7}$$

defined by

$$\chi_q(c) = \int_c \omega_q$$

where we identify $H_1(\tilde{M}_{q_0})^- \cong H_1(\tilde{M}_q)^-$ in a canonical way as discussed in §3.2. We can easily check that $\mathrm{Hom}(H_1(\tilde{M}_{q_0})^-, \mathbb{C})$ is a $\mathbb{C}$-vector space of complex dimension $6g-6+2m$. We shall check the following.

LEMMA 3.1 (Differential of DH). *Let $q_0 \in \mathcal{Q}_{g,m}$ be a generic differential. Let $v \in T_{q_0}\mathcal{Q}_{g,m}$ be a tangent vector associated to a 1-cochain $(\{\phi_i\}_i, \{X_{ij}\}_{i,j})$. Let $\{X_i\}_i$ be a 0-cochain of the sheaf of $C^\infty$-vector fields on $M_{q_0}$ for $\{X_{ij}\}_{i,j}$. Then, the derivative of the mapping (3.7) along $v$ satisfies*

$$\mathrm{DH}_*[v](c) = \int_c \Phi \tag{3.8}$$

*for all $c \in H_1(\tilde{M}_{q_0})^-$, where*

$$\Phi = \left(\frac{\tilde{\phi}_i}{2\omega_{q_0}} - \omega'_{q_0}\tilde{X}_i - \omega_{q_0}(\tilde{X}_i)_z\right)dz - \omega_{q_0}(\tilde{X}_i)_{\overline{z}}d\overline{z}$$

*on $\mathcal{V}_{i,0}$.*

One can easily check that $\Phi$ in Lemma 3.1 is a well-defined closed one-form on $\tilde{M}_{q_0}$ which satisfies $(r_{q_0})^*(\Phi) = -\Phi$. In the definition of $\Phi$, the first term of the coefficient of $dz$ corresponds to the differentiation in the fiber of $\mathcal{Q}_{g,m} \to \mathcal{T}_{g,m}$, and the remaining term comes from the differentiation in the base.

PROOF. We calculate the $\lambda$-derivative at $\lambda = 0$ with the notation in §3.3. Then

$$\alpha_i(\lambda) \circ \alpha_i^{-1}(z_i) = z_i - \lambda\tilde{X}_i(z_i) - \overline{\lambda}\tilde{Y}_i(z_i) + o(|\lambda|)$$
$$\tilde{q}_{\lambda;i}(z_i) = \tilde{q}_i(z_i) + \lambda\tilde{\phi}_i(z_i) + \overline{\lambda}\tilde{\psi}_i(z_i) + o(|\lambda|)$$

on $\mathbb{D} = \alpha_i(\mathcal{V}_{i,0})$ as $\lambda \to 0$, where

$$\tilde{Y}_i = -\left.\frac{\partial\alpha_i(\lambda)}{\partial\overline{\lambda}}\right|_{\lambda=0} \circ \alpha_i^{-1}, \quad \tilde{\psi}_i = \left.\frac{\partial q_{\lambda;i}}{\partial\overline{\lambda}}\right|_{\lambda=0},$$

and the convergence is uniform on any compact sets of $\mathbb{D}$. In our case, $\lambda$-derivative of $D \ni \lambda \mapsto \mathrm{DH}(q_\lambda)(c)$ at $\lambda = 0$ will be the desired formula.

Notice that the partial $\lambda$- derivative of $\omega_\lambda$ at $\lambda = 0$ is

$$\left.\frac{\partial\omega_\lambda}{\partial\lambda}\right|_{\lambda=0} = \frac{\tilde{\phi}_i}{2\omega_{q_0}}$$

on $\mathbb{D} = \alpha_i(\mathcal{V}_{i,0})$. In general, the derivative of a function defined by the integration

$$(-\delta, \delta) \ni s \mapsto \int_{a(s)}^{b(s)} f(s,x)dx$$

at $s = 0$ is equal to

$$\int_{a(0)}^{b(0)} \frac{\partial f}{\partial s}(0,x)dx + f(0,b(0))\frac{db}{ds}(0) - f(0,a(0))\frac{da}{ds}(0)$$

when $a(s)$ and $b(s)$ are of class $C^1$. Hence, by the standard argument, like as the discussion in Proposition 1 of [**4**], one can check the equations (3.8) hold. □

Though the following is well-known (cf. [**25**], [**34**] and [**35**]), we confirm it in the following for the completeness. Indeed, the calculation in the following proof will be a guide in the later discussion.

LEMMA 3.2 (DH defines a local coordinate). *Let $q_0 \in \mathcal{Q}_{g,m}$ be as above. When we take the above neighborhood $U$ of $q_0$ to be sufficiently small, the mapping* (3.7) *defines a holomorphic local coordinate around $q_0$.*

PROOF. Let $v \in T_{q_0}\mathcal{Q}_{g,m}$ be a tangent vector which is represented by the 1 cocycle $(\{\phi_i\}_i, \{X_{ij}\}_{i,j})$ in $C^0(\mathcal{V}_0, \Omega_{M_{q_0}}^{\otimes 2}) \oplus C^1(\mathcal{V}_0, \Theta_{M_{q_0}})$, where $\mathcal{V}_0 = \{\{\pi_{q_0}(\mathcal{V}_{i,0})\}_i\}_i$ is a covering of $M_{q_0}$ (cf. (3.4)).

Suppose that $\mathrm{DH}_*[v] = 0$. Let $\Phi$ be a closed one form defined in Lemma 3.1. Since $(r_0)^*(\Phi) = -\Phi$,

$$\int_c \Phi = \int_{r_0(c)} \Phi = -\int_c \Phi$$

for all $c \in H_1(\tilde{M}_{q_0})^+$, where $H_1(\tilde{M}_{q_0})^+$ is the even part of the homology of $\tilde{M}_q$ with coefficient in $\mathbb{Z}$. Hence, there is a smooth function $f$ on $\tilde{M}_{q_0}$ such that $df = \Phi$ by de Rham's theorem. Namely,

$$\frac{\tilde{\phi}_i}{2\omega_{q_0}} - \omega'_{q_0}\tilde{X}_i - \omega_{q_0} \cdot (\tilde{X}_i)_z = f_z$$
$$-\omega_{q_0} \cdot (\tilde{X}_i)_{\bar{z}} = f_{\bar{z}}$$

hold on $\mathbb{D} = \alpha_i(\mathcal{V}_{i,0})$, where $\tilde{\phi}_i$ and $\tilde{X}_i$ are lifts of $\phi_i$ and $\tilde{X}_i$ to $\tilde{M}_{q_0}$, respectively. Since $\{(\mathcal{V}_{i,\lambda}, \alpha_i(\lambda))\}_i$ is a system of holomorphic local coordinates on $\tilde{M}_{q_\lambda}$ for each $\lambda \in D$, $\dot{\nu} = -(\tilde{X}_i)_{\bar{z}}$ defines a Beltrami differential on $\tilde{M}_{q_0}$ which descends to a Beltrami differential $\dot{\mu}$ on $M_{q_0}$. For any holomorphic quadratic differential $\varphi$ on $M_{q_0}$, the quotient $(\pi_{q_0})^*(\varphi)/\omega_{q_0}$ is a holomorphic 1-form on $\tilde{M}_{q_0}$. Since

$$\begin{aligned}
\int_{M_{q_0}} \dot{\mu}\varphi &= \frac{1}{2}\int_{\tilde{M}_{q_0}} \dot{\nu} \cdot (\pi_{q_0})^*(\varphi) = \frac{1}{2}\int_{\tilde{M}_{q_0}} f_{\bar{z}}\frac{(\pi_{q_0})^*(\varphi)}{\omega_{q_0}} \\
&= \frac{1}{2}\int_{\tilde{M}_{q_0}} \bar{\partial}\left(f(z)\frac{(\pi_{q_0})^*(\varphi)(z)}{\omega_{q_0}(z)}dz\right) \\
&= \frac{1}{2}\int_{\tilde{M}_{q_0}} d\left(f(z)\frac{(\pi_{q_0})^*(\varphi)(z)}{\omega_{q_0}(z)}dz\right) = 0,
\end{aligned}$$

the Beltrami differential $\dot{\mu}$ is infinitesimally trivial on $M_{q_0}$ by Teichmüller's lemma (cf. Lemma 7.6 of [**14**]). This means that the cohomology class

$$[\{X_{ij}\}_{i,j}] \in H^1(\mathcal{V}_0, \Theta_{M_{q_0}}) \cong H^1(M_{q_0}, \Theta_{M_{q_0}})$$

is trivial from (3.5) since $\mathcal{V}_0$ is a Leray covering (see also Theorem 3.5 of [**18**]). Therefore, there is a 0-cochain $\{Z_i\}_i \in C^0(\mathcal{V}_{0,i}, \Theta_{M_{q_0}})$ such that $Z_i - Z_j = X_{ij}$ for all $i, j$.

Let $\zeta_i = L_{Z_i}(q_0) \in \Gamma(\mathcal{V}_{0,i}, \Omega_{M_{q_0}}^{\otimes 2})$ and $\tilde{\zeta}_i = L_{\tilde{Z}_i}(\omega_{q_0}^2)$ be the lift of $\zeta_i$. Then,

$$\tilde{\zeta}_i - \tilde{\zeta}_j = L_{\tilde{X}_{ij}}(\omega_{q_0}^2)$$

and hence $\{\tilde{\phi}_i - \tilde{\zeta}_i\}_i$ defines a holomorphic quadratic differential $\tilde{\varphi}$ on $\tilde{M}_{q_0}$ which descends to a holomorphic quadratic differential $\varphi$ on $M_{q_0}$ associated to $\{\phi_i - \zeta_i\}_i$. Since $\tilde{X}_i - \tilde{Z}_i = \tilde{X}_j - \tilde{Z}_j$ on $\mathcal{V}_{i,0} \cap \mathcal{V}_{j,0}$, $\{\tilde{X}_i - \tilde{Z}_i\}_i$ defines a vector field $\tilde{W}$ on $\tilde{M}_{q_0}$

which satisfies

$$\begin{aligned}
&((\omega_{q_0})'(\tilde{X}_i) + \omega_{q_0} \cdot (\tilde{X}_i)_z)dz + \omega_{q_0} \cdot (\tilde{X}_i)_{\bar{z}} d\bar{z} - \frac{L_{\tilde{Z}_i}(\omega_{q_0}^2)}{2\omega_{q_0}} \\
&= \left((\omega_{q_0})'(\tilde{X}_i - \tilde{Z}_i) + \omega_{q_0} \cdot (\tilde{X}_i - \tilde{Z}_i)_z\right) dz + \omega_{q_0} \cdot (\tilde{X}_i - \tilde{Z}_i)_{\bar{z}} d\bar{z} \\
&= d(\omega_{q_0} \cdot \tilde{W})
\end{aligned}$$

since each $\tilde{Z}_i$ is a holomorphic section. Consequently, we deduce from the assumption that

$$0 = \int_c \Omega = \int_c \left(\Omega - \frac{\tilde{\zeta}_i - L_{\tilde{Z}_i}(\omega_{q_0}^2)}{2\omega_{q_0}}\right) = \int_c \frac{\tilde{\varphi}}{2\omega_{q_0}} - \int_c d(\omega_{q_0} \cdot W) = \int_c \frac{\tilde{\varphi}}{2\omega_{q_0}} \tag{3.9}$$

for all $c \in H_1(\tilde{M}_{q_0})$, and hence $\tilde{\varphi} = 0$ and $\varphi = 0$. Therefore $\phi_i = \zeta_i$ and

$$(\{\phi_i\}_i, \{X_{ij}\}_{i,j}) = (\{L_{Z_i}(q_0)\}_i, \{Z_i - Z_j\}_{i,j}).$$

Thus, the cohomology class of the cochain $(\{\phi_i\}_i, \{X_{ij}\}_{i,j})$ vanishes.

As a consequence, the derivative of the map DH on the tangent space $T_{q_0}\mathcal{Q}_{g,m} \cong \mathbb{H}^1(L^\bullet)$ is injective. Since the dimensions of $\mathcal{Q}_{g,m}$ and $\mathrm{Hom}(H_1(\tilde{M}_{q_0})^-, \mathbb{C})$ are the same, the derivative of DH is an isomorphism from $T_{q_0}\mathcal{Q}_{g,m}$ onto the tangent space to $\mathrm{Hom}(H_1(\tilde{M}_{q_0})^-, \mathbb{C})$ at $\mathrm{DH}(q_0)$. □

**3.5. Homomorphisms on slices.** For $F \in \mathcal{MF} - \{0\}$. we define a slice

$$E^v(F) = \{q \in \mathcal{Q}_{g,m} \mid \mathbf{v}(q) = F\}.$$

for $F$ in $\mathcal{Q}_{g,m}$. Hubbard and Masur showed that the restriction of the projection

$$E^v(F) \ni q \mapsto \pi(q) \in \mathcal{T}_{g,m}$$

is a homeomorphism ([**13**, §2]). The following two lemmas are well-known (cf. [**13**, Lemma 4.3] and [**24**, Proposition 1]. See also [**5**]).

LEMMA 3.3. *Let $q_0 \in \mathcal{Q}_{g,m}$ be a generic differential. When we take $U$ sufficiently small,*

$$E^v(\mathbf{v}(q_0)) \cap U = \{q \in \mathcal{Q}_{g,m} \mid \mathrm{Re}(\mathrm{DH}(q)) = \mathrm{Re}(\mathrm{DH}(q_0))\}$$

*holds.*

LEMMA 3.4. *Let $x_0 \in \mathcal{T}_{g,m}$ and $F \in \mathcal{MF}$. If $q_{F,x_0}$ is generic, then the map*

$$\mathcal{T}_{g,m} \ni x \mapsto q_{F,x} \in \mathcal{Q}_{g,m}$$

*is real analytic around $x_0$.*

## 4. Families of quadratic differentials with prescribed vertical foliation

Fix $F \in \mathcal{MF}$. Consider a family $\{q_\lambda\}_{\lambda \in D}$ of holomorphic quadratic differentials such that $\mathbf{v}(q_\lambda) = F$. Suppose that $q_0$ is generic and $x_\lambda = \pi(q_\lambda) \in \mathcal{T}_{g,m}$ varies holomorphically. Then, the family $\{q_\lambda\}_{\lambda \in D}$ is a real analytic family of quadratic differentials (cf. Lemma 3.4). Take a neighborhood $U$ of $q_0$ as the previous sections.

**4.1. Differentials via the complex conjugate.** We first make a simple observation. Let $\pi\colon E \to N$ be a holomorphic vector bundle over a complex manifold $N$. Let $g\colon \mathbb{D} \to N$ be a holomorphic mapping and $G\colon \mathbb{D} \to E$ be a $C^1$ mapping such that $\pi \circ G = g$ on $\mathbb{D}$. Then,

$$\pi_*\left(G_*\left(\frac{\partial}{\partial\overline{\lambda}}\right)\right) = g_*\left(\frac{\partial}{\partial\overline{\lambda}}\right) = 0$$

since $g$ is holomorphic. Hence, the $(1,0)$-part of the push-forward of a tangent vector $\partial/\partial\overline{\lambda}$ on $\mathbb{D}$ by $G$ satisfies

$$G_*\left(\frac{\partial}{\partial\overline{\lambda}}\right)\Big|_{\lambda} \in \operatorname{Ker}(\pi_*) \cong E_{g(\lambda)} \tag{4.1}$$

for all $\lambda \in \mathbb{D}$ because $\pi$ is holomorphic and the differential $\pi_*$ evaluates only the tangent vectors of type $(1,0)$.

**4.2. Differentials of periods along the slice.** Let $x = (M, f)$ and $v = [\dot\mu] \in T_x\mathcal{T}_{g,m}$. Let $q_0 \in \mathcal{Q}_x$ be a generic differential. A holomorphic quadratic differential $\eta_v \in \mathcal{Q}_x$ is called the $q_0$ *realization of* $v$ if

$$\langle v, \psi\rangle = \int_M \dot\mu\psi = \int_M \frac{\overline{\eta_v}}{|q_0|}\psi \tag{4.2}$$

holds for all $\psi \in \mathcal{Q}_x$. The existence and the uniqueness of $\eta_v$ follows from an argument by Dumas [**5**, Theorem 5.3] since each $q_0$ is generic. Indeed, the pairing

$$(\eta, \psi) \mapsto \int_M \frac{\overline{\eta}\psi}{|q_0|}$$

is a positive definite hermitian inner product on $\mathcal{Q}_x$.

LEMMA 4.1 (First derivatives). *Let $\{q_\lambda\}_{\lambda\in D}$ be the family of holomorphic quadratic differentials defined in the beginning of the section. Let $v_\lambda \in T_{x_\lambda}\mathcal{T}_{g,m}$ be the tangent vector for the holomorphic mapping*

$$D \ni \lambda \mapsto x_\lambda = \pi(q_\lambda) \in \mathcal{T}_{g,m}$$

*at $\lambda$. For any $\lambda \in D$, the $q$ realization $\eta_{v_\lambda} \in \mathcal{Q}_{x_\lambda}$ of $v_\lambda$ satisfies*

$$\frac{\partial\chi_{q_\lambda}(c)}{\partial\lambda} = \int_c \overline{\left(\frac{(\pi_{q_\lambda})^*(\eta_{v_\lambda})}{\omega_{q_\lambda}}\right)}, \quad \frac{\partial\chi_{q_\lambda}(c)}{\partial\overline{\lambda}} = -\int_c \frac{(\pi_{q_\lambda})^*(\eta_{v_\lambda})}{\omega_{q_\lambda}}$$

*for any $c \in H_1(\tilde M_{q_0})$ and $\lambda \in D$.*

PROOF. We first remark that

$$\frac{\partial\chi_{q_\lambda}(c)}{\partial\overline{\lambda}} = 0 = \int_c \frac{(\pi_{q_\lambda})^*(\eta_{v_\lambda})}{\omega_{q_\lambda}}$$

for any $c \in H_1(\tilde M_{q_0})^+$ and $\lambda \in D$. Hence, it suffices to show the case where $\lambda = 0$ and $c \in H_1(\tilde M_{q_0})^-$. Let $\dot\mu$ be the infinitesimal Beltrami differential on $M_{q_0}$ associated to the tangent vector $v_0$. As we discussed around (4.1), from Lemma 3.1 and the proof of Lemma 3.2, there is a holomorphic quadratic differential $\varphi$ on $M_{q_0}$ such that

$$\frac{\partial\chi_{q_\lambda}(c)}{\partial\overline{\lambda}}\Big|_{\lambda=0} = \int_c \frac{(\pi_{q_0})^*(\varphi)}{\omega_{q_0}}$$

for any $c \in H_1(\tilde{M}_{q_0})^-$ since $q \mapsto \chi_q(c)$ is holomorphic near $q_0$ (see (3.9)). Since the vertical foliation $|\mathrm{Re}(\omega_{q_\lambda})|$ of $\omega_{q_\lambda}$ is unchanged on $D$, for all $c \in H_1(\tilde{M}_{q_0})^-$, the real part of $\chi_{q_\lambda}(c)$ is constant on $D$ by Lemma 3.3. Therefore,

$$\frac{\partial \chi_{q_\lambda}(c)}{\partial \overline{\lambda}} = -\overline{\frac{\partial \chi_{q_\lambda}(c)}{\partial \lambda}}$$

holds on $D$. Let $(\{\phi_i\}_i, \{X_{ij}\}_{i,j})$ be the cocycle representing the tangent vector associated to the $\lambda$-derivative of the mapping $D \ni \lambda \mapsto q_\lambda$ at $\lambda = 0$. Let $\{X_i\}_i$ be a 0-cochain of the sheaf of $C^\infty$-vector fields with $X_i - X_j = X_{ij}$. Then,

$$\Omega = \left( \frac{\tilde{\phi}_i}{2\omega_{q_0}} - \omega'_{q_0}\tilde{X}_i - \omega_{q_0}(\tilde{X}_i)_z \right) dz - (\omega_{q_0}(\tilde{X}_i)_{\overline{z}})d\overline{z}$$

on $U_i$ defines a closed 1-form on $\tilde{M}_{q_0}$ satisfies

$$\frac{\partial \chi_{q_\lambda}(c)}{\partial \lambda} = \int_c \Omega$$

for all $c \in H_1(\tilde{M}_{q_0})$. Therefore, there is a $C^\infty$-function $f$ on $\tilde{M}_{q_0}$ such that

$$\Omega = -\overline{\frac{(\pi_{q_0})^*(\varphi)(z)}{\omega_{q_0}(z)}} d\overline{z} + df,$$

that is,

$$\frac{\tilde{\phi}_i}{2\omega_{q_0}} - \omega'_{q_0}\tilde{X}_i - \omega_{q_0}(\tilde{X}_i)_z = f_z \tag{4.3}$$

$$-\omega_{q_0}(\tilde{X}_i)_{\overline{z}} = -\overline{\left( \frac{(\pi_{q_0})^*(\varphi)(z)}{\omega_{q_0}(z)} \right)} + f_{\overline{z}} \tag{4.4}$$

on $\mathcal{V}_{i,0}$. From (4.4), the lift $(\pi_{q_0})^*(\dot{\mu})$ of $\dot{\mu} = -(X_i)_{\overline{z}}$ satisfies

$$(\pi_{q_0})^*(\dot{\mu}) = -\frac{\overline{(\pi_{q_0})^*(\varphi)}}{|\omega_{q_0}|^2} + \frac{f_{\overline{z}}}{\omega_{q_0}} \tag{4.5}$$

on $\tilde{M}_{q_0}$. Let $\psi$ be a holomorphic quadratic differenital on $M_{q_0}$. From (4.5)

$$\begin{aligned}
\int_{\tilde{M}_{q_0}} (\pi_{q_0})^*(\dot{\mu})(\pi_{q_0})^*(\psi) &= -\int_{\tilde{M}_{q_0}} \frac{\overline{(\pi_{q_0})^*(\varphi)}}{|\omega_{q_0}|^2}(\pi_{q_0})^*(\psi) + \int_{\tilde{M}_{q_0}} \frac{f_{\overline{z}}}{\omega_{q_0}}(\pi_{q_0})^*(\psi) \\
&= -\int_{\tilde{M}_{q_0}} \frac{\overline{(\pi_{q_0})^*(\varphi)}}{|\omega_{q_0}|^2}(\pi_{q_0})^*(\psi) + \int_{\tilde{M}_{q_0}} d\left( f\frac{(\pi_{q_0})^*(\psi)}{\omega_{q_0}} \right) \\
&= -\int_{\tilde{M}_{q_0}} \frac{\overline{(\pi_{q_0})^*(\varphi)}}{|\omega_{q_0}|^2}(\pi_{q_0})^*(\psi)
\end{aligned}$$

By descending to $M_{q_0}$, we have

$$\int_{M_{q_0}} \frac{\overline{\eta_{v_0}}}{|q_0|}\psi = \int_{M_{q_0}} \dot{\mu}\psi = -\int_{M_{q_0}} \frac{\overline{\varphi}\psi}{|q_0|}$$

for all $\psi$. Therefore, $\varphi = -\eta_{v_0}$ from [**5**, Theorem 5.3] again. □

**4.3. Laplacian of periods along the slice.** This section is devoted to deriving a formula for the laplacian of the Douady-Hubbard map (3.7) along the slice. The author must confess that for the proof of the plurisubharmonicity of the extremal length funcitons, we need a representation (4.7) of the differential rather than the explicit formula (4.6) of the laplacian given in the following lemma. However, we give the following lemma for its own interests.

LEMMA 4.2 (Laplacian). *Under the assumption in Lemma 4.1, we have*

$$\frac{\partial^2(\chi_{q_\lambda}(c))}{\partial\lambda\partial\overline{\lambda}}\bigg|_{\lambda=0} = -4i\,\mathrm{Im}\int_{\tilde{M}_{q_0}} \frac{|(\pi_{q_0})^*(\eta_{v_0})|^2}{|\omega_{q_0}|^2}\frac{\sigma_c}{\omega_{q_0}} \tag{4.6}$$

*for* $c \in H_1(\tilde{M}_{q_0})$, *where* $\sigma_c = \sigma_c(z)dz$ *is the holomorphic part of the reproducing harmonic differential associated to* $c$.

PROOF. When the homology class $c$ is in the even homology group, $\chi_{q_\lambda}(c) = 0$. The integrand of the right-hand side of (4.6) satisfies

$$r_0^*\left(\frac{|(\pi_{q_0})^*(\eta_{v_0})|^2}{|\omega_{q_0}|^2}\frac{\sigma_c}{\omega_{q_0}}\right) = \frac{|(\pi_{q_0})^*(\eta_{v_0})|^2}{|-\omega_{q_0}|^2}\frac{\sigma_c}{-\omega_{q_0}} = -\frac{|(\pi_{q_0})^*(\eta_{v_0})|^2}{|\omega_{q_0}|^2}\frac{\sigma_c}{\omega_{q_0}}$$

and hence, the integral over $\tilde{M}_{q_0}$ vanishes. This means that (4.6) holds for homology classes in the even homology group.

By applying the same argument as that of Lemma 3.1, one obtain

$$\frac{\partial^2\chi_{q_\lambda}(c)}{\partial\lambda\partial\overline{\lambda}}\bigg|_{\lambda=0} = \int_c \Omega_0$$

for any $c \in H_1(\tilde{M}_{q_0})$, where $\Omega_0$ is a closed differential

$$\left(\dot{\eta}_0 + \left(\frac{(\pi_{q_0})^*(\eta_{v_0})}{\omega_{q_0}}\right)'(\tilde{X}_i) + \left(\frac{(\pi_{q_0})^*(\eta_{v_0})}{\omega_{q_0}}\right)(\tilde{X}_i)_z\right)dz + \left(\frac{(\pi_{q_0})^*(\eta_{v_0})}{\omega_{q_0}}\right)(\tilde{X}_i)_{\overline{z}}d\overline{z}$$

where

$$\dot{\eta}_0 = -\frac{\partial}{\partial\lambda}\left(\frac{(\pi_{q_\lambda})^*(\eta_{v_\lambda})}{\omega_{q_\lambda}}\right)\bigg|_{\lambda=0}.$$

Since a real part of the function $D \ni \lambda \mapsto \chi_{q_\lambda}(c)$ is constant and the Laplacian is the real operator,

$$\int_c \Omega_0 = \frac{\partial^2\chi_{q_\lambda}(c)}{\partial\lambda\partial\overline{\lambda}}\bigg|_{\lambda=0} = -\overline{\frac{\partial^2\chi_{q_\lambda}(c)}{\partial\lambda\partial\overline{\lambda}}\bigg|_{\lambda=0}} = -\overline{\left(\int_c\Omega_0\right)} = -\int_c\overline{\Omega_0}$$

for all $c \in H_1(\tilde{M}_{q_0})$. Hence, there is a $C^\infty$-function $f$ on $\tilde{M}_{q_0}$ such that

$$\Omega_0 = -\overline{\Omega_0} + df.$$

Namely, we have

$$\begin{aligned}\dot{\eta}_0 + \left(\frac{(\pi_{q_0})^*(\eta_{v_0})}{\omega_{q_0}}\right)'(X_i) + \left(\frac{(\pi_{q_0})^*(\eta_{v_0})}{\omega_{q_0}}\right)(\tilde{X}_i)_z &= -\overline{\left(\frac{(\pi_{q_0})^*(\eta_{v_0})}{\omega_{q_0}}\right)(\tilde{X}_i)_{\overline{z}}} + f_z \\ &= \overline{\left(\frac{(\pi_{q_0})^*(\eta_{v_0})}{\omega_{q_0}}\right)(\pi_{q_0})^*(\dot{\mu})} + f_z,\end{aligned}$$

where $\dot{\mu} = -(X_i)_{\bar{z}}$ is the infinitesimal Beltrami differential representing the tangent vector of the mapping $D \ni \mapsto \pi(q_\lambda)$ at $\lambda = 0$ (cf. (3.6)). Applying the above equation, we also obtain

$$\begin{aligned}\Omega_0 &= \left(\overline{\left(\frac{(\pi_{q_0})^*(\eta_{v_0})}{\omega_{q_0}}\right)(\pi_{q_0})^*(\dot{\mu}) + f_z}\right)dz - \left(\frac{(\pi_{q_0})^*(\eta_{v_0})}{\omega_{q_0}}\right)(\pi_{q_0})^*(\dot{\mu})d\bar{z} \\ &= \overline{\left(\frac{(\pi_{q_0})^*(\eta_{v_0})}{\omega_{q_0}}\right)(\pi_{q_0})^*(\dot{\mu})}dz - \left(\frac{(\pi_{q_0})^*(\eta_{v_0})}{\omega_{q_0}}\right)(\pi_{q_0})^*(\dot{\mu})d\bar{z} \\ &\quad + \frac{1}{2}df + \frac{i}{2}{}^*df. \end{aligned} \tag{4.7}$$

Let $\sigma(c) = \sigma_c(z)dz + \overline{\sigma_c(z)}d\bar{z}$ be the reproducing harmonic differential on $\tilde{M}_{q_0}$ associated to $c \in H_1(\tilde{M}_{q_0})$, where $\sigma_c(z)dz$ is a holomorphic 1-form on $\tilde{M}_{q_0}$ (we call this holomorphic 1-form the *holomorphic part* of $\sigma(c)$). Namely, the differential $\sigma(c)$ is a unique harmonic differential on $\tilde{M}_{q_0}$ which satisfies

$$\int_c \omega = \int_{\tilde{M}_{q_0}} \omega \wedge {}^*\overline{\sigma(c)}$$

for all closed 1-form $\omega$ on $\tilde{M}_{q_0}$. Then, by applying the orthogonal decomposition theorem for the space of $L^2$-closed forms, we have

$$\begin{aligned}\left.\frac{\partial^2 \chi_{q_\lambda}(c)}{\partial\lambda\partial\bar{\lambda}}\right|_{\lambda=0} &= \int_c \Omega_0 = \int_{\tilde{M}_{q_0}} \Omega_0 \wedge {}^*\overline{\sigma(c)} \\ &= \int_{\tilde{M}_{q_0}} \left(\overline{\frac{(\pi_{q_0})^*(\eta_{v_0})}{\omega_{q_0}}(\pi_{q_0})^*(\dot{\mu})}\,dz - \frac{(\pi_{q_0})^*(\eta_{v_0})}{\omega_{q_0}}(\pi_{q_0})^*(\dot{\mu})\,d\bar{z}\right) \\ &\qquad \wedge {}^*\overline{(\sigma_c(z)dz + \overline{\sigma_c(z)}d\bar{z})} \\ &= \int_{\tilde{M}_{q_0}} \left(\overline{\frac{(\pi_{q_0})^*(\eta_{v_0})}{\omega_{q_0}}(\pi_{q_0})^*(\dot{\mu})}\,dz - \frac{(\pi_{q_0})^*(\eta_{v_0})}{\omega_{q_0}}(\pi_{q_0})^*(\dot{\mu})\,d\bar{z}\right) \\ &\qquad \wedge (-i\overline{\sigma_c}d\bar{z} + i\sigma_c dz) \\ &= -4i\,\mathrm{Im}\int_{\tilde{M}_{q_0}} \frac{(\pi_{q_0})^*(\eta_{v_0})}{\omega_{q_0}}(\pi_{q_0})^*(\dot{\mu})\sigma_c.\end{aligned}$$

Now, we suppose that the homology class $c$ is in $H_1(\tilde{M}_{q_0})^-$. Since $(r_0)_*(c) = -c$, $r_0^*\sigma(c) = -\sigma(c)$. This means that $\sigma(c)$ has zeroes at each branch points of $\pi_{q_0}\colon \tilde{M}_{q_0} \to M_{q_0}$ and the holomorphic quadratic differential

$$\frac{(\pi_{q_0})^*(\eta_{v_0})}{\omega_{q_0}}\sigma_c = \frac{(\pi_{q_0})^*(\eta_{v_0})(z)}{\omega_{q_0}(z)}\sigma_c(z)dz^2$$

descends to a holomorphic quadratic differential on $M_{q_0}$. Therefore, we obtain from the definition of $\eta_{v_0}$ that

$$\begin{aligned}\left.\frac{\partial^2 \chi_{q_\lambda}(c)}{\partial\lambda\partial\bar{\lambda}}\right|_{\lambda=0} &= -4i\,\mathrm{Im}\int_{\tilde{M}_{q_0}} \frac{(\pi_{q_0})^*(\eta_{v_0})}{\omega_{q_0}}\frac{\overline{(\pi_{q_0})^*(\eta_{v_0})}}{|\omega_{q_0}|^2}\sigma_c \\ &= -4i\,\mathrm{Im}\int_{\tilde{M}_{q_0}} \frac{|(\pi_{q_0})^*(\eta_{v_0})|^2}{|\omega_{q_0}|^2}\frac{\sigma_c}{\omega_{q_0}}\end{aligned} \tag{4.8}$$

(cf. Lemma 4.1). This is what we wanted. □

## 5. Levi forms and Plurisubharmonicity of Extremal length functions

**5.1. Levi forms of extremal length functions.** For a real-valued $C^2$-function $U$ on a complex manifold $N$, the *Levi form* of $U$ is an hermitian inner product on the holomorphic tangent bundle $TN$ of $N$ defined as

$$\mathcal{L}(U)[v,\overline{v}] = \sum_{i,j=1}^{n} \frac{\partial^2 U}{\partial z_i \partial \overline{z}_j} v_i \overline{v_j}$$

for $v = \sum_{i=1}^{n} v_i(\partial/\partial z_i) \in TN$. Let $p \in N$ and $v \in T_pN$. Let $F\colon \mathbb{D} \to N$ be a holomorphic mapping with $f(0) = p$ and $f_*(\partial/\partial\lambda \mid_{\lambda=0}) = v$, then

$$\mathcal{L}(U)[v,\overline{v}] = \left.\frac{\partial^2 (U \circ f)}{\partial\lambda\partial\overline{\lambda}}\right|_{\lambda=0}.$$

THEOREM 5.1 (Levi forms). *Let $x_0 = (M_0, f_0) \in \mathcal{T}_{g,m}$ and $v = [\dot{\mu}] \in T_{x_0}\mathcal{T}_{g,m}$. Let $F \in \mathcal{MF}$. Suppose that the Hubbard-Masur differential $q_0 = q_{F,x_0}$ is generic. Then, the extremal length function is real analytic around $x_0$, and the Levi form of the extremal length function $\mathcal{T}_{g,m} \ni x \mapsto \mathrm{Ext}_x(F)$ at $x_0$ satisfies*

$$\mathcal{L}(\mathrm{Ext}_\bullet(F))[v,\overline{v}] = 2\int_{M_0} \frac{|\eta_v|^2}{|q_0|},$$

*where $\eta_v \in \mathcal{Q}_{x_0}$ is the $q_0$ realization of the tangent vector $v$ defined in §4.2.*

REMARK 5.1. *In §6, we will discuss a geometric interpretation of the anti-complex linear map*

$$T_{x_0}\mathcal{T}_{g,m} \ni v \mapsto \eta_v \in \mathcal{Q}_{x_0}.$$

PROOF OF THEOREM 5.1. We continue to use the symbols defined in the previous sections. Let $\tilde{g} = 4g - 3 + m$ ($\geq 2$) be the genus of $\tilde{M}_{q_0}$. Let $\{\alpha_k, \beta_k\}_{k=1}^{\tilde{g}}$ be a symplectic generator of $H_1(\tilde{M}_{q_0})$. Let $x_\lambda = \pi(q_\lambda)$ for instance. By the bilinear relation, we have

$$\begin{aligned}\mathrm{Ext}_{x_\lambda}(F) &= \frac{1}{2}\|\omega_{q_\lambda}\|^2 = \frac{1}{4}\int_{\tilde{M}_{q_\lambda}} \omega_{q_\lambda} \wedge {}^*\overline{\omega_{q_\lambda}} \\ &= \frac{i}{4}\sum_{k=1}^{\tilde{g}} \left(\chi_{q_\lambda}(\alpha_k)\overline{\chi_{q_\lambda}(\beta_k)} - \chi_{q_\lambda}(\beta_k)\overline{\chi_{q_\lambda}(\alpha_k)}\right)\end{aligned}$$

for $\lambda \in D$ (cf. [**7**, Corollary in §III.2.3]). From Lemmas 3.2 and 3.3, $\chi_{q_{F,x}}(\alpha_k)$ and $\chi_{q_{F,x}}(\beta_k)$ are real analytic around $x_0$ for each $k$, and hence so is the extremal length function. From Lemma 4.2, we have

$$\begin{aligned}
&\frac{\partial^2}{\partial\lambda\partial\overline{\lambda}}\left(\chi_{q_\lambda}(c_1)\overline{\chi_{q_\lambda}(c_2)}\right)\Big|_{\lambda=0} \\
&= \frac{\partial^2\chi_{q_\lambda}(c_1)}{\partial\lambda\partial\overline{\lambda}}\Big|_{\lambda=0}\overline{\chi_{q_0}(c_2)} + \chi_{q_0}(c_1)\frac{\partial^2\overline{\chi_{q_\lambda}(c_2)}}{\partial\lambda\partial\overline{\lambda}}\Big|_{\lambda=0} \\
&\quad + \frac{\partial\chi_{q_\lambda}(c_1)}{\partial\lambda}\Big|_{\lambda=0}\frac{\partial\overline{\chi_{q_\lambda}(c_2)}}{\partial\overline{\lambda}}\Big|_{\lambda=0} + \frac{\partial\chi_{q_\lambda}(c_1)}{\partial\overline{\lambda}}\Big|_{\lambda=0}\frac{\partial\overline{\chi_{q_\lambda}(c_2)}}{\partial\lambda}\Big|_{\lambda=0} \\
&= \frac{\partial^2\chi_{q_\lambda}(c_1)}{\partial\lambda\partial\overline{\lambda}}\Big|_{\lambda=0}\overline{\chi_{q_0}(c_2)} + \chi_{q_0}(c_1)\frac{\partial^2\overline{\chi_{q_\lambda}(c_2)}}{\partial\lambda\partial\overline{\lambda}}\Big|_{\lambda=0} \\
&\quad + \frac{\partial\chi_{q_\lambda}(c_1)}{\partial\lambda}\Big|_{\lambda=0}\overline{\frac{\partial\chi_{q_\lambda}(c_2)}{\partial\lambda}\Big|_{\lambda=0}} + \overline{\frac{\partial\chi_{q_\lambda}(c_1)}{\partial\lambda}\Big|_{\lambda=0}}\frac{\partial\chi_{q_\lambda}(c_2)}{\partial\lambda}\Big|_{\lambda=0}
\end{aligned}$$

for $c_1, c_2 \in H_1(\tilde{M}_{q_0})$, since the real part of $\chi_{q_\lambda}(c)$ is constant for all $c \in H_1(\tilde{M}_{q_0})$ by Lemma 3.3. Since the sum of the last two terms of above calculation is a real number, we obtain

$$\begin{aligned}
&\frac{\partial^2}{\partial\lambda\partial\overline{\lambda}}\left(\chi_{q_\lambda}(\alpha_k)\overline{\chi_{q_\lambda}(\beta_k)} - \chi_{q_\lambda}(\beta_k)\overline{\chi_{q_\lambda}(\alpha_k)}\right)\Big|_{\lambda=0} \\
&= \int_{\alpha_k}\Omega_0\overline{\int_{\beta_k}\omega_{q_0}} + \overline{\int_{\beta_k}\Omega_0}\int_{\alpha_k}\omega_{q_0} - \int_{\beta_k}\Omega_0\overline{\int_{\alpha_k}\omega_{q_0}} - \int_{\alpha_k}\overline{\Omega_0}\int_{\beta_k}\omega_{q_0} \\
&= \int_{\alpha_k}\Omega_0\int_{\beta_k}\overline{\omega_{q_0}} - \int_{\beta_k}\Omega_0\int_{\alpha_k}\overline{\omega_{q_0}} + \int_{\beta_k}\overline{\Omega_0}\int_{\alpha_k}\omega_{q_0} - \int_{\alpha_k}\overline{\Omega_0}\int_{\beta_k}\omega_{q_0}
\end{aligned}$$

where $\Omega_0$ is a closed form defined as (4.7). Since $\omega_{q_0}$ is a holomorphic differential, ${}^*\omega_{q_0}$ and ${}^*\overline{\omega_{q_0}}$ are harmonic differentials and they are perpendicular to the exact and co-exact differentials in the $L^2$-inner product of differential forms (cf. [**7**, §II.3]). From (4.7), we obtain

$$\begin{aligned}
\mathcal{L}(\mathrm{Ext}_\bullet(F))[v,\overline{v}] &= \frac{\partial^2}{\partial\lambda\partial\overline{\lambda}}\mathrm{Ext}_{x_\lambda}(F)\Big|_{\lambda=0} \\
&= \frac{i}{4}\sum_{k=1}^{\tilde{g}}\left(\int_{\alpha_k}\Omega_0\int_{\beta_k}\overline{\omega_{q_0}} - \int_{\beta_k}\Omega_0\int_{\alpha_k}\overline{\omega_{q_0}}\right) \\
&\quad + \frac{i}{4}\sum_{k=1}^{\tilde{g}}\left(\int_{\beta_k}\overline{\Omega_0}\int_{\alpha_k}\omega_{q_0} - \int_{\alpha_k}\overline{\Omega_0}\int_{\beta_k}\omega_{q_0}\right) \\
&= \frac{i}{4}\int_{\tilde{M}_{q_0}}\Omega_0\wedge\overline{\omega_{q_0}} - \frac{i}{4}\int_{\tilde{M}_{q_0}}\overline{\Omega_0}\wedge\omega_{q_0} \\
&= \mathrm{Re}\int_{\tilde{M}_{q_0}}\frac{(\pi_{q_0})^*(\eta_v)}{\omega_{q_0}}(\pi_{q_0})^*(\dot{\mu})\omega_{q_0} \\
&= 2\mathrm{Re}\int_{M_{q_0}}\eta_v\dot{\mu} = 2\int_{M_0}\frac{|\eta_v|^2}{|q_0|}
\end{aligned}$$

from the definition of the differential $\eta_v$ (cf. [**7**, Proposition III.2.3]). This implies what we wanted. □

**5.2. Differential inequality.** From Theorem 5.1, we deduce the following.

PROPOSITION 5.1 (Levi forms and Gradients). *Let $x_0 \in \mathcal{T}_{g,m}$ and $F \in \mathcal{MF}$. Suppose that $q_0 = q_{F,x_0}$ is generic.*

$$\mathrm{Ext}_{x_0}(F) \cdot \mathcal{L}(\mathrm{Ext}_\bullet(F))[v, \overline{v}] - 2\left|(\mathrm{Ext}_\bullet(F))_*[v]\right|^2 \geq 0$$

*for $v = [\dot{\mu}] \in T_{x_0}\mathcal{T}_{g,m}$.*

PROOF. Let $x_0 = (M_0, f_0)$. It follows from Gardiner's formula that

$$(\mathrm{Ext}_\bullet(F))_*[v] = -\int_{M_0} \dot{\mu}\, q_0 = -\int_{M_0} \frac{\overline{\eta_v}}{|q_0|} q_0. \tag{5.1}$$

See [**9**, §11.8] (See also §7.1). In comparing Gardiner's original formula with (5.1), we remind that our differential $q_0 = q_{F,x_0}$ has $F$ as the vertical foliation, while Gardiner considers the quadratic differential with horizontal foliation $F$. Hence, we have to put the minus sign in (5.1). From Cauchy-Schwarz inequality, one can see from Theorem 5.1 that

$$\begin{aligned}
&\mathrm{Ext}_{x_0}(F)\mathcal{L}(\mathrm{Ext}_\bullet(F))[v, \overline{v}] - 2\left|(\mathrm{Ext}_\bullet(F))_*[v]\right|^2 \\
&= \mathrm{Ext}_{x_0}(F) \cdot 2\int_{M_0} \frac{|\eta_v|^2}{|q_0|} - 2\left|-\int_{M_0} \frac{\overline{\eta_v}}{|q_0|} q_0\right|^2 \\
&= 2\left(\int_{M_0} |q_0| \cdot \int_{M_0} \frac{|\eta_v|^2}{|q_0|} - \left|\int_{M_0} \frac{\overline{\eta_v}}{|q_0|^{1/2}} \frac{q_0}{|q_0|^{1/2}}\right|^2\right) \geq 0,
\end{aligned}$$

which implies what we wanted. □

THEOREM 5.2 (Differential inequality). *For any measured foliation $F$, the extremal length function satisfies the differential inequality* (1.1) *holds in the sense of currents. Especially, the log-extremal length function $\mathcal{T}_{g,m} \ni x \mapsto \log \mathrm{Ext}_x(F)$ is a real analytic strictly plurisubharmonic function, when $F$ is essentially complete.*

PROOF. By approximation, we may assume that $F$ is essentially complete (cf. [**17**, §3.3]). Then, the inequality follows from Proposition 5.1. □

**5.3. Plurisubharmonicity for extremal length functions.** A function $U$ on a complex manifold $N$ is said to be *plurisubharmonic* if $U$ is upper semi-continuous, and $U \circ f$ is subharmonic for all holomorphic mapping $f\colon \mathbb{D} \to N$. A $C^2$-function on $N$ is plurisubharmonic if the Levi form is positive semi-definite. We call a $C^2$-function *strictly plurisubharmonic* if the Levi form is positive-definite. When a function $U$ is plurisubharmonic, so is $g \circ U$ for any increasing convex function $g$ on $\mathbb{R}$ such that the limit $\lim_{t\to-\infty} g(t)$ exists.

In the proof of the following theorem, we remind that if $U(z)$ is a real-valued positive $C^2$-function on a domain on $\mathbb{C}$, $\log U$ and $-1/U$ satisfy

$$(\log U)_{\lambda\overline{\lambda}} = \frac{U \cdot U_{\lambda\overline{\lambda}} - |U_\lambda|^2}{U^2} \tag{5.2}$$

$$\left(-\frac{1}{U}\right)_{\lambda\overline{\lambda}} = \frac{U \cdot U_{\lambda\overline{\lambda}} - 2|U_\lambda|^2}{U^3}. \tag{5.3}$$

Especially, for a positive $C^2$-function $U$ on a complex manifold, if $-1/U$ is plurisubharminic, $\log U$ and $U$ are also plurisubharmonic. From (5.2) and (5.3), Theorem 1.1 follows from the following theorem.

THEOREM 5.3 (Plurisuperharmonicity of reciprocals). *Let $F_1, F_2, \cdots, F_n$ be measured foliations. Let $a_k, c \geq 0$ $(k = 1, \cdots, n)$ with $\sum_{k=1}^{n} a_k > 0$. If $n \geq 2$, $a_k, c > 0$ $(k = 1, \cdots, n)$ and some two of them are filling in the sense of* (1.2)*, the function*

$$\rho(x) = -\frac{1}{c + \sum_{k=1}^{n} a_k \mathrm{Ext}_x(F_k)}$$

*is a negative, plurisubharmonic exhaustion function of $\mathcal{T}_{g,m}$ with lower bound. When $n \geq 1$ or $c \geq 0$, $\rho$ is a negative, plurisubharmonic function on $\mathcal{T}_{g,m}$.*

PROOF. It suffices to show that any function in the cone spanned by a finite number of extremal length functions of essential complete measured foliations also satisfies the differential inequality (1.1). Indeed, by applying the continuity of the function (2.2), the general case follows from a standard approximating procedure (cf. [**17**, Theorem 2.6.1]). From an inductive argument, it suffices to show the case $n = 2$. We may also assume that $F_1$ and $F_2$ are filling in the sense of (1.2). Under the assumption, each extremal length function $\mathcal{T}_{g,m} \ni x \mapsto \mathrm{Ext}_x(F_k)$ is real analytic.

Let $x = (M, f) \in \mathcal{T}_{g,m}$ and $v \in T_{x_0}\mathcal{T}_{g,m}$. Let $h\colon \mathbb{D} \to \mathcal{T}_{g,m}$ be a holomorphic mapping with $h(0) = x_0$ and $h_*((\partial/\partial\lambda)\mid_{\lambda=0}) = v$. Set $E^1(\lambda) = \mathrm{Ext}_{h(\lambda)}(F_1)$, $E^2(\lambda) = \mathrm{Ext}_{h(\lambda)}(F_2)$ and $E = a_1E^1 + a_2E^2$ for simplicity.

Then, the Levi form of $\rho$ is

$$\mathcal{L}(\rho)[v, \overline{v}] = \frac{c\, E_{\lambda\overline{\lambda}}\big|_{\lambda=0}}{(c + E(0))^3} + \frac{E(0)\, E_{\lambda\overline{\lambda}}\big|_{\lambda=0} - 2|\, E_{\lambda}|_{\lambda=0}\,|^2}{(c + E(0))^3}.$$

Therefore, we conclude from Proposition 5.1 that

$$\begin{aligned}
&E(0)\, E_{\lambda\overline{\lambda}}\big|_{\lambda=0} - 2|\, E_{\lambda}|_{\lambda=0}\,|^2 \\
&= a_1^2 \left(E^1(0)\, E^1_{\lambda\overline{\lambda}}\big|_{\lambda=0} - 2|\, E^1_{\lambda}\big|_{\lambda=0}\,|^2\right) + a_2^2 \left(E^2(0)\, E^2_{\lambda\overline{\lambda}}\big|_{\lambda=0} - 2|\, E^2_{\lambda}\big|_{\lambda=0}\,|^2\right) \\
&\quad + a_1a_2 \left(E^1(0)\, E^2_{\lambda\overline{\lambda}}\big|_{\lambda=0} + E^2(0)\, E^1_{\lambda\overline{\lambda}}\big|_{\lambda=0} - 4\mathrm{Re}\left(\overline{E^1_{\lambda}\big|_{\lambda=0}}\, E^2_{\lambda}\big|_{\lambda=0}\right)\right) \\
&\geq a_1a_2 \left(E^1(0)\, E^2_{\lambda\overline{\lambda}}\big|_{\lambda=0} + E^2(0)\, E^1_{\lambda\overline{\lambda}}\big|_{\lambda=0} - 4\left|E^1_{\lambda}\big|_{\lambda=0}\right|\left|E^2_{\lambda}\big|_{\lambda=0}\right|\right) \\
&\geq a_1a_2 \Big(E^1(0)\, E^2_{\lambda\overline{\lambda}}\big|_{\lambda=0} + E^2(0)\, E^1_{\lambda\overline{\lambda}}\big|_{\lambda=0} \\
&\qquad -2E^1(0)^{1/2}\left(E^2_{\lambda\overline{\lambda}}\big|_{\lambda=0}\right)^{1/2} E^1(0)^{1/2}\left(E^2_{\lambda\overline{\lambda}}\big|_{\lambda=0}\right)^{1/2}\Big) \\
&= a_1a_2 \left(E^1(0)^{1/2}\left(E^2_{\lambda\overline{\lambda}}\big|_{\lambda=0}\right)^{1/2} - E^1(0)^{1/2}\left(E^2_{\lambda\overline{\lambda}}\big|_{\lambda=0}\right)^{1/2}\right)^2 \geq 0
\end{aligned}$$

and hence

$$\mathcal{L}(\rho)[v, \overline{v}] \geq \frac{c\, E_{\lambda\overline{\lambda}}\big|_{\lambda=0}}{(c + E(0))^3} = c\frac{a_1^2\mathcal{L}(\mathrm{Ext}_{\bullet}(F_1))[v, \overline{v}] + a_2^2\mathcal{L}(\mathrm{Ext}_{\bullet}(F_2))[v, \overline{v}]}{(c + a_1\mathrm{Ext}_{x_0}(F_1) + a_2\mathrm{Ext}_{x_0}(F_2))^3}$$

when $a + b > 0$, $c > 0$. Indeed, since $q_{F,x}$ is generic, the anti-complex linear map

$$T_{x_0}\mathcal{T}_{g,m} \ni v \mapsto \eta_v \in \mathcal{Q}_{x_0}$$

is an isomorphism. Therefore, $\rho$ is (real analytic) strictly plurisubharmonic on $\mathcal{T}_{g,m}$. From the definition, $\rho$ satisfies that $\rho(x) < 0$ for $x \in \mathcal{T}_{g,m}$. Since $F_1$ and $F_2$ are filling, for any $\epsilon > 0$, there is a compact set $K \subset \mathcal{T}_{g,m}$ such that $\rho > -\epsilon$ for all

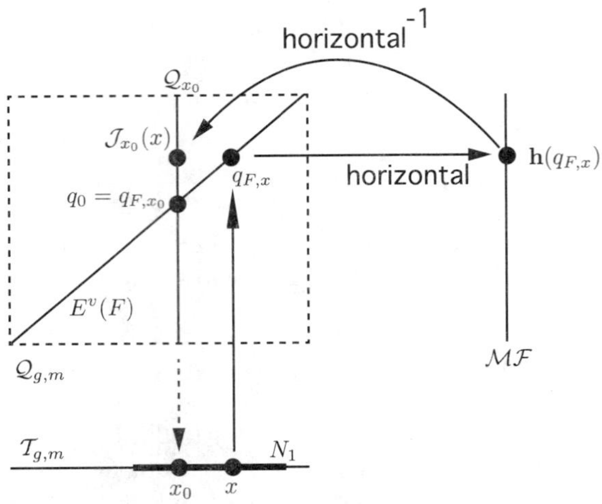


FIGURE 2. A schematic of the map $\mathcal{J}_{x_0}$.

$x \in \mathcal{T}_{g,m} - K$ (cf. §2.4). Since $\rho(x) > -1/a$ for all $x \in \mathcal{T}_{g,m}$, the function $\rho(x)$ satisfies the desired condition. □

## 6. Geometric interpretation of realizations of tangent vectors

**6.1. Motivation.** Let $x_0 = (M_0, f_0) \in \mathcal{T}_{g,m}$. Suppose $q_0 = q_{F,x_0}$ is generic. Let $v = [\dot{\mu}] \in T_{x_0}\mathcal{T}_{g,m}$. In §4.2, we defined the $q_0$ realization $\eta_v \in \mathcal{Q}_{g,m}$ of $v$ as a holomorphic quadratic differential which satisfies

$$\langle v, \psi \rangle = \int_{M_0} \frac{\overline{\eta_v}}{|q_0|}\psi \tag{6.1}$$

for all $\psi \in \mathcal{Q}_{x_0}$. The definition of $\eta_v$ implies as if the infinitesimal Beltrami differential $\dot{\mu}$ were infinitesimally equivalent to the "infinitesimal deformation" along $\overline{\eta_v}/|q_0|$ on $M_{q_0}$ (cf. [**14**, §7.2]). However, $\overline{\eta_v}/|q_0|$ is not a Beltrami differential in general, becauce it may have $1/|z|$-singularities at zeroes of $q_0$ and hence it could happen $\|\overline{\eta_v}/|q_0|\|_\infty = \infty$. Therefore, the above discussion does not interpret correctly the geometry of the anti-complex linear map

$$T_{x_0}\mathcal{T}_{g,m} \ni v \mapsto \eta_v \in \mathcal{Q}_{x_0} \tag{6.2}$$

in general. In this section, we shall try to give a geometric description of the correspondence (6.2).

**6.2. Geometric interpretation.** From [**5**, Theorem 5.8] and [**10**], the map

$$\mathcal{Q}_{g,m} \ni q \mapsto (\mathbf{h}(q), \mathbf{v}(q)) \in \mathcal{MF} \times \mathcal{MF} \tag{6.3}$$

is a real analytic diffeomorphism around $q_0$. Hence, the map

$$\mathcal{J}\colon \mathcal{T}_{g,m} \times \mathcal{T}_{g,m} \ni (x, y) \mapsto -q_{\mathbf{h}(q_{F,x}),y} \in \mathcal{Q}_{g,m} \tag{6.4}$$

is also a real analytic diffeomorphism on a neighborhood $N_1 \times N_1$ of $(x_0, x_0)$ onto the image, where $N_1$ is a neighborhood of $x_0 = \pi(q_0) \in \mathcal{T}_{g,m}$. By definition, $\mathbf{h}(\mathcal{J}(x,y)) = \mathbf{h}(q_{F,x})$, $\mathcal{J}(x,y) \in \mathcal{Q}_y$ and $\mathcal{J}(x,x) = q_{F,x}$ for all $x, y \in N_1$. Let $\mathcal{J}_y(x) = \mathcal{J}(x,y)$ for $(x,y) \in N_1 \times N_1$ (Figure 2).

THEOREM 6.1 (Geometric interpretation). *For $v \in T_{x_0}\mathcal{T}_{g,m}$, we have*

$$(\mathcal{J}_{x_0})_*[v] = -4\eta_v.$$

PROOF. Consider the triangulation $\Delta_q$ and the train track $\tau_{q_0}$ on $M_q$ adapted to the horizontal foliation $|\mathrm{Im}(\sqrt{q})|$ of $q$ (cf. [**5**, Theorem 5.8]). The vertex set consists of the singular points $q$, and edges are saddle connections. We may assume that $\Delta_q$ contains no horizontal edge (cf. [**5**, Lemma 5.7]). Consider the lifts $\tilde{\Delta}_q$ and $\tilde{\tau}_q$ of $\Delta_q$ and $\tau_q$ to $\tilde{M}_q$, respectivey. We orient each edge of $\tilde{\Delta}_q$ so that $\mathrm{Im}(\omega_q)$ is positive along the edge. Each branch $b$ of $\tilde{\tau}_q$ intersects a unique edge $e_b$ of $\tilde{\Delta}_q$. We orient $b$ so that $b \cdot e_b = +1$, where $\cdot$ means the algebraic intersection number. By taking $U$ sufficiently small, we may also assume that the marking $\tilde{M}_{q_0} \to \tilde{M}_q$ induces isomorphisms from $\tilde{\Delta}_{q_0}$ and $\tilde{\tau}_{q_0}$ to $\tilde{\Delta}_q$ and $\tilde{\tau}_q$, respectively.

From [**5**, Theorem 5.8], the tangential map of the map $\mathbf{h}\colon \mathcal{Q}_{x_0} \to \mathcal{MF}$ of the horizontal foliations satisfies

$$\mathrm{Im}\int_{M_{q_0}} \frac{\psi_1\overline{\psi_2}}{4|q_0|} = \omega_{Th}\left(\mathbf{h}_*(\psi_1)\,,\,\mathbf{h}_*(\psi_2)\right)_{\tau_{q_0}} \tag{6.5}$$

for $\psi_1, \psi_2 \in T_{q_0}\mathcal{Q}_{x_0} \cong \mathcal{Q}_{x_0}$, where $\omega_{Th}\left(\cdot\,,\,\cdot\right)_{\tau_{q_0}}$ is the Thurston's symplectic form on $W(\tau_{q_0})$, where $W(\tau_{q_0})$ is the space of real valued functions on the branches of $\tau_{q_0}$ satisfying the switch condition. It is known that the space $W(\tau_{q_0})$ is identified with the tangent space of $\mathcal{MF}$ at $\mathbf{h}(q_0)$ since $\tau_{q_0}$ is maximal (cf. [**30**, §3.2]).

From the definition, $|\mathrm{Im}(\omega_{q_{F,x}})|$ is the lift of the horizontal foliation of $q_{F,x}$ to $\tilde{M}_{q_{F,x}}$. By the definition of the orientation of edges of $\tilde{\Delta}_{q_0}$, the image of the map $N_1 \ni x \mapsto \mathbf{h}(q_{F,x}) \in \mathcal{MF}$ at $x$ is identified with $w_x \in W(\tilde{\tau}_{q_0})$ which is defined by

$$w_x(b) = \int_b |\mathrm{Im}(\omega_{q_{F,x}})| = \int_b \mathrm{Im}(\omega_{q_{F,x}}).$$

From Lemmas 3.1 and 4.1, the image of $v \in T_{x_0}\mathcal{T}_{g,m}$ under the differential of the map $N_1 \ni x \mapsto \mathrm{Im}(\chi_{q_{F,x}})$ is identified with the homomorphism

$$V_v\colon H_1(\tilde{M}_{q_0}) \ni c \mapsto -2\mathrm{Im}\int_c \frac{(\pi_{q_0})^*(\eta_v)}{\omega_{q_0}} \in \mathbb{R}. \tag{6.6}$$

Indeed, when $\{q_\lambda\}_\lambda \in D$ is taken as Lemma 4.1, (6.6) is derived from

$$\begin{aligned}\frac{\partial \mathrm{Im}(\chi_{q_\lambda}(c))}{\partial t} &= -i\left(\frac{\partial \chi_{q_\lambda}(c)}{\partial \lambda} + \frac{\partial \chi_{q_\lambda}(c)}{\partial \overline{\lambda}}\right) \\ &= -i\int_c \overline{\left(\frac{(\pi_{q_\lambda})^*(\eta_\lambda)}{\omega_{q_\lambda}}\right)} - \frac{(\pi_{q_\lambda})^*(\eta_\lambda)}{\omega_{q_\lambda}} \\ &= -2\mathrm{Im}\int_c \frac{(\pi_{q_\lambda})^*(\eta_\lambda)}{\omega_{q_\lambda}}\end{aligned}$$

where $\lambda = t + is$, since the real part of $\chi_{q_\lambda}$ is constant on $D$.

Let $p\colon \tilde{\tau}_{q_0} \to \tau_{q_0}$ be the induced covering. For a weight $w \in W(\tau_{q_0})$, we define $C_w \in H_1(\tilde{M}_{q_0}, \mathbb{R})^-$ as the homology class of a cycle

$$\sum_{b:\text{branches of } \tilde{\tau}_{q_0}} w(p(b))b.$$

Then,

$$w_x(C_w) = \sum_b w(p(b))w_x(b)$$
$$V_v(C_w) = -2\mathrm{Im}\int_{C_w}\frac{(\pi_{q_0})^*(\eta_v)}{\omega_{q_0}} = -2\sum_b w(p(b))\mathrm{Im}\int_b \frac{(\pi_{q_0})^*(\eta_v)}{\omega_{q_0}}.$$

These equations imply that the differential of the map $N_1 \ni x \mapsto w_x \in W(\tilde{\tau}_{q_0})$ at $x = x_0$ is equal to the weight (which we abbreviate as)

$$V_v(b) = -2\,\mathrm{Im}\int_b \frac{(\pi_{q_0})^*(\eta_v)}{\omega_{q_0}}.$$

The weight $V_v$ is invariant under the action of covering transformation $r_0$ of $\tilde{M}_{q_0} \to M_{q_0}$ on $\tilde{\tau}_{q_0}$, because for any branch $b$ of $\tilde{\tau}_{q_0}$, the orientation of $r_0(b)$ is opposite to the orientation induced from $b$. Thus, $V_v$ descends to a weight $p_*(V_v)$ on $\tau_{q_0}$.

Let $\psi_0 = (\mathcal{J}_{x_0})_*[v]$. For all $\psi \in \mathcal{Q}_{x_0} \subset T_{q_0}\mathcal{Q}_{g,m}$, we define $w(\psi) \in W(\tilde{\tau}_{q_0})$ by

$$w(\psi)(b) = \mathrm{Im}\int_b \frac{(\pi_{q_0})^*(\psi)}{2\omega_{q_0}}.$$

Then, we have $\mathbf{h}_*[\psi] = p_*(w(\psi))$ (cf. [**5**, Proof of Theorem 5.8]). By the definition of $\mathcal{J}_{x_0}$, $\mathbf{h}\circ\mathcal{J}_{x_0}(x) = \mathbf{h}(q_{F,x})$ for all $x \in N_1$. Hence, $\mathbf{h}_*[\psi_0] = p_*(V_v)$ as elements of $W(\tilde{\tau}_{q_0}) \cong T_{\mathbf{h}(q_0)}\mathcal{MF}$. From (6.5) and [**5**, (5.4)], the following holds for all $\psi \in \mathcal{Q}_{x_0}$:

$$\begin{aligned}
\mathrm{Im}\int_{M_{q_0}} \frac{\psi_0\overline{\psi}}{4|q_0|} &= \frac{1}{2}\mathrm{Im}\int_{\tilde{M}_{q_0}} \frac{(\pi_{q_0})^*(\psi_0)\overline{(\pi_{q_0})^*(\psi)}}{4|\omega_{q_0}|^2} \\
&= \frac{1}{2}\omega_{Th}\left(V_v\,,\,w(\psi)\right)_{\tilde{\tau}_{q_0}} \\
&= -2\int_{\tilde{M}_{q_0}} \mathrm{Im}\frac{(\pi_{q_0})^*(\eta_v)}{2\omega_{q_0}} \wedge \mathrm{Im}\frac{(\pi_{q_0})^*(\psi)}{2\omega_{q_0}} \\
&= -\mathrm{Re}\int_{\tilde{M}_{q_0}} \frac{(\pi_{q_0})^*(\eta_v)}{2\omega_{q_0}} \wedge \overline{\frac{(\pi_{q_0})^*(\psi)}{2\omega_{q_0}}} \\
&= -2\mathrm{Im}\int_{\tilde{M}_{q_0}} \frac{(\pi_{q_0})^*(\eta_v)\overline{(\pi_{q_0})^*(\psi)}}{4|\omega_{q_0}|^2} = -\mathrm{Im}\int_{M_{q_0}} \frac{\eta_v\overline{\psi}}{|q_0|}.
\end{aligned}$$

Therefore, we have $(\mathcal{J}_{x_0})_*[v] = \psi_0 = -4\eta_v$. □

## 7. Appendix

**7.1. Gardiner's formula by double covers.** Applying our method in Theorem 5.1, one can also get an alternative proof of Gardiner's formula for the derivative of extremal length in some case. Indeed, this is the case we need in the proof of Theorem 1.1.

Let $F \in \mathcal{MF}$ and $x_0 = (M, f) \in \mathcal{T}_{g,m}$. Suppose that the Hubbard-Masur differential $q_0 = q_{F,x_0}$ is generic. Let $v = [\dot{\mu}]$ and $\{q_\lambda\}_{\lambda\in D}$ as the proof of Theorem 5.1. Suppose $x_\lambda = \pi(q_\lambda)$ varies holomorphically. By applying the discussion in the proof of Theorem 5.1, from Lemma 4.1, we have

$$
\begin{aligned}
&\frac{\partial}{\partial\lambda}\mathrm{Ext}_{x_\lambda}(F)\Big|_{\lambda=0}\\
&=\frac{i}{4}\sum_{k=1}^{\tilde{g}}\left(\int_{\alpha_k}\overline{\left(\frac{(\pi_{q_0})^*(\eta_v)}{\omega_{q_0}}\right)}\int_{\beta_k}\overline{\omega_{q_0}}-\int_{\beta_k}\overline{\left(\frac{(\pi_{q_0})^*(\eta_v)}{\omega_{q_0}}\right)}\int_{\alpha_k}\overline{\omega_{q_0}}\right)\\
&\quad+\frac{i}{4}\sum_{k=1}^{\tilde{g}}\left(\int_{\alpha_k}\omega_{q_0}\int_{\beta_k}\overline{\left(-\frac{(\pi_{q_0})^*(\eta_v)}{\omega_{q_0}}\right)}-\int_{\beta_k}\omega_{q_0}\int_{\alpha_k}\overline{\left(-\frac{(\pi_{q_0})^*(\eta_v)}{\omega_{q_0}}\right)}\right)\\
&=\frac{i}{4}\int_{\tilde{M}_{q_0}}\overline{\left(\frac{(\pi_{q_0})^*(\eta_v)}{\omega_{q_0}}\right)}\wedge\overline{\omega_{q_0}}-\frac{i}{4}\int_{\tilde{M}_{q_0}}\omega_{q_0}\wedge\overline{\left(\frac{(\pi_{q_0})^*(\eta_v)}{\omega_{q_0}}\right)}\\
&=-\frac{1}{2}\int_{\tilde{M}_{q_0}}\frac{\overline{(\pi_{q_0})^*(\eta_v)}}{|\omega_{q_0}|^2}\omega_{q_0}^2=-\int_{M_{q_0}}\frac{\overline{\eta_v}}{|q_0|}q_0=-\int_M\dot{\mu}\,q_{F,x_0}
\end{aligned}
$$

because $M=M_{q_0}$ by definition.

**7.2. Examples.** In this section, we treat the exceptional cases in the proofs of pluriharmonicity in the previous sections.

7.2.1. *Levi forms for the case of flat tori.* We start to consider the Levi forms of extremal length functions on the Teichmüller space of flat tori, or once punctured tori, and check that our formula is also valid in this case. Though this case is excluded from our assumption in the theorem, this example assists in understanding our results.

The Teichmüller space of a flat torus is identified with the upper half plane $\mathbb{H}$. A point $\tau\in\mathbb{H}$ corresponds to a marked Riemann surface $M_\tau$ which is defined as the quotient space of $\mathbb{C}$ by a lattice generated by $z+1$ and $z+\tau$. It is known that the space $\mathcal{PMF}=\mathcal{PMF}(M_i)$ of projective measured foliations on $M_i$ ($i=\sqrt{-1}\in\mathbb{H}$) is identified with the projective space $\mathbb{RP}^1=\{[x\colon y]\mid(x,y)\neq(0,0)\}$. A point $[x\colon y]\in\mathbb{RP}^1$ corresponds to a measured foliation whose leaves are parallel to the line of direction $x+yi$.

Let $\alpha$ be a closed curve on $M_i$ corresponding to $[1\colon 0]\in\mathbb{RP}^1$. Then, the Jenkins-Strebel differential $q_{\alpha,\tau}$ on $M_\tau$ for $\alpha$ and the extremal length function of $\alpha$ at $M_\tau$ are obtained as

$$
\begin{aligned}
q_{\alpha,\tau}&=-\frac{1}{\mathrm{Im}(\tau)^2}dz^2\\
\mathrm{Ext}_\tau(\alpha)&=\frac{1}{\mathrm{Im}(\tau)}.
\end{aligned}
$$

Fix $\tau_0\in\mathbb{H}$. For $V\in\mathbb{C}$, let $v$ be the tangent vector at $\tau_0$ corresponding to the infinitesimal quasiconformal deformation from $M_{\tau_0}$ to $M_{\tau_0+\lambda V}$ as $\lambda\to 0$. The Levi form (Laplacian) of the extremal length function is

$$
\mathcal{L}(\mathrm{Ext}_\bullet(\alpha))[v,\overline{v}]=\frac{\partial^2}{\partial\lambda\partial\overline{\lambda}}\mathrm{Ext}_{\tau_0+\lambda V}(\lambda)\Big|_{\lambda=0}=\frac{|V|^2}{2\mathrm{Im}(\tau_0)^3}\tag{7.1}
$$

The Beltrami differential $\mu(\lambda)$ of the quasiconformal deformation from $M_{\tau_0}$ to $M_{\tau_0+\lambda V}$ behaves

$$
\mu(\lambda)=\lambda\cdot\dot{\mu}+o(|\lambda|)=\lambda\cdot\frac{iV}{2\mathrm{Im}(\tau_0)}\frac{d\overline{z}}{dz}+o(|\lambda|)
$$

as $\lambda \to 0$. By definition, $v = [\dot{\mu}]$. We define a holomorphic quadratic differential $\eta_v$ on $M_{\tau_0}$ by

$$\eta_v = -\frac{i\overline{V}}{2\mathrm{Im}(\tau_0)^3}dz^2. \tag{7.2}$$

This differential satisfies that

$$\langle v, \psi\rangle = \int_{M_{\tau_0}} \dot{\mu}\psi = \int_{M_{\tau_0}} \frac{\overline{\eta_v}}{|q_{\alpha,\tau_0}|}\psi$$

for all holomorphic quadratic differential $\psi$ on $M_{\tau_0}$. Therefore, the right-hand side of the formula in Theorem 5.1 is equal to

$$2\int_{M_{\tau_0}} \frac{|\eta_v|^2}{|q_{\alpha,\tau_0}|} = \frac{|V|^2}{2\mathrm{Im}(\tau_0)^3},$$

which coincides with the right-hand side of (7.1).

Let $F$ be a measured foliation corresponding to $[a\colon b] \in \mathbb{RP}^1$ with $b \neq 0$. We normalize $F$ with $i(\alpha, F) = 1$. Then the Hubbard-Masur differential $q_{F,\tau}$ on $M_\tau$ and the extremal length function for $F$ are

$$q_{F,\tau} = -\frac{(a+b\overline{\tau})^2}{b^2\mathrm{Im}(\tau)^2}dz^2$$

$$\mathrm{Ext}_\tau(F) = \frac{|a+b\tau|^2}{b^2\mathrm{Im}(\tau)}.$$

Let $v$ be the tangent vector at $\tau_0$ corresponding to $\dot{\mu} = \frac{iV}{2\mathrm{Im}(\tau_0)}\frac{d\overline{z}}{dz}$. Then,

$$\mathcal{L}(\mathrm{Ext}_\bullet(\alpha))[v,\overline{v}] = \frac{\partial^2}{\partial\lambda\partial\overline{\lambda}}\mathrm{Ext}_{\tau_0+\lambda V}(\lambda)\Big|_{\lambda=0} = \frac{|a+b\tau_0|^2}{2b^2\mathrm{Im}(\tau_0)^3}|V|^2 \tag{7.3}$$

In this case, the differential $\eta_v \in \mathcal{Q}_{\tau_0}$ is

$$\eta_v = -\frac{i|a+b\tau_0|^2\overline{V}}{2b^2\mathrm{Im}(\tau_0)^3}dz^2. \tag{7.4}$$

and hence

$$2\int_{M_{\tau_0}} \frac{|\eta_v|^2}{|q_{F,\tau_0}|} = \frac{|a+b\tau_0|^2}{2b^2\mathrm{Im}(\tau_0)^3}|V|^2,$$

which coincides with the right-hand side of (7.3).

7.2.2. *The geometric interpretation of the realization in the case of tori.* We check Theorem 6.1 in the case of flat tori, or once punctured tori. We use the notation defined in §7.2.1 frequently.

First we consider the case $\alpha = [1\colon 0]$. Fix $\tau_0 \in \mathbb{H}$. Since the projective class of the horizontal foliation of $q_{\alpha,\tau}$ corresponds to a projective class $[-\mathrm{Re}(\tau)\colon 1] \in \mathbb{RP}^1$, the underlying foliation of the horizontal foliation $J_{\tau_0}(\tau)$ is foliated by the lines (leaves) directed to $-(\mathrm{Re}(\tau))+\tau_0$. Since the horizontal foliations of $J_{\tau_0}(\tau)$ and $q_{\alpha,\tau}$ are the same,

$$\int_\alpha |\mathrm{Im}\sqrt{J_{\tau_0}(\tau)}| = i(\mathbf{h}(J_{\tau_0}(\tau)),\alpha) = i(\mathbf{h}(q_{\alpha,\tau}),\alpha) = \int_\alpha |\mathrm{Im}\sqrt{q_{\alpha,\tau}}|$$

holds. Therefore, we can check that

$$J_{\tau_0}(\tau) = \left(\frac{-\mathrm{Re}(\tau)+\overline{\tau_0}}{\mathrm{Im}(\tau)\mathrm{Im}(\tau_0)}\right)^2 dz^2 \in \mathcal{Q}_{\tau_0}$$

for $\tau \in \mathbb{H}$. Let $v = [\dot{\mu}]$ be a tangent vector at $\tau_0$, where $\dot{\mu} = \frac{iV}{2\mathrm{Im}(\tau_0)}\frac{d\bar{z}}{dz}$ with $V \in \mathbb{C}$ as §7.2.1. Then, from the above calculation, we obtain

$$(J_{\tau_0})_*[v] = \left(2\frac{i\overline{V}}{\mathrm{Im}(\tau_0)^3}\right) dz^2 \in \mathcal{Q}_{\tau_0}.$$

Comparing with (7.2), we get $(J_{\tau_0})_*[v] = -4\eta_v$ as in Theorem 6.1.

Let $F \in \mathcal{PMF}$ be a measured foliation corresponding to $[a\colon b] \in \mathbb{RP}^1$ with $b \neq 0$ and $i(\alpha, F) = 1$. Then, the underlying foliation of the horizontal foliation of $q_{F,\tau}$ is foliated by the lines of direction $[a\mathrm{Re}(\tau) + b|\tau|^2 \colon a + b\mathrm{Re}(\tau)]$. By a similar argument as above, one can check that

$$J_{\tau_0}(\tau) = \left(\frac{-(a\mathrm{Re}(\tau) + b|\tau|^2) + (a + b\mathrm{Re}(\tau))\overline{\tau_0}}{b\mathrm{Im}(\tau)\mathrm{Im}(\tau_0)}\right)^2 dz^2$$

$$(J_{\tau_0})_*[v] = \left(2\frac{i|a + b\tau_0|^2\overline{V}}{b^2\mathrm{Im}(\tau_0)^3}\right) dz^2 \in \mathcal{Q}_{\tau_0}.$$

From (7.4), we also have $(J_{\tau_0})_*[v] = -4\eta_v$.

7.2.3. *A comment on the case of* $3g - 3 + m = 1$. The case of once punctured tori is treated in the same way as that in the case of flat tori. Indeed, the canonical completion at the puncture gives an isomorphism between the Teichmüller space of once punctured tori and that of flat tori. The extremal length functions on both spaces are identified under the isomorphism. The case of four puncture sphere is treated in a similar way.

## References


[1] G. Berg, *Hyperconvexity and the Carathéodory metric*, Arch. Math. (Basel) **32** (1979), no. 2, 189–191, DOI 10.1007/BF01238489. MR534932

[2] L. Bers and L. Ehrenpreis, *Holomorphic convexity of Teichmüller spaces*, Bull. Amer. Math. Soc. **70** (1964), 761–764, DOI 10.1090/S0002-9904-1964-11230-1. MR0168800

[3] F. Bonahon, *Bouts des variétés hyperboliques de dimension* 3 (French), Ann. of Math. (2) **124** (1986), no. 1, 71–158, DOI 10.2307/1971388. MR847953

[4] A. Douady and J. Hubbard, *On the density of Strebel differentials*, Invent. Math. **30** (1975), no. 2, 175–179, DOI 10.1007/BF01425507. MR0396936

[5] D. Dumas, *Skinning maps are finite-to-one*, Acta Math. **215** (2015), no. 1, 55–126, DOI 10.1007/s11511-015-0129-6. MR3413977

[6] C. J. Earle, *On the Carathéodory metric in Teichmüller spaces*, Discontinuous groups and Riemann surfaces (Proc. Conf., Univ. Maryland, College Park, Md., 1973), Princeton Univ. Press, Princeton, N.J., 1974, pp. 99–103. Ann. of Math. Studies, No. 79. MR0352450

[7] H. M. Farkas and I. Kra, *Riemann surfaces*, 2nd ed., Graduate Texts in Mathematics, vol. 71, Springer-Verlag, New York, 1992. MR1139765

[8] A. Fathi, F. Laudenbach, and V. Poénaru, *Travaux de Thurston sur les surfaces* (French), Astérisque, vol. 66, Société Mathématique de France, Paris, 1979. Séminaire Orsay; With an English summary. MR568308

[9] F. P. Gardiner, *Teichmüller theory and quadratic differentials*, Pure and Applied Mathematics (New York), John Wiley & Sons, Inc., New York, 1987. A Wiley-Interscience Publication. MR903027

[10] F. P. Gardiner and H. Masur, *Extremal length geometry of Teichmüller space*, Complex Variables Theory Appl. **16** (1991), no. 2-3, 209–237. MR1099913

[11] R. Godement, *Topologie algébrique et théorie des faisceaux* (French), Actualit'es Sci. Ind. No. 1252. Publ. Math. Univ. Strasbourg. No. 13, Hermann, Paris, 1958. MR0102797

[12] P. Griffiths and J. Harris, *Principles of algebraic geometry*, Wiley-Interscience [John Wiley & Sons], New York, 1978. Pure and Applied Mathematics. MR507725

[13] J. Hubbard and H. Masur, *Quadratic differentials and foliations*, Acta Math. **142** (1979), no. 3-4, 221–274, DOI 10.1007/BF02395062. MR523212

[14] Y. Imayoshi and M. Taniguchi, *An introduction to Teichmüller spaces*, Springer-Verlag, Tokyo, 1992. Translated and revised from the Japanese by the authors. MR1215481
[15] Y. Kawamata, *On deformations of compactifiable complex manifolds*, Math. Ann. **235** (1978), no. 3, 247–265, DOI 10.1007/BF01420124. MR499279
[16] S. P. Kerckhoff, *The asymptotic geometry of Teichmüller space*, Topology **19** (1980), no. 1, 23–41, DOI 10.1016/0040-9383(80)90029-4. MR559474
[17] M. Klimek, *Pluripotential theory*, London Mathematical Society Monographs. New Series, vol. 6, The Clarendon Press, Oxford University Press, New York, 1991. Oxford Science Publications. MR1150978
[18] K. Kodaira, *Complex manifolds and deformation of complex structures*, Grundlehren der Mathematischen Wissenschaften [Fundamental Principles of Mathematical Sciences], vol. 283, Springer-Verlag, New York, 1986. Translated from the Japanese by Kazuo Akao; With an appendix by Daisuke Fujiwara. MR815922
[19] S. L. Krushkal′, *Strengthening pseudoconvexity of finite-dimensional Teichmüller spaces*, Math. Ann. **290** (1991), no. 4, 681–687, DOI 10.1007/BF01459267. MR1119946
[20] S. L. Krushkal, *The Green function of Teichmüller spaces with applications*, Bull. Amer. Math. Soc. (N.S.) **27** (1992), no. 1, 143–147, DOI 10.1090/S0273-0979-1992-00294-X. MR1142683
[21] P. Lelong, *Fonctions plurisousharmoniques et formes différentielles positives* (French), Gordon & Breach, Paris-London-New York (Distributed by Dunod éditeur, Paris), 1968. MR0243112
[22] A. Lenzhen and K. Rafi, *Length of a curve is quasi-convex along a Teichmüller geodesic*, J. Differential Geom. **88** (2011), no. 2, 267–295. MR2838267
[23] L. Liu and W. Su, Variations of extremal length function on Teichmüller space, ArXiv e-prints, October 2012.
[24] H. Masur, *The Teichmüller flow is Hamiltonian*, Proc. Amer. Math. Soc. **123** (1995), no. 12, 3739–3747, DOI 10.2307/2161902. MR1283559
[25] H. Masur and J. Smillie, *Hausdorff dimension of sets of nonergodic measured foliations*, Ann. of Math. (2) **134** (1991), no. 3, 455–543, DOI 10.2307/2944356. MR1135877
[26] C. T. McMullen, *Complex earthquakes and Teichmüller theory*, J. Amer. Math. Soc. **11** (1998), no. 2, 283–320, DOI 10.1090/S0894-0347-98-00259-8. MR1478844
[27] Y. N. Minsky, *Teichmüller geodesics and ends of hyperbolic 3-manifolds*, Topology **32** (1993), no. 3, 625–647, DOI 10.1016/0040-9383(93)90013-L. MR1231968
[28] T. Ohsawa, *On the Bergman kernel of hyperconvex domains*, Nagoya Math. J. **129** (1993), 43–52. MR1210002
[29] A. Papadopoulos and W. Su, *On the Finsler structure of Teichmüller's metric and Thurston's metric*, Expo. Math. **33** (2015), no. 1, 30–47, DOI 10.1016/j.exmath.2013.12.007. MR3310926
[30] R. C. Penner and J. L. Harer, *Combinatorics of train tracks*, Annals of Mathematics Studies, vol. 125, Princeton University Press, Princeton, NJ, 1992. MR1144770
[31] H. Shiga, *On analytic and geometric properties of Teichmüller spaces*, J. Math. Kyoto Univ. **24** (1984), no. 3, 441–452, DOI 10.1215/kjm/1250521274. MR766636
[32] J.-L. Stehlé, *Fonctions plurisousharmoniques et convexité holomorphe de certains fibrés analytiques* (French), C. R. Acad. Sci. Paris Sér. A **279** (1974), 235–238. MR0374493
[33] W. Thurston, *The Geometry and Topology of Three-Manifolds*, Lecture Note at Princeton University, 1980. Available at `http://library.msri.org/nonmsri/gt3m/`.
[34] W. A. Veech, *The Teichmüller geodesic flow*, Ann. of Math. (2) **124** (1986), no. 3, 441–530, DOI 10.2307/2007091. MR866707
[35] W. A. Veech, *Moduli spaces of quadratic differentials*, J. Analyse Math. **55** (1990), 117–171, DOI 10.1007/BF02789200. MR1094714
[36] S. A. Wolpert, *Geodesic length functions and the Nielsen problem*, J. Differential Geom. **25** (1987), no. 2, 275–296. MR880186
[37] S.-K. Yeung, *Bounded smooth strictly plurisubharmonic exhaustion functions on Teichmüller spaces*, Math. Res. Lett. **10** (2003), no. 2-3, 391–400, DOI 10.4310/MRL.2003.v10.n3.a8. MR1983031

Department of Mathematics, Graduate School of Science, Osaka University, Machikaneyama 1-1, Toyonaka, Osaka 560-0043, Japan
*E-mail address*: `miyachi@math.sci.osaka-u.ac.jp`

## Selected Published Titles in This Series

696 **Ara S. Basmajian, Yair N. Minsky, and Alan W. Reid, Editors,** In the Tradition of Ahlfors–Bers, VII, 2017

692 **Anatole Katok, Yakov Pesin, and Federico Rodriguez Hertz, Editors,** Modern Theory of Dynamical Systems, 2017

691 **Farrell Brumley, Maria Paula Gómez Aparicio, and Alberto Mínguez, Editors,** Around Langlands Correspondences, 2017

690 **Andrés Eduardo Caicedo, James Cummings, Peter Koellner, and Paul B. Larson, Editors,** Foundations of Mathematics, 2017

689 **Erica Flapan, Allison Henrich, Aaron Kaestner, and Sam Nelson, Editors,** Knots, Links, Spatial Graphs, and Algebraic Invariants, 2017

688 **Jeffrey Bergen, Stefan Catoiu, and William Chin, Editors,** Groups, Rings, Group Rings, and Hopf Algebras, 2017

687 **Fernanda Botelho, Raena King, and T. S. S. R. K. Rao, Editors,** Problems and Recent Methods in Operator Theory, 2017

686 **Alp Bassa, Alain Couvreur, and David Kohel, Editors,** Arithmetic, Geometry, Cryptography and Coding Theory, 2017

685 **Heather A. Harrington, Mohamed Omar, and Matthew Wright, Editors,** Algebraic and Geometric Methods in Discrete Mathematics, 2017

684 **Anna Beliakova and Aaron D. Lauda, Editors,** Categorification in Geometry, Topology, and Physics, 2017

683 **Anna Beliakova and Aaron D. Lauda, Editors,** Categorification and Higher Representation Theory, 2017

682 **Gregory Arone, Brenda Johnson, Pascal Lambrechts, Brian A. Munson, and Ismar Volić, Editors,** Manifolds and $K$-Theory, 2017

681 **Shiferaw Berhanu, Nordine Mir, and Emil J. Straube, Editors,** Analysis and Geometry in Several Complex Variables, 2017

680 **Sergei Gukov, Mikhail Khovanov, and Johannes Walcher, Editors,** Physics and Mathematics of Link Homology, 2016

679 **Catherine Bénéteau, Alberto A. Condori, Constanze Liaw, William T. Ross, and Alan A. Sola, Editors,** Recent Progress on Operator Theory and Approximation in Spaces of Analytic Functions, 2016

678 **Joseph Auslander, Aimee Johnson, and Cesar E. Silva, Editors,** Ergodic Theory, Dynamical Systems, and the Continuing Influence of John C. Oxtoby, 2016

677 **Delaram Kahrobaei, Bren Cavallo, and David Garber, Editors,** Algebra and Computer Science, 2016

676 **Pierre Martinetti and Jean-Christophe Wallet, Editors,** Noncommutative Geometry and Optimal Transport, 2016

675 **Ana Claudia Nabarro, Juan J. Nuño-Ballesteros, Raúl Oset Sinha, and Maria Aparecida Soares Ruas, Editors,** Real and Complex Singularities, 2016

674 **Bogdan D. Suceavă, Alfonso Carriazo, Yun Myung Oh, and Joeri Van der Veken, Editors,** Recent Advances in the Geometry of Submanifolds, 2016

673 **Alex Martsinkovsky, Gordana Todorov, and Kiyoshi Igusa, Editors,** Recent Developments in Representation Theory, 2016

672 **Bernard Russo, Asuman Güven Aksoy, Ravshan Ashurov, and Shavkat Ayupov, Editors,** Topics in Functional Analysis and Algebra, 2016

671 **Robert S. Doran and Efton Park, Editors,** Operator Algebras and Their Applications, 2016

670 **Krishnendu Gongopadhyay and Rama Mishra, Editors,** Knot Theory and Its Applications, 2016

For a complete list of titles in this series, visit the
AMS Bookstore at **www.ams.org/bookstore/conmseries/**.